AF361810

Sensors and Communications for the Social Good 2022

Sensors and Communications for the Social Good 2022

Editors

Claudio Palazzi
Ombretta Gaggi
Pietro Manzoni

MDPI • Basel • Beijing • Wuhan • Barcelona • Belgrade • Manchester • Tokyo • Cluj • Tianjin

Editors

Claudio Palazzi
Università degli Studi
di Padova,
Padova, Italy

Ombretta Gaggi
University of Padua,
Padova, Italy

Pietro Manzoni
Universitat Politècnica
de València,
Valencia, Spain

Editorial Office
MDPI
St. Alban-Anlage 66
4052 Basel, Switzerland

This is a reprint of articles from the Topical Collection published online in the open access journal *Sensors* (ISSN 1424-8220) (available at: https://www.mdpi.com/journal/sensors/topical_collections/topical_collection_goodit).

For citation purposes, cite each article independently as indicated on the article page online and as indicated below:

LastName, A.A.; LastName, B.B.; LastName, C.C. Article Title. *Journal Name* **Year**, *Volume Number*, Page Range.

ISBN 978-3-0365-7098-3 (Hbk)
ISBN 978-3-0365-7099-0 (PDF)

Contents

Claudio Palazzi, Ombretta Gaggi and Pietro Manzoni
Sensors and Communications for the Social Good
Reprinted from: *Sensors* **2023**, *23*, 2448, doi:10.3390/s23052448 1

Lulu Gao and Shin'ichi Konomi
Indoor Spatiotemporal Contact Analytics Using Landmark-Aided Pedestrian Dead Reckoning
on Smartphones
Reprinted from: *Sensors* **2023**, *23*, 113, doi:10.3390/s23010113 5

Mirko Zichichi, Stefano Ferretti, Víctor Rodríguez Doncel
Decentralized Personal Data Marketplaces: How Participation in a DAO Can Support the
Production of Citizen-Generated Data
Reprinted from: *Sensors* **2022**, *22*, 6260, doi:10.3390/s22166260 29

Andrea Masciadri, Changhong Lin, Sara Comai and Fabio Salice
A Multi-Resident Number Estimation Method for Smart Homes
Reprinted from: *Sensors* **2022**, *22*, 4823, doi:10.3390/s22134823 61

Jonathan Lacanlale, Paruyr Isayan, Katya Mkrtchyan and Ani Nahapetian
Sensoring the Neck: Classifying Movements and Actions with a Neck-Mounted Wearable
Device [†]
Reprinted from: *Sensors* **2022**, *22*, 4313, doi:10.3390/s22124313 81

**Sonia Bergamaschi, Stefania De Nardis, Riccardo Martoglia, Federico Ruozzi, Luca Sala,
Matteo Vanzini and Riccardo Amerigo Vigliermo**
Novel Perspectives for the Management of Multilingual and Multialphabetic Heritages through
Automatic Knowledge Extraction: The DigitalMaktaba Approach [†]
Reprinted from: *Sensors* **2022**, *22*, 3995, doi:10.3390/s22113995 95

**Kévin Chapron, Florentin Thullier, Patrick Lapointe, Julien Maître, Kévin Bouchard and
Sébastien Gaboury**
LIPSHOK: LIARA Portable Smart Home Kit
Reprinted from: *Sensors* **2022**, *22*, 2829, doi:10.3390/s22082829 115

Armir Bujari, Alessandro Calvio, Luca Foschini, Andrea Sabbioni and Antonio Corradi
A Digital Twin Decision Support System for the Urban Facility Management Process [†]
Reprinted from: *Sensors* **2021**, *21*, 8460, doi:10.3390/s21248460 131

**Edoardo Longo, Fatih Alperen Sahin, Alessandro E. C. Redondi, Patrizia Bolzan, Massimo
Bianchini and Stefano Maffei**
A 5G-Enabled Smart Waste Management System for University Campus [†]
Reprinted from: *Sensors* **2021**, *21*, 8278, doi:10.3390/s21248278 145

Giovanni Andreatta, Carla De Francescoand Luigi De Giovanni
Algorithms for Smooth, Safe and Quick Routing on Sensor-Equipped Grid Networks
Reprinted from: *Sensors* **2021**, *21*, 8188, doi:10.3390/s21248188 167

Emilia Corina Corbu and Eduard Edelhauser
Responsive Dashboard as a Component of Learning Analytics System for Evaluation in
Emergency Remote Teaching Situations
Reprinted from: *Sensors* **2021**, *21*, 7998, doi:10.3390/s21237998 187

Shirin Hajahmadi and Gustavo Marfia
Effects of the Uncertainty of Interpersonal Communications on Behavioral Responses of the
Participants in an Immersive Virtual Reality Experience: A Usability Study
Reprinted from: *Sensors* **2023**, *23*, 2148, doi:10.3390/s23042148 . . . . **209**

Editorial

Sensors and Communications for the Social Good

Claudio Palazzi [1], Ombretta Gaggi [1] and Pietro Manzoni [2,]*

1 Department of Mathematics, University of Padua, Via Trieste, 35131 Padova, Italy
2 Department of Computer Engineering (DISCA), Universitat Politècnica de València, 46022 Valencia, Spain
* Correspondence: pmanzoni@disca.upv.es

Citation: Palazzi, C.; Gaggi, O.;
Manzoni, P. Sensors and
Communications for the Social Good.
Sensors **2023**, *23*, 2448. https://
doi.org/10.3390/s23052448

Received: 7 February 2023
Accepted: 7 February 2023
Published: 22 February 2023

This topical collection focuses on applying sensors and communications technologies for social good. Social good is typically defined as an action that provides some sort of benefit to the general public. In this case, the Internet connection, education, and health care are all excellent examples of social good. However, the development of new media and the explosion of online communities have added a new meaning to the term. Social good is now about global citizens uniting to unlock the potential of individuals, technology, and collaboration to create a positive societal impact.

This reprint includes the recent advances and novel contributions from academic researchers and industry practitioners in this growing area. Each article was assigned and reviewed by at least three experts in the field during the review process, with a rigorous multi-round review process. Consequent to the great support and dedicated work of numerous reviewers, we accepted 10 excellent articles covering various topics under the umbrella of "Sensors and Communications for the Social Good". We will introduce these articles and highlight their main contributions in what follows.

The first paper focuses on how the need to provide safe environments and reduce the risks of virus exposure plays a crucial role in our daily lives. Contact tracing is a well-established and widely used approach to tracking and suppressing the spread of viruses. Most digital contact tracing systems can detect direct face-to-face contact based on estimated proximity without quantifying the exposed virus concentration. In particular, they rarely allow for the quantitative analysis of indirect environmental exposure due to virus survival time in the air and constant airborne transmission. The authors of [1] proposed an indoor spatiotemporal contact awareness framework (iSTCA), which explicitly considers the self-containing quantitative contact analytics approach with spatiotemporal information to provide accurate awareness of the quanta concentration of virus in different origins at various times. Smartphone-based pedestrian dead reckoning (PDR) was employed to precisely detect the locations and trajectories for distance estimation and time assessment without additional infrastructure. The PDR technique calibrates the accumulative error by automatically identifying spatial landmarks. They utilized a custom deep-learning model composed of bidirectional long short-term memory (Bi-LSTM) and multi-head convolutional neural networks (CNNs) to extract the local correlation and long-term dependency to recognize landmarks. By considering the spatial distance and time difference in an integrated manner, they quantified the virus quanta concentration in the indoor environment at any time with all contributed virus particles. They conducted an extensive experiment based on practical scenarios to evaluate the performance of the proposed system. They showed that the average positioning error is reduced to less than 0.7 m.

Big Tech companies operating in a data-driven economy offer services that rely on their users' data and usually store this personal information in "data silos", which prevent transparency about their use and opportunities for data sharing for the public interest. In [2], the authors presented a solution that promotes the development of decentralized personal data marketplaces, exploiting the use of distributed ledger technologies (DLTs), decentralized file storages (DFSs), and smart contracts to store personal data and manage access control in a decentralized way. Moreover, they focused on the lack of efficient decentralized mechanisms in DLTs and DFSs for querying a specific data type. For this reason, the authors proposed using a hypercube-structured distributed hash table (DHT) on top of DLTs, organized for the efficient processing of multiple keyword-based queries on the ledger data. They tested their approach by implementing a use case for creating citizen-generated data based on direct participation and decentralized autonomous organization (DAO) participation. Performance evaluation demonstrated the viability of their approach for decentralized data searches, distributed authorization mechanisms, and smart contract exploitation.

In [3], the authors focused on how population aging requires innovative solutions to increase the quality of life and preserve autonomous and independent living at home. A need of particular significance is the identification of behavioral drifts. A relevant behavioral drift concerns sociality: Older people tend to isolate themselves. There is a need to find methodologies to identify if, when, and for how long the person is in the company of other people (possibly also considering the number). The challenge is to address this task in poorly sensorized apartments, with non-intrusive sensors that are typically wireless and can only provide local and simple information. The proposed method addresses technological issues, such as PIR (passive infrared) blind times, topological issues, such as sensor interference due to the inability to separate detection areas, and algorithmic issues. A house was modeled as a graph to constrain the transitions between adjacent rooms. Each room was associated with a set of values for each identified person. These values would decay over time and represent the probability that each person is still in the room. Because the used sensors cannot determine the number of people, this approach is based on a multi-branch inference that, over time, differentiates the movements in the apartment and estimates the number of people.

Sensor technology that captures information from the user's neck region can enable a variety of new possibilities, including less intrusive mobile software interfaces. In [4], the authors investigated the feasibility of using a single inexpensive flex sensor mounted at the neck to capture information about head gestures, mouth movements, and the presence of audible speech. Different sensor sizes and various sensor positions on the neck were experimentally evaluated. With data collected from experiments carried out on the finalized prototype, a classification accuracy of 91% was achieved to differentiate common head gestures, a classification accuracy of 63% was achieved to differentiate mouth movements, and a classification accuracy of 83% was achieved in speech detection.

The work in [5] highlighted the fact that the linguistic and social impact of multiculturalism cannot be ignored in any sector, creating the urgent need to create systems and procedures for managing and sharing cultural heritage in both supranational and multi-literate contexts. Text sensing is one of the most crucial research areas to achieve this goal. The long-term objective of the DigitalMaktaba project, born from an interdisciplinary collaboration between computer scientists, historians, librarians, engineers, and linguists, is to establish procedures for creating, managing, and cataloging archival heritage in non-Latin alphabets. In this paper, the authors discussed the currently ongoing design of an innovative workflow and tool in the area of text sensing, for the automatic extraction of knowledge and cataloging of documents written in non-Latin languages (Arabic, Persian, and Azerbaijani). The current prototype leverages different OCR, text processing, and information extraction techniques to provide highly accurate extracted text and rich metadata content (including automatically identified cataloging metadata), overcoming the typical limitations of current state-of-the-art approaches. The initial tests revealed promising results. The paper included a discussion of future steps (e.g., AI-based techniques further leveraging the extracted data/metadata and making the system learn from user feedback) and of the many foreseen advantages of this research, both from a technical and a broader cultural preservation and sharing points of view.

In [6], the authors focused on several smart home architecture implementations proposed in the last decade. These architectures are mostly deployed in laboratories or inside real habitations built for research purposes to enable ambient intelligence using various sensors, actuators, and machine learning algorithms. However, the major issues for most related smart home architectures are their price, proprietary hardware requirements, and the need for highly specialized personnel to deploy such systems. To address these challenges, lighter forms of smart home architectures known as smart homes in a box (SHiB) have been proposed. While SHiB remains an encouraging first step towards lightweight yet affordable solutions, they still suffer a few drawbacks. Indeed, some of these kits lack hardware support for some technologies, and others do not include enough sensors and actuators to cover the requirements of most smart homes. Thus, this paper introduced the LIARA Portable Smart Home Kit (LIPSHOK), designed to provide an affordable SHiB solution that anyone can install in an existing home. Moreover, LIPSHOK is a generic kit that includes four specialized sensor modules that have been independently introduced since the authors' laboratory has been working on their development over the last few years. This paper first summarized these modules and their respective benefits within a smart home context. Then, it mainly focused on introducing the LIPSHOK architecture, which provides a framework to unify the use of the proposed sensors owing to a common modular infrastructure capable of managing heterogeneous technologies. Finally, the authors compared their work to existing SHiB kit solutions and revealed that it offers a more affordable, extensible, and scalable solution with resources distributed under an open-source license.

The ever-increasing pace of IoT deployment is opening the door to the concrete implementations of smart city applications, enabling the large-scale sensing and modeling of (near) real-time digital replicas of physical processes and environments. This digital replica could serve as the basis of a decision support system, providing insight into the possible optimizations of resources in a smart city scenario. In [7], the authors discussed an extension of prior work, presenting a detailed proof-of-concept implementation of a digital twin solution for the urban facility management (UFM) process. The Interactive Planning Platform for Adaptive Maintenance Operations for the City District (IPPODAMO) is a distributed geographical system fed and ingested heterogeneous data sources from different urban data providers. The data are subject to continuous refinements and algorithmic processes used to quantify and build synthetic indexes measuring the activity level inside an area of interest. IPPODAMO considers the potential interference from other stakeholders in the urban environment, enabling informed operations scheduling to minimize interference and operating costs.

The authors of [8] revealed that future university campuses will be characterized by a series of novel services enabled by the vision of the Internet of Things, such as smart parking and smart libraries. In this paper, the authors proposed a complete solution for a smart waste management system to increase the recycling rate on campus and better manage the entire waste cycle. The system is based on a prototype of a smart waste bin, which can accurately classify pieces of trash typically produced on campus premises with a hybrid sensor/image classification algorithm and automatically separate the different waste materials. The authors discussed the system prototype's entire design, from the analysis of the requirements to the implementation details, and evaluated its performance in different scenarios. Finally, they discussed advanced application functionalities built around the smart waste bin, such as optimized maintenance scheduling.

Automation plays an important role in modern transportation and handling systems, e.g., controlling aircraft and ground service equipment routes in airport aprons, automated guided vehicles in port terminals or public transportation, handling robots in automated factories, drones in warehouse picking operations, etc. Information technology provides hardware and software (e.g., collision detection sensors, routing, and collision avoidance logic) that contribute to safe and efficient operations, with relevant social benefits in terms of improved system performance and reduced accident rates. In this context, the authors of [9] addressed the design of efficient collision-free routes in a minimum-sized routing network. They considered a grid and a set of vehicles, each moving from the bottom of the origin column to the top of the destination column. Smooth nonstop paths are required, without collisions or deviations from shortest paths, and they investigated the minimum number of horizontal lanes allowing for such routing. The problem is known as the fleet's quickest routing problem on grids. They proposed a mathematical formulation solved for small instances using standard solvers. For larger instances, they devised heuristics that define priorities based on known combinatorial properties and design collision-free routes. Their experiments on random instances showed that their algorithms could quickly provide good-quality solutions.

Finally, in [10], the authors considered how the pandemic crisis has forced the development of teaching and evaluation activities exclusively online. In this context, the emergency remote teaching (ERT) process, which raised many problems for institutions, teachers, and students, led the authors to consider it important to design a model to evaluate teaching and evaluation processes. The study objective presented in this paper was to develop a model for the evaluation system called the learning analytics and evaluation model (LAEM). The authors also validated a software instrument they designed called the EvalMathI system, which is to be used in the evaluation system and was developed and tested during the pandemic. The evaluation process was optimized by including and integrating the dashboard model in a responsive panel. The EvalMathI dashboard monitored six online courses in the 2019/2020 and 2020/2021 academic years. For each of the six monitored courses, the curricula were evaluated through the analyzed parameters by highlighting the percentage achieved by each course on various components, such as content, adaptability, skills, and involvement. In addition, after collecting the data through interview guides, the authors determined the extent to which online education during the COVID-19 pandemic has influenced the educational process. Through the developed model, the authors also found software tools to solve problems raised by teaching and evaluation in the ERT environment.

The editors and authors express their thanks to the publisher and staff members for their ongoing dedication and valuable advice and encouragement, which helped to improve the quality of this reprint.

Conflicts of Interest: The authors declare no conflict of interest.

References

1. Gao, L.; Konomi, S. Indoor Spatiotemporal Contact Analytics Using Landmark-Aided Pedestrian Dead Reckoning on Smartphones. *Sensors* **2023**, *23*, 113. [CrossRef] [PubMed]
2. Zichichi, M.; Ferretti, S.; Rodríguez-Doncel, V. Decentralized Personal Data Marketplaces: How Participation in a DAO Can Support the Production of Citizen-Generated Data. *Sensors* **2022**, *22*, 6260. [CrossRef] [PubMed]
3. Masciadri, A.; Lin, C.; Comai, S.; Salice, F. A Multi-Resident Number Estimation Method for Smart Homes. *Sensors* **2022**, *22*, 4823. [CrossRef] [PubMed]
4. Lacanlale, J.; Isayan, P.; Mkrtchyan, K.; Nahapetian, A. Sensoring the Neck: Classifying Movements and Actions with a Neck-Mounted Wearable Device. *Sensors* **2022**, *22*, 4313. [CrossRef] [PubMed]
5. Bergamaschi, S.; De Nardis, S.; Martoglia, R.; Ruozzi, F.; Sala, L.; Vanzini, M.; Vigliermo, R.A. Novel Perspectives for the Management of Multilingual and Multialphabetic Heritages through Automatic Knowledge Extraction: The DigitalMaktaba Approach. *Sensors* **2022**, *22*, 3995. [CrossRef] [PubMed]
6. Chapron, K.; Thullier, F.; Lapointe, P.; Maître, J.; Bouchard, K.; Gaboury, S. LIPSHOK: LIARA Portable Smart Home Kit. *Sensors* **2022**, *22*, 2829. [CrossRef] [PubMed]
7. Bujari, A.; Calvio, A.; Foschini, L.; Sabbioni, A.; Corradi, A. A Digital Twin Decision Support System for the Urban Facility Management Process. *Sensors* **2021**, *21*, 8460. [CrossRef] [PubMed]
8. Longo, E.; Sahin, F.A.; Redondi, A.E.C.; Bolzan, P.; Bianchini, M.; Maffei, S. A 5G-Enabled Smart Waste Management System for University Campus. *Sensors* **2021**, *21*, 8278. [CrossRef] [PubMed]
9. Andreatta, G.; De Francesco, C.; De Giovanni, L. Algorithms for Smooth, Safe and Quick Routing on Sensor-Equipped Grid Networks. *Sensors* **2021**, *21*, 8188. [CrossRef] [PubMed]
10. Corbu, E.C.; Edelhauser, E. Responsive Dashboard as a Component of Learning Analytics System for Evaluation in Emergency Remote Teaching Situations. *Sensors* **2021**, *21*, 7998. [CrossRef] [PubMed]

Article

Indoor Spatiotemporal Contact Analytics Using Landmark-Aided Pedestrian Dead Reckoning on Smartphones

Lulu Gao [1] and Shin'ichi Konomi [2,*]

1   Graduate School of Information Science and Electrical Engineering, Kyushu University, Fukuoka 819-0395, Japan
2   Faculty of Arts and Science, Kyushu University, Fukuoka 819-0395, Japan
*   Correspondence: konomi@artsci.kyushu-u.ac.jp

Abstract: Due to the prevalence of COVID-19, providing safe environments and reducing the risks of virus exposure play pivotal roles in our daily lives. Contact tracing is a well-established and widely-used approach to track and suppress the spread of viruses. Most digital contact tracing systems can detect direct face-to-face contact based on estimated proximity, without quantifying the exposed virus concentration. In particular, they rarely allow for quantitative analysis of indirect environmental exposure due to virus survival time in the air and constant airborne transmission. In this work, we propose an indoor spatiotemporal contact awareness framework (iSTCA), which explicitly considers the self-containing quantitative contact analytics approach with spatiotemporal information to provide accurate awareness of the virus quanta concentration in different origins at various times. Smartphone-based pedestrian dead reckoning (PDR) is employed to precisely detect the locations and trajectories for distance estimation and time assessment without the need to deploy extra infrastructure. The PDR technique we employ calibrates the accumulative error by identifying spatial landmarks automatically. We utilized a custom deep learning model composed of bidirectional long short-term memory (Bi-LSTM) and multi-head convolutional neural networks (CNNs) for extracting the local correlation and long-term dependency to recognize landmarks. By considering the spatial distance and time difference in an integrated manner, we can quantify the virus quanta concentration of the entire indoor environment at any time with all contributed virus particles. We conducted an extensive experiment based on practical scenarios to evaluate the performance of the proposed system, showing that the average positioning error is reduced to less than 0.7 m with high confidence and demonstrating the validity of our system for the virus quanta concentration quantification involving virus movement in a complex indoor environment.

Keywords: COVID-19; contact awareness; spatiotemporal analytics; indoor positioning; pedestrian dead reckoning; landmark identification

Citation: Gao, L.; Konomi, S. Indoor Spatiotemporal Contact Analytics Using Landmark-Aided Pedestrian Dead Reckoning on Smartphones. *Sensors* **2023**, *23*, 113. https://doi.org/10.3390/s23010113

Academic Editors: Pietro Manzoni, Claudio Palazzi and Ombretta Gaggi

Received: 15 November 2022
Revised: 16 December 2022
Accepted: 19 December 2022
Published: 22 December 2022

1. Introduction

The worldwide COVID-19 pandemic has brought about many changes in our daily lives and struck a devastating blow to the global economy. It is widely recognized that airborne transmission serves as the primary pathway for the spread of COVID-19 via expiratory droplets, especially in indoor environments [1]. During the viral outbreak, many people were infected due to exposure to virus droplets generated by human exhalation activities [2–4]. Some infected patients spread the virus unknowingly without properly being examined because there is an incubation period that varies for different mutations and asymptomatic patients who never experience apparent symptoms [5]. Reliable and efficient tracing and quarantining have become more important than ever to alert individuals to take actions to interrupt the transmission between people and further curb the spread of the disease. Contact tracing involves identifying, assessing, and managing people who are at risk of the infection, and tracking subsequent victims as recorded by the public health department [6]; contact tracing can be performed via manual or digital methods. Since the manual contact tracing is labor-intensive and time-consuming and may be incomplete and inaccurate due to forgetfulness; automatic digital contact tracing has been widely researched in recent years [7]. Usually, digital contact tracing applications are installed on portal devices, typically smartphones, to conveniently and intelligently realize tracing with the help of existing sensors based on various technologies, such as a global navigation satellite system (GNSS), Bluetooth, and Wi-Fi.

Contact tracing in indoor environments can complement the ones used in outdoor environments to enable comprehensive digital contact tracing. However, indoor contact tracing imposes unique technical challenges due to virus concentrations and unreliable GNSS signals in indoor environments [8]. The virus concentration, which plays a critical role in calculating the amount of a virus we are exposed to and further assesses the infection risk, should be explicitly considered in indoor contact tracing applications [3,9]. The quantitative infection risk for a susceptible person is significantly associated with the quantity of the pathogen inhaled in the surrounding ambient air, from the respiratory droplets exhaled by infected individuals [10]. Thus, inhaling a large amount of the virus in a short period, i.e., under the 15 min time mark, can greatly increase the infection risk, especially for so-called "superspreading events", which invariably occur indoors [11]. Moreover, the majority of time has to be spent by people in indoor contexts with plenty of daily activities performed. However, GNSS-based approaches do not work well in indoor environments due to signal attenuation. Phone-to-phone pairing-based methods using Bluetooth low energy (BLE) work only for direct face-to-face contact tracing scenarios and are inapplicable to indirect virus exposure in ambient aerosols. The expelled pathogen-containing particles can remain active in the air for hours without sufficient sanitization, especially in indoor environments, constructing a significant fraction of the virus concentration [3]. Recently, vContact was proposed as a means to detect exposure to the virus with the consideration of asynchronous contacts by leveraging Wi-Fi networks, while the spatiotemporal dynamism in the virus concentration is not fully being considered [8].

Although the virus concentration will gradually decrease due to inactivation, deposition, and air purification after the virus-laden droplets are exhaled, the poor air exchange rate, superspreaders, and more virulent variants will keep it at a relatively high concentration for a long time in an indoor environment [9,12]. The viral particles are continuously ejected by infected people at different locations, relying on human movement. Moreover, due to the initial motion state and environmental airflow, these droplets maintain a ceaseless transmission before they are removed and meet somewhere (at some time), which leads to constant changes in the virus concentration within the control volume [13]. To accurately estimate the concentration, investigating the airborne transmission of these ejected particles is, thus, of fundamental importance in a closed environment because of the assemblage, in which human movement is implicitly involved to achieve the initial motion state of droplets [13]. The qualitatively location-specific assessment of the viral concentration is proposed with the dual use of computational fluid dynamic simulations and surrogate aerosol measurements for different real-world settings [14]. Moreover, the transmission of the virus brings about changes in the viral concentration of a specific location in an overall space, as well as the movements of people. Z. Li et al. analyzed the dispersion of cough-generated droplets in the wake of a walking person [4].

To be precisely aware of the amount of the virus one is exposed to and to detect both direct and indirect contacts, an indoor spatiotemporal contact awareness (iSTCA) framework is proposed. Since the virus concentration (at different times in the same area) is not the same because of the dispersion and diffusion of the virus and human movements, we employed a self-contained PDR technique to calculate the human trajectory with accuracy and further achieve the location and time of the expelled virus droplets for the quantitative measurement of the concentration at any time in different spots. Moreover, based on the acquired changing virus concentration and reliable trajectories, the exposure time and distance of both direct and indirect contacts can be derived via cross-examination to realize quantitative spatiotemporal contact awareness.

Our main contributions are as follows:

1. To accurately present the virus concentrations at different times, we established quantitative virus concentration changes in various areas of indoor environments at different times by infected individuals. The viral-laden droplets were continuously released during the expiratory activities, moving forward. During the movements of viral-loaded droplets exhaled by infectious individuals at different locations and times, the virus instances met in certain spots at certain times and contributed to the calculation of the concentration. Finally, the concentration of each virus instance was integrated.
2. We properly selected PDR for the acquisition of the trajectory to conduct contact awareness without requiring extra infrastructure or being affected by coverage limitations compared with other indoor positioning techniques.
3. We considered various landmarks to calibrate the accumulative error for trajectory achievement by using PDR. A custom deep neural network using bidirectional long short-term memory (Bi-LSTM) and multi-head convolutional neural networks (CNNs) with residual concatenations were designed and implemented to extract temporal information in forward and backward directions and spatial features at various resolutions from built-in sensor readings for landmark identification.
4. Additionally, we demonstrate the effectiveness of the proposed Bi-LSTM-CNN classification model for landmark identification through empirical experiments, as well as the performance of our proposed iSTCA system for quantitative spatiotemporal contact analytics.

The remainder of this paper is organized as follows. The related work about contact awareness and indoor localization techniques, including PDR, is reviewed in Section 2. Definitions and preliminaries about virus concentrations and different contact types are introduced in Section 3. Section 4 introduces the theoretical methodology and the architecture of the proposed iSTCA. The experimental methodology and results based on the collected datasets are presented in Section 5. Section 6 reveals the limitations of this work. Finally, we present the conclusion and future work in Section 7.

2. Related Work

Contact tracing is used to identify and track people who may have been exposed to a virus due to the prevalence of many infectious diseases in our society. To conduct contact tracing, it is necessary for the infected individuals to provide their visited locations and people whom they encountered based on the specific definitions of meetups for different diseases. Instead of interviews and questionnaires via traditional manual tracing, technology-aided contact tracing can track people at risk conveniently and intelligently. To reduce the spread of COVID-19 effectively, digital contact tracing, which generally depends on applications installed on smartphones, has been developed in both academia and industry, using various technologies, such as GNSS, Bluetooth, and Wi-Fi.

There are typically two approaches for encounter determinations, peer-to-peer proximity detection-based and geolocation-based. Peer-to-peer proximity can be estimated by the received signal strength (RSS) of wireless signals, such as Bluetooth and ultra-wideband (UWB), and the distance between two devices in geolocation-based approaches can be precisely derived from the cross-examination after obtaining the accurate location and trajectory with the help of localization techniques using various technologies, such as global positioning system (GPS), Wi-Fi, and PDR.

Some systems based on peer-to-peer proximity using Bluetooth or BLE have been implemented, and part of them are deployed by the governments of various countries, such as Australia (COVIDSafe), Singapore (Trace together), and the United Kingdom (NHS COVID-19 App) due to their ubiquitous embedding in mobile phones [15]. Among these systems, the most representative protocols are Blue Trace and ROBERT [11,16]. The data from Bluetooth device-to-device communications are stored and checked against the data uploaded by the infector. In Blue Trace, the health authority contacts individuals who had a high probability of virus exposure, whereas ROBERT users need to periodically probe the server for their infection risk scores. In addition, Google and Apple provide a broadly used toolkit based on Bluetooth, named Google and Apple Exposure Notification (GAEN), to facilitate a contact tracing system in Android and iOS and curb the spread of COVID-19 [17]. Despite some minor differences in implementation and efficiency, these schemes are all independently designed and very similar. When exposure is detected, the RSS in the communication data frame is utilized to estimate the distance between two devices and notify the user. However, it has been demonstrated that the signal strengths can only provide very rough estimations of the actual distances between devices, as they are affected by device orientation, shadowing, shading effects, and multipath losses in different environments [18,19]. Although it is difficult to measure the distances among users accurately by using Bluetooth and other technologies, the UWB radio technology has the capacity to measure distances at the accuracy level of a few centimeters, which is significantly bettering than Bluetooth [20]. The use of UWB, however, has some significant drawbacks, including the fact that UWB is not widely supported by mobile devices, requires extra infrastructure, and is not energy efficient, which makes UWB less useful in practice [21]. All of the above works that are based on calculated proximity using RSS do not consider the user's specific physical location, resulting in unsatisfactory tracing results. Moreover, these approaches cannot be applied to the detection of temporal contact due to the dispersion and lifespan of the virus.

To achieve accurate geolocation in contact tracing, plenty of localization systems have been researched with the joint efforts of researchers and engineers in the past based on GNSS, cellular technology, radio frequency identification (RFID), and quick response (QR) code [7]. GNSS can be used for contact tracing as the exact position of a person can be located and it is available globally. Many countries, including Israel (HaMagen 2.0) and Cyprus (CovTracer), use GPS-based contact tracing approaches [15] as well. GNSS signals are usually weak in indoor environments due to the absence of the line of sight and the attenuation of satellite signals, as well as the noisiness of the environment. Many people may spend most of their time in indoor environments, which can result in limited contact coverage. It is difficult to detect contact based on cellular data due to the large coverage of cell towers and high location errors [8]. RFID was used to reveal the spread of infectious diseases and detect face-to-face contact in [22,23]. QR codes for contact tracing require users to check in at various venues by scanning the placed QR codes manually to record their locations and times, which are deployed in some countries, such as New Zealand (NZ COVID Tracer) [15]. However, special devices or codes have to be deployed at scale for data collection. Recently, some protocols were proposed for Wi-Fi-based contact tracing with the pre-installed Wi-Fi Access Point. WiFiTrace was proposed by proposed in [24]. WiFiTrace is a network-centric contact tracing approach with passive Wi-Fi sensing and without client-side involvement, in which the locations visited are reconstructed by network logs; graph-based model and graph algorithms are employed to efficiently perform contact tracing. Wi-Fi association logs were also investigated in [25] to infer the social intersections with coarse collocation behaviors. Li et al. utilized active Wi-Fi sensing for data collection; they leveraged signal processing approaches and similarity metrics to align and detect virus exposure with temporally indirect contact [8]. As the changes in virus concentrations over time (due to the transmission of aerosols and environmental factors) are not considered, their results are in relatively low spatiotemporal resolutions. The approach presented in [26] divides contact tracing into two separate parts, duration and distance of exposure. The duration is captured from the Wi-Fi network logs and the distance is calculated by the PDR positioning trajectory, calibrated by recognized landmarks with the help of a CNN, ensuring the performance of contact tracing. Although integration with the existing infrastructure is beneficial in mitigating the deployment costs, it may not fully satisfy the requirements of contact tracing with the high spatiotemporal resolution because of the absent coverage [27]. The trajectory obtained by the PDR technique, without requiring special infrastructure, can improve the coarse-grained duration and make it fine-grained. This can enable the development of a contact-tracing environment that considers the virus lifespan in detail.

One of the ultimate goals of contact awareness systems is to estimate the risk based on the recorded encounter data [28]. Moreover, with the exposure duration and distance obtained, the virus concentration is significant to determine the exposed viral load, which is closely associated with the infection risk [29]. Typically, the virus concentration in a given space depends on the total amount of viral load contained in the viable virus-laden droplets in the air and maintains a downward trend because of the self-inactivation and environmental factors. Researchers presented the qualitative location-specific assessment of viral concentration with the dual use of computational fluid dynamic simulations and surrogate aerosol measurements for different real-world settings [14]. The practical viral loads emitted by contagious subjects based on the viral loads in the mouth (or sputum) with various types of respiratory activities and activity levels are presented in [29]. Furthermore, to quantitatively shape the virus concentration in a targeted environment at different times, the constant viral load emission rate is adopted with the virus removal rate, including the air exchange rate, particle sediment, and viral inactivation rate in [30].

The aforementioned contact tracing research usually only considers the static virus concentration without considering the exposure to the environmental virus and dynamism in the virus concentration. Moreover, in contrast to the qualitative estimation of exposure risks that can be achieved in previous works, there is a lack of sufficient quantitative awareness about the concentrations of contracted viruses. Such awareness would be useful in our daily lives to protect ourselves from virus infections.

3. Definitions and Preliminaries

Virus-encapsulating secretions are continuously exhaled and aerosolized into airborne virus-laden particles with infectivity from daily expository activities. There is a great difference between the size and number of droplets expelled, depending on their origin locations in the respiratory tract [4]. The time and distances of these droplets traveling in indoor environments largely depend on the expiration air jet, particle weight, and ambient factors. The movements and the viral loads of virus-containing particles are directly associated with the virus concentrations in different regions. To quantitatively become aware of the exposure of the virus, the quanta concentration as a medical virus concentration indicator, virus airborne pattern, and various contact types are present.

3.1. Quanta Concentration

The viral loads of virus-containing droplets change after leaving the human expiratory tract with airborne transmission and a combination of environmental factors. In particular, the viral load emitted is expressed in terms of the quanta emission rate (ER_q, quanta $\cdot$ h^{-1}), in which a quantum is defined as the dose of airborne droplet nuclei that infect 63% of susceptible persons with exposure [30]. The quanta concentration in an indoor area at time t, $q(t)$ is measured by:

$$q(t, ER_q) = N_I \cdot \frac{ER_q}{RR_{iv} \cdot V} + \left(q_0 + N_I \cdot \frac{ER_q}{RR_{iv}} \right) \cdot \frac{e^{-RR_{iv} \cdot t}}{V} \left(\text{quanta} \cdot \text{m}^{-3} \right) \qquad (1)$$

where ER_q is the quanta emission rate of the infector (measure in quanta $\cdot$ h^{-1}), q_0 is a constant declaring the initial number of quanta in the space, V $\left(m^3 \right)$ is the target indoor volume, N_I represents the number of infected individuals in the investigated volume, RR_{iv} (h^{-1}) is the removal rate for the infectious virus in the considered spaces [30]. RR_{iv} consists of three contributions, the air exchange rate (AER) via ventilation, the deposition on surface rate (k) caused by gravitational sedimentation and turbulent eddy impaction, and the viral inactivation rate (λ). The typical k is 0.24 h^{-1} and the inactivation rate λ of viable COVID-19 particles in a typical indoor environment without sunlight is generally $0.63\ \text{h}^{-1}$, as indicated in [30,31].

The ER_q is determined by the viral load in sputum, the volume of signal droplets, and the quantity of all expelled droplets per exhalation. Thus, the quanta concentration ER_q is modeled as:

$$ER_q = c_v \cdot c_i \cdot IR \cdot \int N_d(D) \cdot dV_d(D)\ (\text{quanta} \cdot \text{h}^{-1}) \tag{2}$$

where c_v represents the viral load in the sputum of the infector (RNA copies $\cdot\text{mL}^{-1}$), IR is the inhalation/exhalation rate produced by the breathing rate and tidal volume, N_d is the droplet concentrations in different expiratory activities of the infected person (particles $\cdot\ \text{cm}^{-3}$), V_d is the volume of a single droplet (cm^3) with the function of particle diameters D, and c_i is the conversion factor, presenting the ratio between one infectious quantum and the infectious dose expressed in the viral RNA copies [29]. There is a wide range of variations in the quanta emission estimation via Equation (2), depending on these and other factors, such as virus concentration in the mouth, activity level, and the type of coughing or exhaling. With light exercise and speaking, a quanta concentration of 142 (quanta $\cdot\ \text{h}^{-1}$) can be obtained, which was widely adopted in many works [29].

3.2. Spatial–Temporal Contact

COVID-19 contained in expiratory droplets and expelled from the infector is transported and dispersed in the ambient airflow before finally being removed, inactivated, and inhaled by a susceptible. There are a number of factors that contribute to the droplet's movement, such as the horizontally emitted velocity, the particle weight and the external environment. Occasionally, coughing and sneezing generate more particles with higher initial velocities ($11.7\ \text{m} \cdot \text{s}^{-1}$ for coughing) and virus quanta concentrations, while constantly performed breathing and speaking ($3.9\ \text{m} \cdot \text{s}^{-1}$ for speaking) produce fewer particles with relatively lower initial velocity and virus quanta concentrations [29]. Large droplets usually settle quickly in a few seconds or minutes owing to gravitational sedimentation and are evaporated into small nuclei in indoor environments, where the particle can disperse for a long distance in the vaporization process. Tiny particles, including ones that are evaporated and originally expelled, are trapped and carried continuously forward within a moist, warm, turbulent cloud of gas, with the help of airflow movement. To facilitate the calculation, the movement of each virus-laden droplet expelled at each moment is independent and divided into two stages, maintaining a uniform motion with the initial horizontal velocity (e.g., $3.9\ \text{m} \cdot \text{s}^{-1}$), being well-mixed within the moved space in the first phase (e.g., 1 s), and then instantaneously and evenly distributed in the overall considered space.

The contact in COVID-19 contact tracing is originally equivalent to direct face-to-face contact, while due to the transmission of the virus and survival time in the air, more cases of indirect contact have emerged [2]. Here, indirect contact mainly represents the asynchronous time contact, called temporal contact. Direct and indirect contacts are types of spatiotemporal contacts. If there is no time difference between two people, and only a spatial distance is presented, it is called spatial contact. Similarly, if there is no space difference between two people, and only a temporal distance is presented, it is called temporal contact. There are time and space gaps, a mixture of two single cases, called spatiotemporal contacts. Since both the time and space differences would decrease the virus quanta concentration, it is necessary to obtain the accurate value for the precise awareness of the virus quanta concentration.

4. Methodology

This work utilizes the trajectories, including the spatial position coordinates and time obtained by the PDR technique to quantitatively estimate the time-dependent changes in the virus quanta concentration derived from the movement and lifespan of the virus in various places of the considered indoor environment. The overview of the proposed scheme is systematically introduced in Section 4.1. In Section 4.2, we provided the data processing approaches utilized in PDR-based trajectory construction and the estimation of droplet exhalation. The PDR technique (with a calibration of the landmark recognized by a landmark identification model based on a residual Bi-LSTM and CNN structure) is discussed in Section 4.3. Further, the contact awareness model relying on the precisely constructed pedestrian trajectory is detailed in Section 4.4.

4.1. System Overview

An overview of the proposed iSTCA system is presented in Figure 1. More precisely, the data flow of various sensors for the analysis was primarily collected from the existing sensors in handhold smartphones, which record the changes in the environment and body motion. The signals need to be processed, including data filtering and scaling, to reduce the noise for a better state of motion estimation before training the landmark identification model and performing the PDR. The trajectory can be achieved based on the PDR technique and properly corrected with the assistance of the identified landmark distinguished by the trained landmark recognition model. The trajectory is defined as a set of points consisting of the time and position, $\{(t_0, x_0, y_0), (t_1, x_1, y_1), \dots (t_n, x_n, y_n)\}$ where (x_i, y_i) represents the location coordinates and t_i is the moment when the individual passes the location. The virus quanta concentrations in different spaces at various moments can be measured quantitatively to achieve sufficient awareness with the help of the estimated spatial distance, temporal distance, and infectivity model, as shown in Equation (1).

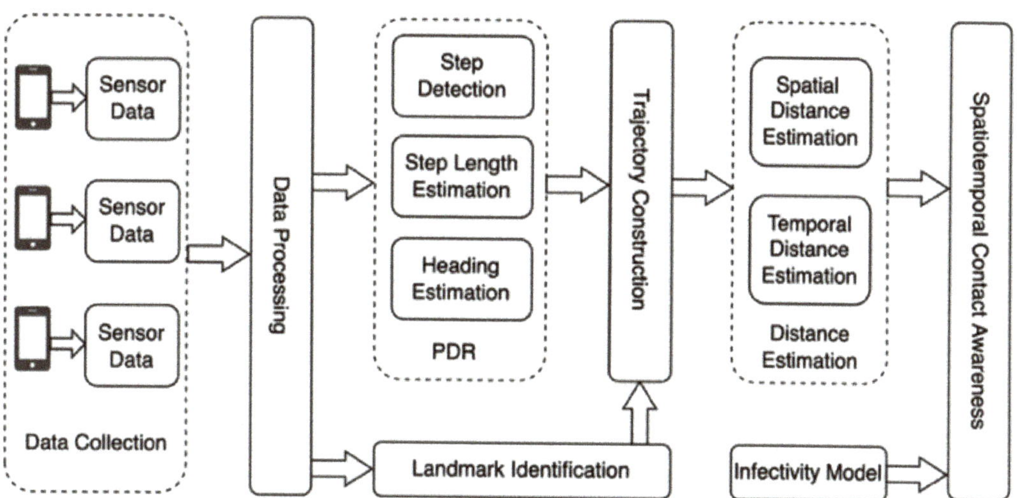

Figure 1. Overview of iSTCA.

4.2. Data Preprocessing

Data alignment. The same sampling rate for data collection is set to 50 Hz due to the low frequency of human movements [32]. Although the constant rate is defined, the time interval between the recorded adjacent readings of each sensor is not always the same because of the observational error and random error, and it oscillates within a certain range in practice. To acquire the same number of samples for conveniently performing the subsequent procedures, we take the timestamp of the first data collected as the starting time to align the sensor readings at the same time interval with the help of data interpolation.

Data interpolation. During the practical data collection using smartphone sensors, some data points in the acquired dataset are lost due to malfunctioning; such data points are typically replaced by 0, NaN, or none [33]. To fill in the missing values, the data interpolation technique was developed, in which the new data point is estimated based on the known information. Linear interpolation, as the prevalent type of interpolation approach, was adopted in this paper, using linear polynomials to construct new data points [34]. Generally, the strategy for linear interpolation is to use a straight line to connect the known data points on either side of the unknown point and, thus, it is defined as the concatenation of linear interpolation between each pair of data points on a set of samples.

Data filtering. Due to the environmental noise and interference caused by the unconscious jittering of the human body, there are many undesirable components in the obtained signals that need to be dealt with [34]. This usually means removing some frequencies to suppress interfering signals and reduce the background noise. A low-pass filter is a type of electronic filter that attempts to pass low-frequency signals through the filter unchanged while reducing the amplitude of signals with a frequency above what is known as the cutoff frequency. A Butterworth low-pass filter with a cutoff frequency of 3 Hz is applied to denoise and smooth the raw signals.

Data scaling. The difference in the scale of each input variable increases the difficulty of the problem being modeled. If one of the features has a broad range of values, the objective functions of THE established model will be highly probably governed by the particular feature without normalization, suffering from poor performance during learning and sensitivity to input values and further resulting in a higher generalization error [35]. Therefore, the range of all data should be normalized so that each feature contributes approximately proportionately to the final result. Standardization makes the values of each feature in the data have zero means by subtracting from the mean in the numerator and unit variance, as shown in Equation (3):

$$X'_i = \frac{X_i - \mu}{\sigma} \ (i = 1, 2, 3 \cdots , n) \tag{3}$$

where the X'_i is the standardized data, n represents the number of data channels, and μ and σ are the mean and standard deviations of the i-th channel of the samples [35]. This method is widely used for normalization in many machine learning algorithms and is also adopted in this work to normalize the range of data we obtained.

Data segmentation. A sensor-based landmark recognition model is typically fed with a short sequence of continuously recorded sensor readings since only a single data point cannot reflect the characteristics of landmarks. The sequence consists of all the channels of selected sensors. To preserve the temporal relationship between the acquired data points with the aligned times, we partition the multivariate time-series sensor signals into sequences or segments leveraging the sliding operation, which consists of 128 samples (corresponding to 2.56 s for the sampling frequency at 50 Hz) [34,36]. It is noteworthy that the length of the window is picked empirically to achieve the segments for all considered landmarks, in which the features of the landmarks can be precisely captured to promote the landmark identification model training [32,37].

4.3. PDR-Based Trajectory Construction Model

For the quantitative evaluation of the virus quanta concentration, the precise spatial distance and temporal distance between two individuals should be efficiently estimated. To reach this objective, a variety of indoor positioning techniques have been proposed for various scenarios. The widely studied fingerprinting-based method relies on the latest fingerprint database that needs to be precisely updated in time. In addition to the time-consuming and labor-intensive collection and re-establishment, the instability of RSS due to environmental uncertainties poses another challenge to the accuracy [38]. Moreover, coverage and distribution are also not satisfied in countries with poor ICT infrastructure [39]. Therefore, the self-contained PDR algorithm without extra requirements and coverage limitations is employed in this work, and its accuracy is improved by the identified landmark.

4.3.1. PDR

Since PDR does not need additional equipment or a pre-survey, it has a wide range of potential applications for the indoor positioning of pedestrians. It relies on the inertial sensors extensively existing in mobile devices, e.g., smartphones, to acquire information about the user's movements, which are then combined with the user's previous location to estimate the present position and further achieve complete trajectory. The equation utilized for location estimation is as follows:

$$\begin{cases} x_t = x_{t-1} + SL_t sin\theta_t \\ y_t = y_{t-1} + SL_t cos\theta_t \end{cases} \tag{4}$$

where (x_t, y_t) is the pedestrian position at time t, SL_t is the step length, and θ_t details the heading direction of the pedestrian [40].

As mobile technology continues to evolve, a growing number of physical sensors are being installed in smartphones and, thus, various combinations of sensors can provide increasingly rich information, which makes PDR more feasible and accessible. A typical PDR consists of three main components: step detection, step-length estimation, and heading estimation [41].

Step detection. As the most popular method for accurate step detection, peak detection is employed in this paper, which relies on the repeating fluctuation patterns during human movement. Using the smartphone's accelerometer to determine whether the pedestrian is stationary, or walking is straightforward as it directly reflects the moving acceleration. The magnitude of acceleration on three dimensions (a_x, a_y, a_z) instead of the vertical part is employed as the input for peak findings to improve the accuracy, which can be expressed as:

$$a = \sqrt{a_x^2 + a_y^2 + a_z^2} \tag{5}$$

where a_x, a_y, a_z denote the three-axis accelerometer values in the smartphone [42]. A peak is detected when a is greater than the given threshold. To further enhance the performance, the low-pass filter is further applied to the magnitude to reduce the signal noise. Due to the acceleration jitter, the incumbently detected peak points need to be eliminated. Hence, an adaptive threshold technique of the maximum and minimum acceleration is adopted to fit different motion states with a time interval limitation between adjacent detected steps.

Stride length estimation. Various linear and nonlinear methods are proposed to estimate the step length, which varies from person to person because of different walking postures determined by various factors, including height, weight, and step frequency. Therefore, it is not easy to precisely construct the same step-length estimation model. Some researchers assume that the step length is a static value affected by the individual characteristics of different users. On the contrary, the empirical Weinberg model estimates the stride length according to the dynamic movement state, which is closer to reality [43]. The model is given by:

$$SL = k\sqrt[4]{a_{max} - a_{min}} \tag{6}$$

where k is the dynamic value concerned with the acceleration of each step and a_{max}, a_{min} are the maximum and minimum accelerations for each step [44].

Heading estimation. Heading information is a critical component for the entire PDR implementation, which seriously affects localization accuracy. To avoid the accumulative error in the direction estimation based on the gyroscope, and short-term direction disturbances based on the magnetometer, the combination of the gyroscope and magnetometer is typically adopted for heading estimation [42]. The current magnetometer heading signals, current gyroscope readings, and previously fused headings are weight-averaged to form the fused heading. The weighting factor is adaptive and is based on the magnetometer's stability as well as the correlation between the magnetometer and the gyroscope [44]. As they are already fused in the rotation vector achieved from the rotation sensor in the smartphone, the heading change can be calculated by a rotation matrix transformed from the rotation vector [45]. The rotation vector is defined as: $[x, y, z, w]$, and the matrix is defined as M, $M \in R^{3 \times 3}$. The heading direction on three dimensions can be evaluated by:

$$M = \begin{bmatrix} M_{11} & M_{12} & M_{13} \\ M_{21} & M_{22} & M_{23} \\ M_{31} & M_{32} & M_{33} \end{bmatrix} = \begin{bmatrix} 1 - 2y^2 - 2z^2 & 2xy - 2zw & 2xz + 2yw \\ 2xy + 2zw & 1 - 2x^2 - 2z^2 & 2yz - 2xw \\ 2xz - 2yw & 2yz + 2xw & 1 - 2x^2 - 2y^2 \end{bmatrix} \tag{7}$$

$$\theta = \begin{bmatrix} arctan2(M_{12}, M_{22}) \\ arcsin(-M_{32}) \\ arctan2(-M_{31}, M_{33}) \end{bmatrix} = \begin{bmatrix} arctan2(2xy - 2zw, 1 - 2x^2 - 2z^2) \\ arcsin(-2yz - 2xw) \\ arctan2(2yw - 2xz, 1 - 2x^2 - 2y^2) \end{bmatrix} \tag{8}$$

4.3.2. Landmark Identification Model

Although PDR methods can estimate the location and trajectory of pedestrians, low-cost inertial sensors built into smartphones provide poor-quality measurements, resulting in accuracy degradation. Moreover, the cumulative error, including the heading estimation caused by the gyroscope and step-length estimation error caused by an accelerometer, could be produced in the long-term positioning using PDR, increasing the challenge of precise localization collection. Therefore, it is necessary to prepare the reference points with the correct positions known during the movement to reduce the accumulated errors when the user passes. Spatial contexts, such as landmarks, can be properly chosen to calibrate the localization error based on the inherent spatial information without additional deployment costs. Landmark is defined as a spatial point with salient features and semantic characteristics from its near environment in indoor positioning systems, such as corners, stairs, and elevators [27]. These features can be observed for identification in one or a combination of different sensors as people pass through the landmark. The locations of these landmarks are presented by geographical coordinates or the relationships with other locations/areas, where people perform specific and predictable activities. Changes in motion are reflected in sensor readings, and different motions present different patterns. The specific activities that people perform when passing landmarks are also reflected in at least one sensor. Using the data of one sensor or the combination of data from multiple sensors, the changing pattern of a specific activity can be identified, and then the landmark can be recognized [46]. The identified landmark can be used as an anchor point to correct the path we obtained and improve the performance of the calculated trajectory.

Landmark identification involves classifying the sequences of various sensor data recorded at regular intervals by sensing devices, usually smartphones, into a well-defined landmark, which has been extensively regarded as a problem of multivariate time series classification. To address this issue, it is critical to extract and learn the features comprehensively to determine the relationship between sensing information and movement patterns. In recent years, numerous features have been attained in many studies on certain raw signal statistical aspects, such as variance, mean, entropy, kurtosis, correlation coefficients, or frequency domains via the integration of cross-formal codings, such as signals with Fourier transform and wavelet transform [47]. Moreover, the special thresholds of different features for various kinds of landmark recognition are specifically analyzed. For instance, the threshold of angular velocity produced by a gyroscope is usually used to detect the corner landmark, the acceleration changes can recognize the stairs. The combinations of different thresholds of various sensors forming the decision tree can detect the standing motion state to further distinguish common landmarks, such as corners, stairs, and elevators [48,49]. However, despite high accuracy, the calculation, extraction, and selection of features of different sensors for various landmarks are heuristic (with professional knowledge and expertise of the specific domain), time-consuming, and laborious [47].

To facilitate feature engineering and improve performance, artificial neural networks based on deep learning techniques have been employed to conduct activity identification without hand-crafted extraction. Deep learning techniques have been applied in many fields to solve practical problems with remarkable performance, such as image processing, speech recognition, and natural language processing, to solve practical problems [50,51]. Many kinds of deep neural networks have been introduced and investigated to handle landmark identification based on the complexity and unsureness of human movements. Additionally, CNN and LSTM are widely adopted with high accuracy rate activity recognition among the applied networks. CNN is commonly separated into numerous learning stages, each of which consists of a mix of convolutional operation and nonlinear processing units, as follows:

$$h^k = \sigma(\sum_{l \in L} g(x^l, w^k) + b^k) \tag{9}$$

where h^k reveals the latent representation of the k-th feature map of the current layer, σ is the activation function, g denotes the convolution operation, x^l indicates the l-th feature map of the group of the feature maps L achieved from the upper layer, w^k and b^k express the weights matrix and the bias of the k-th feature map of the current layer, receptively [52]. In our model, the rectified linear units (ReLU) were employed as the activation functions to subsequently conduct the non-linear transformation to obtain the feature maps, denoted by:

$$\sigma(x) = \max(0, \ x) \tag{10}$$

More importantly, the convolution operation in CNN can efficiently capture the local spatial correlation features by limiting the hidden unit's receptive field to be local [53]. CNN considers each frame of sensor data as independent and extracts the features for these isolated portions of data without considering the temporal contexts beyond the boundaries of the frame. Due to the continuity of sensor data flow produced by the user's behavior, local spatial correlations and temporally long-term connections are both important to identify the landmark [52]. LSTMs with learnable gates, which modulate the flow of information and control when to forget previous hidden states, as variants of vanilla recurrent neural networks (RNNs), allow the neural network to effectively extract the long-range dependencies of time-series sensor data [54]. The hidden state for the LSTM at time t is represented by:

$$h_t = \sigma(w_{i,h} \cdot x_t + w_{h,h} \cdot h_{t-1} + b) \tag{11}$$

where h_t and h_{t-1} are the hidden state at time t and $t-1$, respectively, σ is the activation function, $w_{i,h}$ and $w_{h,h}$ are the weight matrices between the parts, and b symbolizes the hidden bias vector. The standard LSTM cells barely extract the features from the past movements, ignoring the future part. To comprehensively capture the information for landmark identification, the Bi-LSTM is applied to access the context in both the forward and backward directions [55].

Therefore, both Bi-LSTM and CNN are involved in capturing the spatial and temporal features of signals for landmark identification. The architecture of the proposed landmark identification is shown in Figure 2. It performs the function of landmark recognition using the residual concatenation for classification, followed by Bi-LSTM and multi-head CNN. When preprocessed data segmentations of multiple sensors come, the inherent temporal relationship is extracted sequentially by two Bi-LSTM blocks that consist of a Bi-LSTM layer, a batch normalization (BN) layer, an activation layer, and a dropout layer. BN is a method used to improve training speed and accuracy with the mitigation of the internal covariate shift through normalization of the layer inputs by recentering and re-scaling [34]. Next, multi-head CNN blocks with varying kernels size are followed to learn the spatial features at various resolutions. Each convolutional block is made of four layers: a one-dimensional (1D) convolutional layer, a BN layer, an activation layer, and a dropout layer. To accommodate the three-dimensional input shape (samples, time steps, input channels) of the 1D convolutional layer, we retain the output of the hidden state in the Bi-LSTM layer. Then the acquired spatial and temporal features are combined, namely the concatenations of the outputs of the multi-head CNNs and Bi-LSTMs. To reduce the parameters and avoid overfitting, the global average pooling layer (GAP) with no parameter to optimize rather than the traditional fully connected layer is applied before combining the outputs [32]. Finally, the concatenated features are transmitted into a BN layer to re-normalize before being fed into a dense layer with a softmax classifier to generate the probability distribution over classes.

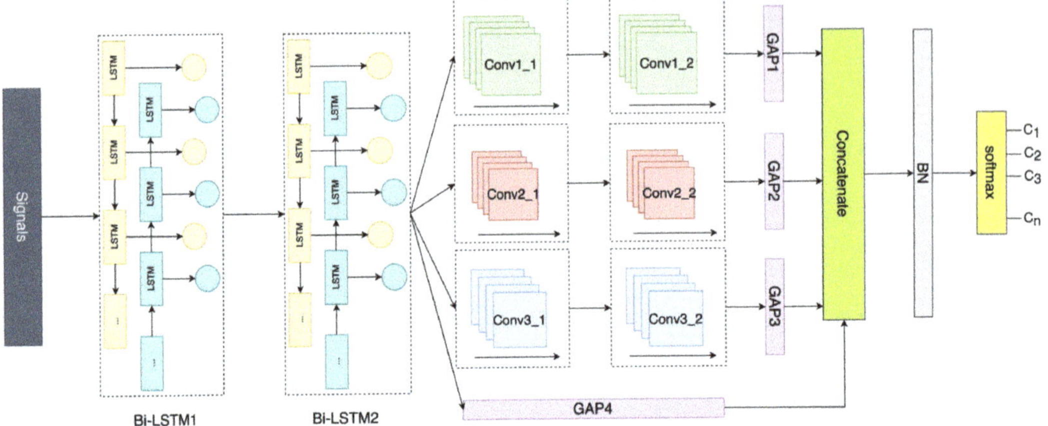

Figure 2. Architecture of the landmark identification model.

4.4. Contact Awareness with Trajectory

Exhalation and inhalation respiratory activities are constantly alternating (e.g., each breath consists of 2.5 s of continuous exhalation and 2.5 s of continuous inhalation), and droplets are continuously being released from the respiratory tract with a horizontal velocity during the process of exhalation with the same direction as the movement of the human. The particles exhaled at each moment will continue to move forward, starting from the user positions when they are expelled. The viral droplets exhaled from the infectious host are transported and dispersed into the ambient airflow before finally being inhaled by a susceptible person. Each exhalation lasts several seconds (e.g., 2.5 s), in which a long distance can be traveled for those who are in motion, and the initial position of droplets expelled cannot be accurately estimated in an indoor environment. Therefore, once complete, the exhalation period is divided into many short-term (e.g., 0.1 s) particle ejections. Because the interval is short, the continuous virus exhalation process can be converted into an instantaneous process, i.e., the virus is released instantly at the beginning of each interval. The virus-laden droplets expelled at different intervals maintain independent and identical motion patterns and the initial positions of the particles released in each interval can be regarded as the locations of the people at the initial moments. The virus-containing particles maintain a uniform motion of initially horizontal velocity (e.g., $3.9 \text{ m} \cdot \text{s}^{-1}$) in the first second and then instantaneously will mix in the overall considered space. Meanwhile, the droplets are evenly distributed within the moved space. In the first movement phase of the exhaled droplets in each interval, the virus moves in the same direction as the people travel, which is called forward transmission. As for the backward transmission, in general, the initial velocity of the virus is faster than the speed of movement and the speed of airflow, so in the first phase, very few virus particles move in the opposite direction.

The movements of all viral-loaded droplets exhaled by infectious people at different locations will meet somewhere at some time and contribute to the calculation of concentration. To precisely present the virus quanta concentrations, the transmissions of all virus particles per exhalation sources from different origins and in different states are assumed to follow the same patterns, in which the particles keep constant initial velocity in the first second and then will instantly mix in the overall space. The time it takes for the virus to move to the current point and the contribution to the virus quanta in the present are estimated with the help of spatial distance and velocity. Thus, the quanta concentration in an indoor area at time t, $q\left(t, ER_q\right)$ is measured by:

$$q\left(t, ER_q\right) = \sum_{i}^{i=N_v} \left(\frac{ER_q^i}{RR_{iv} \cdot V\left(t^i\right)} \cdot \left(1 + e^{-RR_{iv} \cdot t^i}\right) + \left(q_0 \cdot \frac{e^{-RR_{iv} \cdot T}}{V} + q_0^i \cdot \frac{e^{-RR_{iv} \cdot t^i}}{V}\right) \right) \tag{12}$$

where RR_{iv} is the virus removal rate of the target space, N_v represents the virus generated in different places at different moments, ER_q^i is the of the quanta emission rate of the infector at which the virus (i-th) is expelled, T is the time difference from the start of the experiment to present, t^i is the time difference between the current time and the originating time of the virus (i-th), $V\left(t^i\right)$ is the volume of the space that the i-th virus had passed since it was expelled to the present, q_0 is the environmental virus quanta number, q_0^i is the virus exhaled by the infector that has evenly spread to the overall investigated space with the volume of V. Exhaled virus particles eventually become the environmentally well-mixed virus quanta, while different initial states induce different decays.

4.5. Spatiotemporal Contact Awareness

The algorithm of the proposed iSTCA with the landmark-calibrated PDR technology based on a smartphone is detailed in Algorithm 1. The detailed procedures are as follows,

Firstly, the raw signals are acquired via the developed collection application and preprocessed to create the dataset for the landmark identification model training by utilizing the data preprocessing method introduced in Section 4.2.

Algorithm 1 Indoor spatiotemporal contact awareness algorithm

Input:	raw sensor signals of infector's smartphone,
	trained landmark identification model $\mathcal{M}_{lm}$
	target time $\mathcal{T}$
	target position $\mathcal{P}$
	Infectivity model $\mathcal{M}_I$
Output:	quantitative virus quanta concentration in $\mathcal{P}$ at $\mathcal{T}$.

1.   time interval initialized to τ,
2.   quanta concentration in $\mathcal{P}$ at $\mathcal{T}$ ($q_{\mathcal{P}}^{\mathcal{T}}$) initialized to 0,
3.   construct the processed signals $\mathcal{D}$,
4.   achieve the trajectories $\mathcal{S}$ from $\mathcal{D}$, landmark-calibrated via $\mathcal{M}_{lm}$,
5.   establish the initial state set $\left\{Q_0^i\right\}$ of all viruses expelled at different intervals, where $Q_0^i \leftarrow \left(t_0^i, V_0^i, q_0^i\right)$, i represent the number of time intervals,
6.   for each Q_0^i do:
7.   for j in $0, 1, 2, \ldots \left|\frac{T - t_0^i}{\tau}\right|$ do:
8.   achieve the $Q_j^i \leftarrow \left(t_j^i, V_j^i, q_j^i\right)$ based on movement pattern ($\mathcal{M}_I$) itself
9.   if $\mathcal{P}$ in V_j^i then:
10.   update $q_{\mathcal{P}}^{\mathcal{T}}$, $q_{\mathcal{P}}^{\mathcal{T}} \leftarrow q_{\mathcal{P}}^{\mathcal{T}} + q_j^i$,
11.   end if
12.   end for
13.   end for
14.   return $q_{\mathcal{P}}^{\mathcal{T}}$

Secondly, the landmark recognition model designed in Section 4.3.2 would be trained and stored based on the dataset generated in the first step to further the PDR algorithm.

Thirdly, the target trajectory S is constructed by performing the landmark-calibrated PDR technique, including step detection, stride length estimation, heading determination, and landmark identification.

Fourthly, we obtain the initial state set $\left\{Q_0^i\right\}$ of the expelled particles in the i-th ($i = 1, 2, 3\ldots$) short-term period with the help of the calculated human movement trajectory S and the preset viral particle ejection interval τ. Q_0^i defines the state of all i-th emitted particles in interval τ and consists of three parts t, V, q, where t represents the elapsed time after being exhaled, V represents the spread coverage of droplets due to airborne dispersion, and q represents the quanta concentration.

Fifthly, the state set $\left\{Q_j^i\right\}$ at the j-th interval for any Q_0^i after being expelled is acquired by employing the defined movement pattern of the considered particles.

Finally, the virus quanta concentration $q_{\mathcal{P}}^{\mathcal{T}}$ in the target position $\mathcal{P}$ at the target time $\mathcal{T}$ is reached. The virus quanta concentration presented within $\mathcal{P}$ at $\mathcal{T}$ by particles expelled in the various intervals is summed to estimate $q_{\mathcal{P}}^{\mathcal{T}}$. Moreover, the virus quanta concentrations presented in different locations at various times can be further evaluated.

5. Experiments

In this section, we evaluate the performances of the proposed methods through experiments with the dataset we collected in a university building. We introduce the experimental scenario and data collection in Section 5.1 and the results are presented in Section 5.2; we analyzed the performances related to the landmark identification, PDR, and virus quanta concentration.

5.1. Experimental Scenario and Data Acquisition

We collected our experimental data on the fourth floor of the Center-Zone-1 building of Kyushu University's Ito campus. We assume that there is no exchange of virus particles with the room space. Figure 3 shows the floor plan of the experimental area. Based on the practical scenario, the Manhattan distance is applied to measure the virus movement. Since the width (measured as 2 m) and height (assumed to be 3 m based on the practical scenario) of the hallway are generally the same, the volume of the virus coverage can be determined by the virus movement distance for the calculation of the virus quanta concentration. When the virus encounters a corner, its direction changes, leading to a shift in the virus quanta concentration to varying degrees. For a corner with two branches, the concentration is assumed to decrease by half due to the inertia effect while these viral particles continue the forward transmission. If it is a corner with three or more branches, we assume that the virus quanta would be distributed evenly in all other directions.

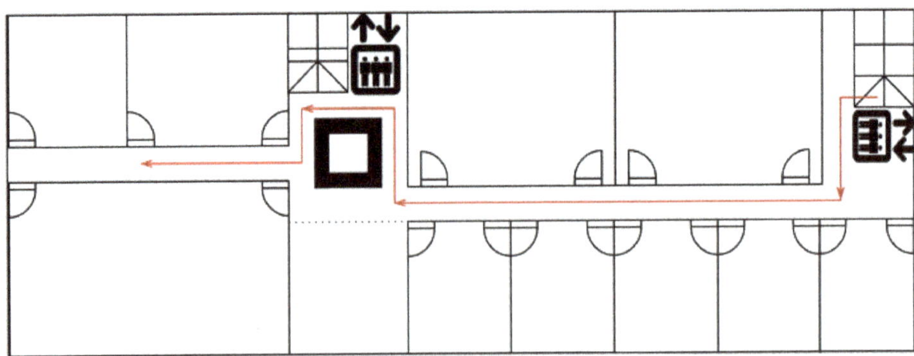

Figure 3. The floor plan of our experiment.

In data collection, five recruited participants held Pixel 4a smartphones with the required sensors integrated (e.g., accelerometer, gyroscope, and rotation sensor) and an Android application installed. The application can periodically read and store the readings of 11 channels (3 for the accelerometer, 3 for the gyroscope, and 5 for the rotation sensor) as the user walks along the prescribed routes at a normal speed in the experimental area. Moreover, participants are required to hold their smartphones at chest level, which is a reasonable position where participants can record extra information to facilitate data processing. Indeed, it is recommended that they record the timestamp and the identification of the passing landmark to construct the dataset for the landmark identification model training.

5.2. Analysis and Discussion

5.2.1. Landmark Identification

The proposed landmark recognition model was extensively evaluated by a series of experiments and implemented using the Keras framework with the TensorFlow backend to minimize the cross-entropy loss. The model was performed using the collected data with 3863 samples. The dataset was divided into training (70%) and testing (30%) sets, randomly, without overlapping. There were a total of 11 landmarks, including 7 corners, 2 stairs, and 2 elevators.

Table 1 details the network configuration considered in our study. Since there were many combinations of parameters, to reduce the selection space, we let all of the Bi-LSTM neurons share the same value, with the 1D convolution filter and kernel sizes accessing the same setting, respectively. To achieve stable performances of different model settings, a grid search with the 10-fold cross-validation method was adopted. It worked through all of the combinations of parameters to find the best settings. It should be noted that the following study uses the bold value for each parameter when it is not otherwise specified.

Table 1. Landmark identification neural network configuration.

Layers	Parameter	Value
Input	shape	(None, 128, 11)
Bi-LSTM1	neurons	32, 64, **128**, 256
Bi-LSTM2	neurons	32, 64, **128**, 256
	kernel size	3, **5**, 9, 11
Conv1_1	filters	32, 64, **128**
	stride	1
	kernel size	3, **5**, 9, 11
Conv2_1	filters	32, 64, **128**
	stride	1
	kernel size	3, 5, **9**, 11
Conv3_1	filters	32, 64, **128**
	stride	1
Conv1_2,	kernel size	3, **5**, 9, 11
Conv2_2,	filters	32, **64**, 128
Conv3_2	stride	1
Dropout	drop rate	0.2, **0.3**, 0.5, 0.8

Moreover, the model configuration (the Adam optimization algorithm) was selected as the optimizer during the gradient descent. Other training hyperparameters were also evaluated and their recognition accuracies are presented in Figure 4. More specifically, the experiment was conducted with the learning rates of 0.00001, 0.00002, 0.00005, 0.0001, 0.0002, 0.0005, 0.001, 0.002, 0.005, 0.01, 0.02, and 0.05, as presented in Figure 4a. The mini-batch size was tested with 16, 20, 32, 50, 64, 100, 128, 150, 200, and 256, as shown in Figure 4b. The model configured as Table 1 achieved the highest identification accuracy of 98.4% when the learning rate was 0.0002 and the mini-batch size was 256. Additionally, early stopping criteria and a learning rate reduction strategy were applied during the model training process in order to reduce the issue of over-fitting and to improve the model performance. The learning rate decreased with a factor of 0.5 when the accuracy was not improved for 10 epochs and the training ended if the accuracy without enhancement on the validation was set after 15 iterations. Detailed training hyperparameter settings are revealed in Table 2.

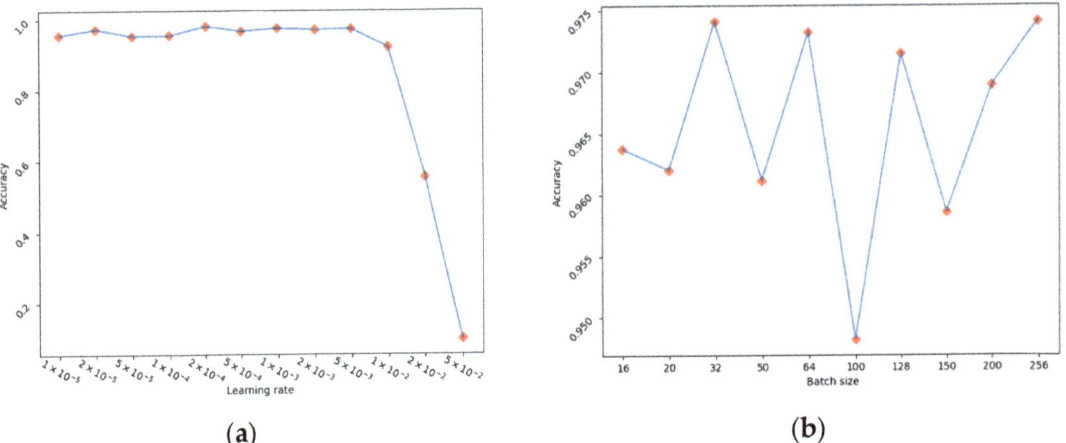

Figure 4. Model accuracy on various learning rates (**a**) and batch sizes (**b**), shown as the red square.

Table 2. Training hyperparameters.

Hyperparameters	Value
Optimizer	Adam
Activation function	ReLU
Batch size	256
Learning rate	0.0002
Epochs	800

Following the considered model configuration and optimal training hyperparameters, the accuracy curve and loss curve of the training and testing processes are illustrated in Figure 5a,b. The recognition results on 11 selected landmarks of the experiment-conducted floor are presented by the confusion matrix in Figure 6.

Moreover, to evaluate the proposed network more comprehensively, further comparisons were conducted on other deep neural networks (CNN, LSTM, and LSTM-CNN without residual connections) with the same depth and training hyperparameters as shown in Table 2. Table 3 presents the obtained experimental results of accuracy, precision, recall, and F1-score using different networks. It can be seen that the proposed Bi-LSTM-CNN network achieved the highest performance in all four metrics thanks to the elaborately extracted spatial and temporal features. Therefore, the effectiveness of the proposed Bi-LSTM-CNN classification model for the landmark identification task is demonstrated with the experimental evaluation.

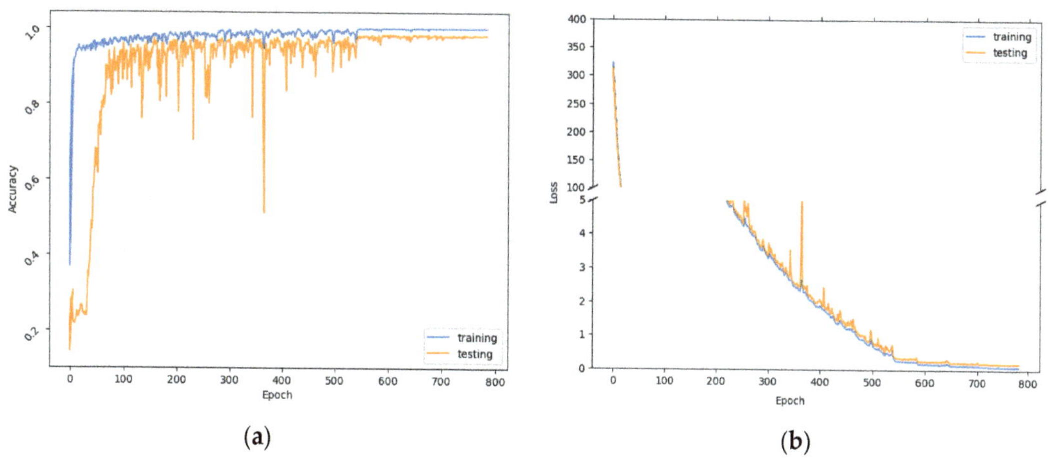

Figure 5. Accuracy (**a**) and loss (**b**) curves of the model on the selected parameters.

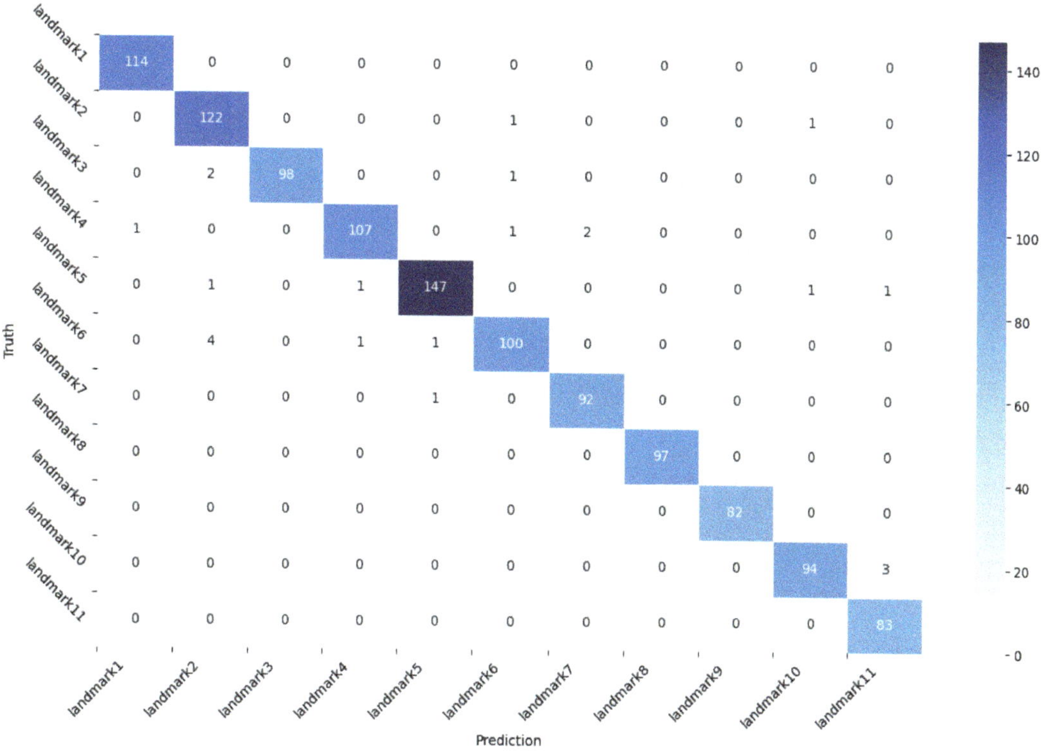

Figure 6. Confusion matrix for landmark identification.

Table 3. Landmark identification performances of different models in the collected dataset.

Method	Accuracy	Precision	Recall	F1-Score
CNN	0.9327	0.9404	0.9116	0.9258
LSTM	0.9637	0.9636	0.9600	0.9618
LSTM-CNN	0.9706	0.9750	0.9705	0.9727
Bi-LSTM-CNN (our)	0.9836	0.9849	0.9850	0.9849

5.2.2. Trajectory Tracing

The path shown as the red line in Figure 3 is designed to evaluate the performance of PDR with landmark calibration and the results are presented in Figure 7a. To quantitatively evaluate the positioning accuracy, we show the accumulative error distribution in Figure 7b. It can be seen from the left figure that the original PDR has an increasing error due to the initial wrong direction, although the information of many short segments can be described relatively accurately. Due to the significant error in the heading estimation without the landmark correction, the cumulative error distribution is not displayed in the right picture. The performance of PDR with the landmark calibration is well examined, nearly 80% of the positioning errors are less than 0.4 m, and the error probability within 0.7 m is higher than 90%. From the conducted experiments, the performance of the PDR-fused landmark calibration was evaluated with a lower positioning error, as compared to the PDR without calibration.

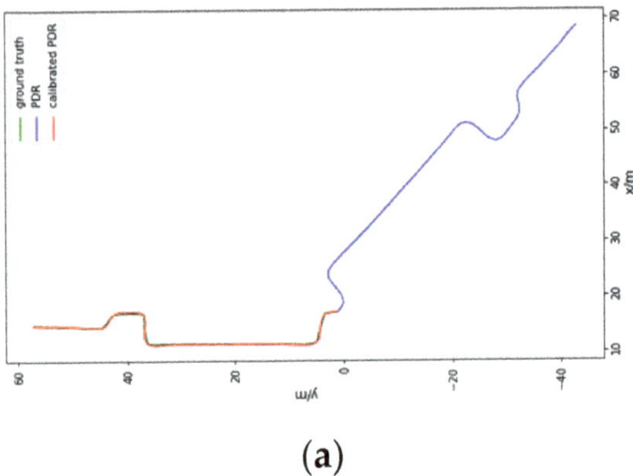

(a)

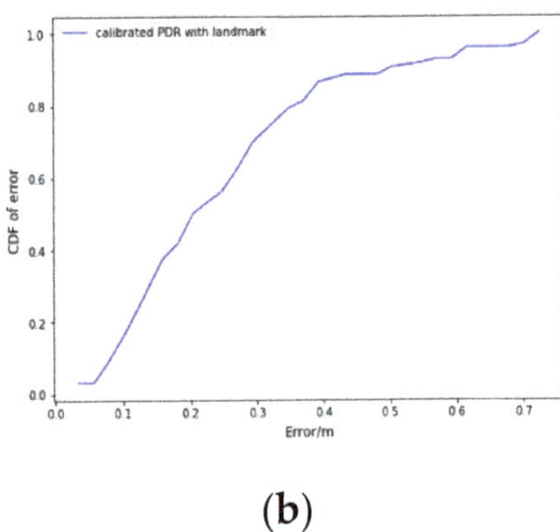

(b)

Figure 7. The performance (**a**) and the accumulative error distribution (**b**) of the proposed landmark-calibrated PDR.

5.2.3. Virus Quanta Concentration

As mentioned above, we regard all virus particles exhaled every 0.1 s during exhalation as virus instances. There will be many virus instances expelled during the entire movement of an infector. During the transmission of each instance, a uniform motion with a velocity of 3.9 ($m \cdot s^{-1}$) is maintained in the first second after exhalation, and the virus quanta are evenly distributed in the space that is passed by. The initial number of quanta ($q_0 = 0$), the virus quanta emission rate (ER_q), and the removal rate of infectious viral-laden particles (RR_{iv}) are 142 and 1.37, respectively, and remain the same within the experiment [29,30].

The virus-laden particles released in each interval follow the same moving pattern, leading to the same trend in the change of the quanta concentration. We chose the instantaneous concentration at the end of each shorter interval with a length of 0.1 s to represent the concentrations at all times during the entire interval, as presented in Figure 8. As can be seen in Figure 8, the overall change in concentration presents an exponentially decreasing trend, from above 88 in the first interval (0~0.1 s) to close to 0 one second later. The sharp decrease one second later is because of an instantaneous expansion of the viral aerosol coverage to the entire considered space.

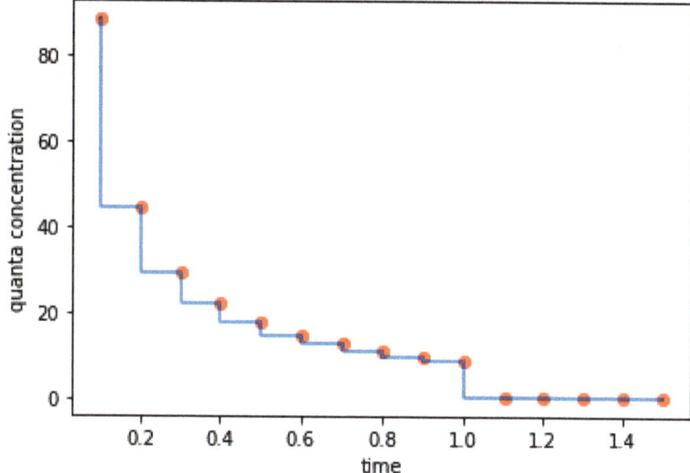

Figure 8. Quanta concentration of viral particles changes over time (first 1.5 s) after being released. Red points represent the instantaneous concentration at the end of each shorter interval.

The time when people started moving can be seen at time 0 of the experiment. Figure 9 presents the virus concentration in the current environment at the time of 0 s, 0.5 s, and 5 s from left to the right (using lines to represent the considered corridor spaces). Among them, at t = 0 s, only the virus concentration near the point start can be seen to exceed 80, while most of the other parts are not covered by viral particles. At t = 0.5 s, under the combined movements of virus droplets and humans, the relatively high quanta concentrations covered more. In addition, after another 0.5 s, the particles initially expelled at t = 0 s will spread to the overall space. At t = 5 s, the area with higher quanta concentration gradually moves forward with the movement of people. Moreover, due to the accumulated particles that diffuse into the entire environment, the quanta concentration in the overall space is increased, gradually reaching a non-negligible level compared with the concentration of the newly expelled virus instance.

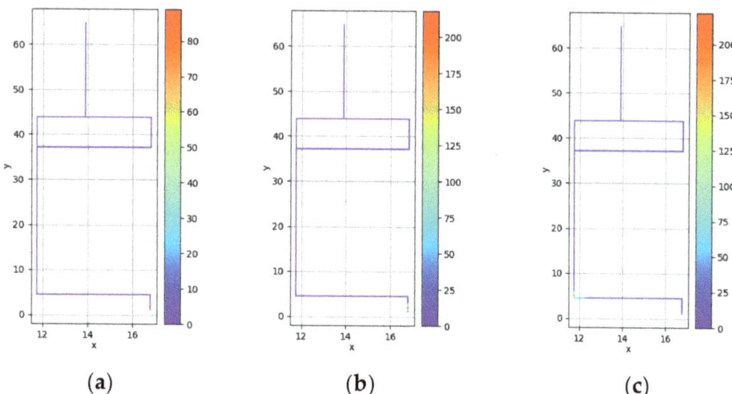

Figure 9. Indoor virus quanta concentrations at 0 s (**a**), 0.5 s (**b**), and 5 s (**c**), respectively, from the start of the movement.

6. Limitations

Although the proposed iSTCA system realizes quantitative representation for exposed virus concentrations with the help of the landmark-calibrated PDR technique, there are some challenges that need to be overcome. First of all, there are some strict restrictions in the data acquisition process. The participants are required to hold the smartphone, specifically the Pixel 4a, at chest level. As a result, except for the diversity of users considered, other factors that affect the motion sensor readings are not seriously taken into account, such as the mobile device heterogeneity (e.g., different types or various vendors) and the device's status variation (e.g., putting in a pocket or handbag). In addition, a large amount of power of the smartphone is consumed during the indoor positioning process, resulting in the smartphone being overheated.

7. Conclusions and Future Work

Technology-assisted virus exposure tracking approaches are increasingly being adopted to mitigate and tame the epidemic. In view of the complexity of quantifying virus exposure due to human movement and airborne dispersion of virus particles, we propose iSTCA, a self-containing contact awareness approach that exploits PDR-based techniques. Quantitative information support directly concerned with risk assessment is provided for self-protection and epidemic control. More precisely, to reduce and calibrate the accumulative errors of trajectories based on landmarks, we apply Bi-LSTM and multi-head CNN with residual concatenation to long-term dependency in forward and backward directions and extract local correlations at various resolutions for landmark identification. The proposed method exploits the trajectories of people with viral-laden droplets exhaled and the transmission and attenuation of viruses in the air to quantify the virus quanta concentration in an indoor environment via spatiotemporal analytics for prevention and sanitization. In future work, we will continue studying the landmark identification model with different devices and various attitudes of the device and conduct further research on the exploration of other advanced deep neural networks and fusion algorithms. We will consider employing wearable devices, such as smartwatches and smart bands, to replace mobile phones for power saving in indoor positioning. Moreover, we plan to apply the proposed techniques for the development of services in developing communities without reliable digital infrastructure.

Author Contributions: Conceptualization, L.G. and S.K.; methodology, L.G. and S.K.; software, L.G.; validation, L.G.; formal analysis, L.G.; investigation, L.G.; resources, L.G. and S.K.; data curation, L.G.; writing—original draft preparation, L.G.; writing—review and editing, S.K.; visualization, L.G.; supervision, S.K.; project administration, S.K.; funding acquisition, S.K. All authors have read and agreed to the published version of the manuscript.

Funding: This work was funded by Japan Society for the Promotion of Science (JSPS), Grants-in-Aid for Scientific Research (KAKENHI), Japan: Grant Number JP20H00622 and supported in part by China Scholarship Council (CSC), China: Grant Number 202008050086.

Institutional Review Board Statement: Not applicable.

Informed Consent Statement: Informed consent was obtained from all subjects involved in the study.

Data Availability Statement: The data presented in this study are available upon request from the corresponding author. The data are not publicly available due to privacy reasons.

Conflicts of Interest: The authors declare no conflict of interest.

References

1. Shah, Y.; Kurelek, J.W.; Peterson, S.D.; Yarusevych, S. Experimental Investigation of Indoor Aerosol Dispersion and Accumulation in the Context of COVID-19: Effects of Masks and Ventilation. *Phys. Fluids* **2021**, *33*, 073315. [CrossRef] [PubMed]
2. Marzoli, F.; Bortolami, A.; Pezzuto, A.; Mazzetto, E.; Piro, R.; Terregino, C.; Bonfante, F.; Belluco, S. A Systematic Review of Human Coronaviruses Survival on Environmental Surfaces. *Sci. Total Environ.* **2021**, *778*, 146191. [CrossRef] [PubMed]
3. Wang, C.C.; Prather, K.A.; Sznitman, J.; Jimenez, J.L.; Lakdawala, S.S.; Tufekci, Z.; Marr, L.C. Airborne Transmission of Respiratory Viruses. *Science* **2021**, *373*, eabd9149. [CrossRef] [PubMed]
4. Li, Z.; Wang, H.; Zhang, X.; Wu, T.; Yang, X. Effects of Space Sizes on the Dispersion of Cough-Generated Droplets from a Walking Person. *Phys. Fluids* **2020**, *32*, 121705. [CrossRef]
5. Ferretti, L.; Wymant, C.; Kendall, M.; Zhao, L.; Nurtay, A.; Abeler-Dörner, L.; Parker, M.; Bonsall, D.; Fraser, C. Quantifying SARS-CoV-2 Transmission Suggests Epidemic Control with Digital Contact Tracing. *Science* **2020**, *368*, eabb6936. [CrossRef]
6. Alo, U.R.; Nkwo, F.O.; Nweke, H.F.; Achi, I.I.; Okemiri, H.A. Non-Pharmaceutical Interventions against COVID-19 Pandemic: Review of Contact Tracing and Social Distancing Technologies, Protocols, Apps, Security and Open Research Directions. *Sensors* **2021**, *22*, 280. [CrossRef]
7. Nguyen, T.D.; Miettinen, M.; Dmitrienko, A.; Sadeghi, A.-R.; Visconti, I. Digital Contact Tracing Solutions: Promises, Pitfalls and Challenges. *arXiv* **2022**, arXiv:2202.06698. [CrossRef]
8. Li, G.; Hu, S.; Zhong, S.; Tsui, W.L.; Chan, S.-H.G. VContact: Private WiFi-Based IoT Contact Tracing with Virus Lifespan. *IEEE Internet Things J.* **2022**, *9*, 3465–3480. [CrossRef]
9. Faust, J.S.; Du, C.; Liang, C.; Mayes, K.D.; Renton, B.; Panthagani, K.; Krumholz, H.M. Excess Mortality in Massachusetts During the Delta and Omicron Waves of COVID-19. *JAMA* **2022**, *328*, 74–76. [CrossRef]
10. Bazant, M.Z.; Bush, J.W.M. A Guideline to Limit Indoor Airborne Transmission of COVID-19. *Proc. Natl. Acad. Sci. USA* **2021**, *118*, e2018995118. [CrossRef]
11. Castelluccia, C.; Bielova, N.; Boutet, A.; Cunche, M.; Lauradoux, C.; Métayer, D.L.; Roca, V. ROBERT: ROBust and Privacy-PresERving Proximity Tracing. Available online: https://hal.inria.fr/hal-02611265/document (accessed on 14 November 2021).
12. Brüssow, H. COVID-19: Omicron—The Latest, the Least Virulent, but Probably Not the Last Variant of Concern of SARS-CoV-2. *Microb. Biotechnol.* **2022**, *15*, 1927–1939. [CrossRef] [PubMed]
13. Wang, J.; Dalla Barba, F.; Roccon, A.; Sardina, G.; Soldati, A.; Picano, F. Modelling the Direct Virus Exposure Risk Associated with Respiratory Events. *J. R. Soc. Interface* **2022**, *19*, 20210819. [CrossRef] [PubMed]
14. Ooi, C.C.; Suwardi, A.; Ou Yang, Z.L.; Xu, G.; Tan, C.K.I.; Daniel, D.; Li, H.; Ge, Z.; Leong, F.Y.; Marimuthu, K.; et al. Risk Assessment of Airborne COVID-19 Exposure in Social Settings. *Phys. Fluids* **2021**, *33*, 087118. [CrossRef] [PubMed]
15. Shahroz, M.; Ahmad, F.; Younis, M.S.; Ahmad, N.; Kamel Boulos, M.N.; Vinuesa, R.; Qadir, J. COVID-19 Digital Contact Tracing Applications and Techniques: A Review Post Initial Deployments. *Transp. Eng.* **2021**, *5*, 100072. [CrossRef]
16. Bay, J.; Kek, J.; Tan, A.; Hau, C.S.; Yongquan, L.; Tan, J.; Quy, T.A. *BlueTrace: A Privacy-Preserving Protocol for Community-Driven Contact Tracing across Borders*; Tech. Rep.; Government Technology Agency: Singapore, 2020.
17. Leith, D.J.; Farrell, S. GAEN Due Diligence: Verifying the Google/Apple COVID Exposure Notification API. In Proceedings of the CoronaDef21, NDSS '21, San Diego, CA, USA, 23–26 February 2020.
18. Leith, D.J.; Farrell, S. Measurement-Based Evaluation of Google/Apple Exposure Notification API for Proximity Detection in a Light-Rail Tram. *PLoS ONE* **2020**, *15*, e0239943. [CrossRef] [PubMed]
19. Leith, D.J.; Farrell, S. Measurement-Based Evaluation of Google/Apple Exposure Notification API for Proximity Detection in a Commuter Bus. *PLoS ONE* **2021**, *16*, e0250826. [CrossRef] [PubMed]
20. Istomin, T.; Leoni, E.; Molteni, D.; Murphy, A.L.; Picco, G.P.; Griva, M. Janus: Dual-Radio Accurate and Energy-Efficient Proximity Detection. *Proc. ACM Interact. Mob. Wearable Ubiquitous Technol.* **2022**, *5*, 1–33. [CrossRef]
21. Biri, A.; Jackson, N.; Thiele, L.; Pannuto, P.; Dutta, P. SociTrack: Infrastructure-Free Interaction Tracking through Mobile Sensor Networks. In Proceedings of the 26th Annual International Conference on Mobile Computing and Networking, London, UK, 21–25 September 2020; pp. 1–14.
22. Salathé, M.; Kazandjieva, M.; Lee, J.W.; Levis, P.; Feldman, M.W.; Jones, J.H. A High-Resolution Human Contact Network for Infectious Disease Transmission. *Proc. Natl. Acad. Sci. USA* **2010**, *107*, 22020–22025. [CrossRef]
23. Isella, L.; Romano, M.; Barrat, A.; Cattuto, C.; Colizza, V.; den Broeck, W.V.; Gesualdo, F.; Pandolfi, E.; Ravà, L.; Rizzo, C.; et al. Close Encounters in a Pediatric Ward: Measuring Face-to-Face Proximity and Mixing Patterns with Wearable Sensors. *PLoS ONE* **2011**, *6*, e17144. [CrossRef]
24. Trivedi, A.; Zakaria, C.; Balan, R.; Becker, A.; Corey, G.; Shenoy, P. WiFiTrace: Network-Based Contact Tracing for Infectious Diseases Using Passive WiFi Sensing. *Proc. ACM Interact. Mob. Wearable Ubiquitous Technol.* **2021**, *5*, 1–26. [CrossRef]
25. Swain, V.D.; Kwon, H.; Sargolzaei, S.; Saket, B.; Morshed, M.B.; Tran, K.; Patel, D.; Tian, Y.; Philipose, J.; Cui, Y.; et al. Leveraging WiFi Network Logs to Infer Student Collocation and Its Relationship with Academic Performance. *arXiv* **2020**, arXiv:2005.11228.
26. Tu, P.; Li, J.; Wang, H.; Wang, K.; Yuan, Y. Epidemic Contact Tracing with Campus WiFi Network and Smartphone-Based Pedestrian Dead Reckoning. *IEEE Sens. J.* **2021**, *21*, 19255–19267. [CrossRef]

27. Gao, L.; Konomi, S. Mapless Indoor Navigation Based on Landmarks. In *Proceedings of the Distributed, Ambient and Pervasive Interactions. Smart Living, Learning, Well-Being and Health, Art and Creativity: 10th International Conference, DAPI 2022, Held as Part of the 24th HCI International Conference, HCII 2022, Virtual Event, 26 June–1 July 2022, Proceedings, Part II*; Springer: Berlin/Heidelberg, Germany, 2022; pp. 53–68.

28. Kindt, P.H.; Chakraborty, T.; Chakraborty, S. How Reliable Is Smartphone-Based Electronic Contact Tracing for COVID-19? *Commun. ACM* **2021**, *65*, 56–67. [CrossRef]

29. Shen, J.; Kong, M.; Dong, B.; Birnkrant, M.J.; Zhang, J. Airborne Transmission of SARS-CoV-2 in Indoor Environments: A Comprehensive Review. *Sci. Technol. Built Environ.* **2021**, *27*, 1331–1367. [CrossRef]

30. Buonanno, G.; Stabile, L.; Morawska, L. Estimation of Airborne Viral Emission: Quanta Emission Rate of SARS-CoV-2 for Infection Risk Assessment. *Environ. Int* **2020**, *141*, 105794. [CrossRef]

31. van Doremalen, N.; Bushmaker, T.; Morris, D.H.; Holbrook, M.G.; Gamble, A.; Williamson, B.N.; Tamin, A.; Harcourt, J.L.; Thornburg, N.J.; Gerber, S.I.; et al. Aerosol and Surface Stability of SARS-CoV-2 as Compared with SARS-CoV-1. *N. Engl. J. Med.* **2020**, *382*, 1564–1567. [CrossRef]

32. Tong, L.; Ma, H.; Lin, Q.; He, J.; Peng, L. A Novel Deep Learning Bi-GRU-I Model for Real-Time Human Activity Recognition Using Inertial Sensors. *IEEE Sens. J.* **2022**, *22*, 6164–6174. [CrossRef]

33. Khatun, M.A.; Yousuf, M.A.; Ahmed, S.; Uddin, M.Z.; Alyami, S.A.; Al-Ashhab, S.; Akhdar, H.F.; Khan, A.; Azad, A.; Moni, M.A. Deep CNN-LSTM with Self-Attention Model for Human Activity Recognition Using Wearable Sensor. *IEEE J. Transl. Eng. Health Med.* **2022**, *10*, 1–16. [CrossRef]

34. Xia, K.; Huang, J.; Wang, H. LSTM-CNN Architecture for Human Activity Recognition. *IEEE Access* **2020**, *8*, 56855–56866. [CrossRef]

35. Demrozi, F.; Turetta, C.; Pravadelli, G. B-HAR: An Open-Source Baseline Framework for in Depth Study of Human Activity Recognition Datasets and Workflows. *arXiv* **2021**, arXiv:2101.10870.

36. Gu, F.; Khoshelham, K.; Valaee, S.; Shang, J.; Zhang, R. Locomotion Activity Recognition Using Stacked Denoising Autoencoders. *IEEE Internet Things J.* **2018**, *5*, 2085–2093. [CrossRef]

37. Zhao, Y.; Yang, R.; Chevalier, G.; Xu, X.; Zhang, Z. Deep Residual Bidir-LSTM for Human Activity Recognition Using Wearable Sensors. *Math. Probl. Eng.* **2018**, *2018*, 1–13. [CrossRef]

38. Subedi, S.; Pyun, J.-Y. A Survey of Smartphone-Based Indoor Positioning System Using RF-Based Wireless Technologies. *Sensors* **2020**, *20*, 7230. [CrossRef] [PubMed]

39. Konomi, S.; Gao, L.; Mushi, D. An Intelligent Platform for Offline Learners Based on Model-Driven Crowdsensing Over Intermittent Networks. In *Proceedings of the Cross-Cultural Design. Applications in Health, Learning, Communication, and Creativity: 12th International Conference, CCD 2020, Held as Part of the 22nd HCI International Conference, HCII 2020, Copenhagen, Denmark, 19–24 July 2020, Proceedings, Part II*; Springer: Berlin/Heidelberg, Germany, 2020; pp. 300–314.

40. Li, W.; Chen, R.; Yu, Y.; Wu, Y.; Zhou, H. Pedestrian Dead Reckoning with Novel Heading Estimation under Magnetic Interference and Multiple Smartphone Postures. *Measurement* **2021**, *182*, 109610. [CrossRef]

41. Liu, T.; Zhang, X.; Li, Q.; Fang, Z. Modeling of Structure Landmark for Indoor Pedestrian Localization. *IEEE Access* **2019**, *7*, 15654–15668. [CrossRef]

42. Yao, H.; Shu, H.; Sun, H.; Mousa, B.G.; Jiao, Z.; Suo, Y. An Integrity Monitoring Algorithm for WiFi/PDR/Smartphone-Integrated Indoor Positioning System Based on Unscented Kalman Filter. *EURASIP J. Wirel. Commun. Netw.* **2020**, *2020*, 246. [CrossRef]

43. Weinberg, H. Using the ADXL202 in Pedometer and Personal Navigation Applications. *Analog Devices 602 Appl. Note* **2002**, *2*, 1–6.

44. De Cock, C.; Joseph, W.; Martens, L.; Trogh, J.; Plets, D. Multi-Floor Indoor Pedestrian Dead Reckoning with a Backtracking Particle Filter and Viterbi-Based Floor Number Detection. *Sensors* **2021**, *21*, 4565. [CrossRef]

45. Yoon, J.; Kim, S. Practical and Accurate Indoor Localization System Using Deep Learning. *Sensors* **2022**, *22*, 6764. [CrossRef]

46. Gu, F.; Valaee, S.; Khoshelham, K.; Shang, J.; Zhang, R. Landmark Graph-Based Indoor Localization. *IEEE Internet Things J.* **2020**, *7*, 8343–8355. [CrossRef]

47. Nafea, O.; Abdul, W.; Muhammad, G.; Alsulaiman, M. Sensor-Based Human Activity Recognition with Spatio-Temporal Deep Learning. *Sensors* **2021**, *21*, 2141. [CrossRef] [PubMed]

48. Zhou, B.; Li, Q.; Mao, Q.; Tu, W.; Zhang, X.; Chen, L. ALIMC: Activity Landmark-Based Indoor Mapping via Crowdsourcing. *IEEE Trans. Intell. Transp. Syst.* **2015**, *16*, 2774–2785. [CrossRef]

49. Wang, X.; Jiang, M.; Guo, Z.; Hu, N.; Sun, Z.; Liu, J. An Indoor Positioning Method for Smartphones Using Landmarks and PDR. *Sensors* **2016**, *16*, 2135. [CrossRef]

50. Kong, T.; Yao, A.; Chen, Y.; Sun, F. HyperNet: Towards Accurate Region Proposal Generation and Joint Object Detection. In Proceedings of the 2016 IEEE Conference on Computer Vision and Pattern Recognition (CVPR), Las Vegas, NV, USA, 27–30 June 2016; pp. 845–853.

51. Le-Hong, P.; Le, A.-C. A Comparative Study of Neural Network Models for Sentence Classification. In Proceedings of the 2018 5th NAFOSTED Conference on Information and Computer Science (NICS), Ho Chi Minh City, Vietnam, 23–24 November 2018; pp. 360–365.

52. Thakur, D.; Biswas, S.; Ho, E.; Chattopadhyay, S. ConvAE-LSTM: Convolutional Autoencoder Long Short-Term Memory Network for Smartphone-Based Human Activity Recognition. *IEEE Access* **2022**, *10*, 4137–4156. [CrossRef]

53. Wang, L.; Xu, Y.; Cheng, J.; Xia, H.; Yin, J.; Wu, J. Human Action Recognition by Learning Spatio-Temporal Features with Deep Neural Networks. *IEEE Access* **2018**, *6*, 17913–17922. [CrossRef]
54. Dua, N.; Singh, S.N.; Semwal, V.B. Multi-Input CNN-GRU Based Human Activity Recognition Using Wearable Sensors. *Computing* **2021**, *103*, 1461–1478. [CrossRef]
55. Ding, X.; Jiang, T.; Zhong, Y.; Huang, Y.; Li, Z. Wi-Fi-Based Location-Independent Human Activity Recognition via Meta Learning. *Sensors* **2021**, *21*, 2654. [CrossRef]

Article

Decentralized Personal Data Marketplaces: How Participation in a DAO Can Support the Production of Citizen-Generated Data

Mirko Zichichi [1,*], Stefano Ferretti [2] and Víctor Rodríguez-Doncel [1]

[1] Ontology Engineering Group, Universidad Politécnica de Madrid, 28040 Madrid, Spain
[2] Department of Pure and Applied Sciences, University of Urbino "Carlo Bo", 61029 Urbino, Italy
* Correspondence: mirko.zichichi@upm.es; Tel.: +34-910-672-914

Abstract: Big Tech companies operating in a data-driven economy offer services that rely on their users' personal data and usually store this personal information in "data silos" that prevent transparency about their use and opportunities for data sharing for public interest. In this paper, we present a solution that promotes the development of decentralized personal data marketplaces, exploiting the use of Distributed Ledger Technologies (DLTs), Decentralized File Storages (DFS) and smart contracts for storing personal data and managing access control in a decentralized way. Moreover, we focus on the issue of a lack of efficient decentralized mechanisms in DLTs and DFSs for querying a certain type of data. For this reason, we propose the use of a hypercube-structured Distributed Hash Table (DHT) on top of DLTs, organized for efficient processing of multiple keyword-based queries on the ledger data. We test our approach with the implementation of a use case regarding the creation of citizen-generated data based on direct participation and the involvement of a Decentralized Autonomous Organization (DAO). The performance evaluation demonstrates the viability of our approach for decentralized data searches, distributed authorization mechanisms and smart contract exploitation.

Keywords: distributed ledger technology; decentralized file storage; distributed hash table; data marketplace; keyword-based search; citizen-generated data

Citation: Zichichi, M.; Ferretti, S.; Rodríguez-Doncel, V. Decentralized Personal Data Marketplaces: How Participation in a DAO Can Support the Production of Citizen-Generated Data. *Sensors* **2022**, *22*, 6260. https://doi.org/10.3390/s22166260

Academic Editors: Pietro Manzoni, Claudio Palazzi and Ombretta Gaggi

Received: 25 May 2022
Accepted: 16 August 2022
Published: 20 August 2022

Publisher's Note: MDPI stays neutral with regard to jurisdictional claims in published maps and institutional affiliations.

1. Introduction

Recent scandals have shown the harm that current data collection, storage and sharing practices can cause with regard to the misuse of personal data [1,2]. As the world is becoming more "smart", so-called smart environments, of which smart cities [3] stand out the most, have in common the ability to transform data (in particular, personal data) into meaningful information needed by the liveness of the ecosystem they generate. Based on this transformation, indeed, they provide services that are becoming more and more targeted towards individuals. For instance, it is commonly known that personal information is used to recommend opportunities to individuals and to make their life easier. However, entities that control these data might not always operate with the aim of social good [4]. Many Big Tech companies rely on data collected about their users, usually storing this personal information in corporate databases, i.e., data silos, and transacting it to third parties with not enough transparency for individuals.

Meanwhile, among the many technologies used for general-purpose data management and storage, Distributed Ledger Technologies (DLTs) are rising up as powerful tools for avoiding control centralization. DLT and the realm of decentralized systems, such as Decentralized File Storages (DFS), that are emerging as solutions able to tackle the issue of obtaining large amounts of data that are not of dubious or of false origin, while providing more disintermediated processes [5,6]. DLTs, in this context, provide a new way of handling personal data, such as recording, storage and transfer. This can be carried out in combination with cryptographic schemes to ensure data confidentiality. By their decentralized nature, indeed, these technologies have the potential to make processes more democratic, transparent and efficient [7]. DLTs and DFS can support the creation of a Personal Information Management System (PIMS) based on decentralized data processing and Personal Data Stores (PDS) [8,9]. In PIMS, data access is granted in line with user policies and these ones, in a decentralized scenario, can be determined by the user via DLTs and smart contracts [8]. PIMS have been proposed by scholars [9–11] or companies [12] and are increasingly gaining attention from policymakers who currently consider mechanisms for regulating and advancing data intermediation services in general [7,8,13,14]. In the context of the European Union's General Data Protection Regulation (GDPR) [15], PIMSs enforce the right of individuals to know the data collected about them and the right to transfer data to other service providers, i.e., data portability. Such features enable the process of moving the data sovereignty towards users, i.e., Self-Sovereign Identity [16], and of providing them more influence over access control, while allowing anyone else to be able to consume this data with transparency. All of this paves the way towards the use of personal data for open data markets and for social good. The ability to easily obtain personal data has the potential to create a marketplace where users are consumers and providers at the same time. By creating a common, decentralized and trustless infrastructure, such as a decentralized personal data marketplace, it will be possible for data owners and consumers to interact and collaborate in Peer-to-Peer (P2P) transactions [17,18]. This means facilitating the transactions of data between owners and consumers without the need for a trusted third-party broker, enabling liquid data markets [19].

The underlying research questions we aim to explore in this work are as follows:

i   Are decentralized personal data marketplaces able to optimally support individuals' personal data protection and portability?

ii   How can decentralized technologies foster a convergence between the protection of individuals' personal data and the development of data aggregation solutions?

Contributions

The contributions of this work include the description of a high-level solution for a decentralized personal data marketplace involving the use of DLTs, DFS and smart contracts for the creation of a PIMS: (i) data owners store their personal data in PDSs implemented using a DFS, such as the InterPlanetary File System (IPFS) [20]; (ii) this storage is complemented by the use of a DLT, which enables data integrity validation and indexing of the data, i.e., the IOTA DLT [21]; (iii) a distributed authorization mechanism enables access control of this data, thanks to the decentralized execution of immutable instructions of smart contracts implemented using the Ethereum protocol [22]; our authorization blockchain network executes access control to enact data owners' preferences and to verify the authenticity of the claims; the nodes of such a network employ a cryptographic schema that enables the protection of data owner's personal data (i.e., based on the use of a Threshold Proxy Re-Encryption [23]); (iv) finally, we provide a system for the search for data according to their content or meaning, relying on the use of a Distributed Hash Table (DHT) as a layer placed over the DLTs; indeed, data inserted into the DLTs and DFS is usually unstructured and no efficient decentralized mechanisms are present to query a certain kind of data; in our solution, data stored in DLT can be searched in a decentralized way through keywords stored in the DHT; the distinctive feature of our DHT network is that it is essentially a hypercube overlay structure [24], in which each node indexes objects representing specific indexes and addresses of a DLT using keywords.

Moreover, in this paper, we address a use case that assists us in describing the implementation of the decentralized personal data marketplace. Since our main driving force is facilitating the use of private data for the social good, we present a use case where citizens can use and give access to their personal data to produce new data, creating value for governments, businesses and other citizens as well. Through practices that promote a collaborative and co-creative approach to working together on the management, control and governance of the use of data, people and society can influence and shape data governance processes and can support greater social and economic equity [2,25]. By implementing this use case, we demonstrate the ability of our system to support these processes. In particular, our system allows citizens to take part in an organization for the creation of new datasets and to steer the developmental decisions through a Decentralized Autonomous Organization (DAO) [26]. Our system provides the elements of a framework for participatory data stewardship [2], being transparent, i.e., informing individuals about their data, and collaboration, i.e., enabling individuals to take action [25], at its core.

The original contributions and novelties of our work are summarized as follows:

i First, we provide a description of the implementation of a decentralized personal data marketplace, where

- Our PDS is implemented using a DFS for storing personal data;
- We use a DLT for providing data integrity, validation and indexing;
- Our smart contract-based authorization system executes distributed data access control;
- Our hypercube DHT enables a decentralized way of searching for data in DLTs.

In particular, we provide a detailed description of the protocols behind the authorization blockchain and the hypercube DHT.

ii Second, we provide the implementation of a use case for the architecture through the description of citizen-generated data creation based on direct participation. This consists of the development of a data aggregation solution through the use of a DAO, where members are citizens.

iii Third, we evaluate the implementation's performance by means of an experimental evaluation. More specifically, (i) we simulate a P2P network executing the hypercube DHT for decentralized search of data, (ii) we test the distributed data access control execution for the use case and (iii) we evaluate the smart contract implementation in terms of gas usage.

The remainder of the paper is organized as follows: Section 2 provides a background on the main concepts and technologies used, while Section 3 focuses on related work. Section 4 presents a description of the decentralized personal data marketplace architecture. Section 5 provides the description of a citizen-generated data creation use case with the intent to present our marketplace implementation. In Section 6, the evaluation of our proposal is shown and the results are discussed. Finally, Section 7 provides the concluding remarks.

2. Background

In this section, we introduce the main concepts and technologies involved in our work.

2.1. Distributed Hash Table (DHT)

A Distributed Hash Table (DHT) is a distributed infrastructure and storage system that provides the functionalities of a hash table, i.e., a data structure that efficiently maps "keys" into "values". It consists of a P2P network of nodes that are supplied with the table data and on a routing mechanism that allows for searching for objects in the network [24]. Each node in the DHT network is responsible for part of the entire system's keys and allows the objects mapped to the keys to be reached. In addition, each node stores a partial view of the entire network, with which it communicates for routing information. To reach nodes from one part of the network to another, a routing procedure typically traverses several nodes, approaching the destination at each hop. This type of infrastructure has been used as a key element to implement complex and decentralized services, such as Content-Addressable Networks (CANs) [27], Decentralized File Storage (DFS) [20], cooperative web caching, multicast and domain name services.

2.2. Decentralized File Storage (DFS)

Decentralized File Storage (DFS) is a solution for storing files as in Cloud Storage [28] but retaining the benefits of decentralization [9]. They offer higher data availability and resilience thanks to data replication. A DFS comprises a P2P network of nodes that provide storage and follow the same protocol for content storing and retrieval. In content-based addressing, contents are directly queried through the network rather than establishing a connection with a server. In order to know which DFS node in the network owns the requested contents, it is possible to rely on a DHT in charge of mapping the contents, i.e., files and directories, to the addresses of the peers owning such data. A principal example of DFS is the InterPlanetary File System (IPFS) [20], a protocol that builds a distributed file system over a P2P network. IPFS is a DFS and a protocol created for distributed environments with a focus on data resilience. The IPFS P2P network stores and shares files and directories in the form of IPFS objects that are identified by a CID (Content Identifier). The CID acts as an immutable universal identifier used to retrieve an object in the network. Only the file digest is needed, i.e., the result of a hash function applied on the data. Users that want to locate that object use this identifier as a handle. When an IPFS object is shared in the network, it is identified by the CID retrieved from the object hash, for instance a directory with a CID equal to *QmbWqxBEKC3P8tqsKc98xmWNzrzDtRLMiMPL8wBuTGsMnR*. Even if other nodes in the network try to share the same exact directory, the CID will always be the same.

2.3. Distributed Ledger Technology (DLT)

Distributed Ledger Technologies (DLTs) consist of networks of nodes that maintain a single ledger and follow the same protocol, including a consensus mechanism, for appending information to it. The blockchain is a type of DLT where the ledger is organized into blocks and where each block is sequentially linked to the previous one. The execution of the same protocol, i.e., source code, guarantees (most of the time) the property of being tamper-proof and not forgeable. This allows for a trust mechanism to be created without the need for third-party intermediaries [29,30].

There are different implementations of DLTs, each one with its pros and cons. In permissionless ones, anyone can take part in the consensus mechanism, while this is not true in permissioned ones. Another distinction lies in the support of smart contracts, e.g., Ethereum [22]. This feature is quite often in contrast with other key features related to the level of scalability and responsiveness of the system [31]. Conversely, some implementations are thought to provide better scalability at the expense of lacking some features. IOTA [21], for instance, implements a more scalable solution for distributing the ledger. It consists of a Layer-1 solution, while, on the other hand, Layer-2 solutions are technologies that operate on top of an underlying DLT to improve its scalability [32].

2.4. Smart Contract and Decentralized Autonomous Organization (DAO)

A smart contract is a new paradigm of contracts that does not completely embody the same features of a legal contract but can act as a self-managed structure able to execute code that forces agreements between two or more parts. A smart contract consists of instructions that, once distributed on the ledger, cannot be altered. Thus, the result of its execution will always be the same for all DLT nodes running the same protocol. When a smart contract is deployed on the DLT and the issuer is confident that the code embodies the intended and proper behavior (e.g., by reviewing the code), then transactions originating from that contract do not require the presence of a third party to have value [33].

Smart contracts are fundamental components of Ethereum that reside on the blockchain and are triggered by specific transactions [34]. Moreover, smart contracts can communicate with other contracts and even create new ones. The use of these contracts grants permission to build Decentralized Applications (dApps) and Decentralized Autonomous Organizations (DAOs) [6,32,35–37]. A DAO is a virtual entity managed by a set of interconnected smart contracts, where various actors maintain the organization state by a consensus system and are able to implement transactions, currency flows, rules and rights within the organization. Members of a DAO are able to propose options for decisions in the organization and to discuss about and vote on those through transparent mechanisms [26].

2.5. IOTA and Streams

In this work, we specifically refer to the IOTA DLT as a technology that uses a different paradigm for managing the ledger; however, there are many other alternatives such as Radix [38] or Nano [39]. IOTA is a DLT that allows hosts in a network to transfer immutable data among each other. In the IOTA ledger, i.e., the Tangle [21], is based on a Directed Acyclic Graph (DAG) where the vertices represent transactions and edges represent validations to previous transactions. The validation approach is thought to address two major issues of traditional blockchain-based DLTs, i.e., latency and fees. IOTA has been designed to offer fast validation, and no fees are required to add a transaction to the Tangle [40]. When a new transaction is to be issued, two previous transactions must be referenced as valid (i.e., tips selection), and then, a small amount of Proof-of-Work is performed.

An important feature offered by IOTA are the Streams [41]. Streams consist of a communication protocol that adds the functionality to emit and access encrypted message streams over the Tangle [40]. Message streams assume the form of channels, i.e., a linked list of ordered messages stored in transactions. Once a stream channel is created, only the channel author can publish encrypted messages on it. Subscribers that possess the channel encryption key (or set of keys, since each message can be encrypted using a different key) are enabled to decode messages. A channel is addressed using an "announcement link". In other words, IOTA Streams enable users to subscribe and follow a messages stream channel, generated by some device. From a logical point of view, channels are an ordered set of messages; in fact, a channel is referenced through the link of a "starting" message.

2.6. Proxy Re-Encryption (PRE) and Cryptographic Threshold Schemes

Distributed systems usually store data as they are received, without further processing for confidentiality. Therefore, data can be accessed by any network participant. In order to deal with the protection of personal data, we employ in this work two cryptographic schemes, which are described in the following. Proxy Re-Encryption (PRE) is a cryptographic protocol where it is not necessary to know the recipient of the data in advance [42]. PRE is a type of public key encryption based on the figure of a proxy. A sender encrypts a plaintext with a specific public key obtaining a ciphertext. Then, the untrusted proxy transforms the ciphertext into a new ciphertext decryptable with the recipient private key, which does not have anything to do with the first public key. This operation is performed without learning anything about the underlying plaintext. This is possible using a re-encryption key generated by the sender using the recipient public key and shared with the proxy.

A Threshold Proxy Re-Encryption (TPRE) adds a layer of complexity [23]. A (t, n)-threshold scheme can be employed to share a secret among a set of n participants, allowing the secret to being reconstructed using any subset of t (with $t \leq n$) or more fragments, but no subset of less than t. In a network where more than one node keeps secret fragments, a mutual consensus can be reached when t nodes provide the shares to a secret recipient, enabling the secret to be known by the latter. This can be used by a sender to share the re-encryption key in fragments with a network of proxies; none of the latter can obtain the whole key without the help of other $t - 1$ proxies.

3. Related Works

In this section, we described the related work based on the different topics we have addressed in this work. To the best of our knowledge, no other works have developed a personal data marketplace using the same set of technologies and techniques; thus, we subdivided this section in work related to each part or parts of our proposed solution.

3.1. Decentralized Data Marketplace

The use of DLTs has been proposed for the implementation of data marketplaces to take advantage of the following advantages [43,44]: (i) no need to rely on third party platforms, (ii) better resilience against network partitioning and single point of failure, and (iii) privacy-preserving mechanisms [45]. Most of the related work investigated the data distribution through DLTs, focusing in particular on the use of off-chain storage based on DFS with data links referenced in DLTs [6,35,45]. Data exchange with such technologies can lead to a transparent market, where transactions between data owners and data consumers are recorded on DLTs and where smart contracts enable the self-enforcement of fair exchanges between participants and the automatic resolution of disputes [46]. In [5], the authors provided the implementation of a data marketplace based on the use of DFS for storing data and a payment protocol that exploits Ethereum smart contracts. Similarly, in [17,18], the proposed systems were based on P2P interactions and smart contracts to reach an agreement while also integrating other components such as the IOTA DLT. Lopez and Farooq [47] presented a framework for Smart Mobility Data Market in which the participants shared their data and could transact this information with another participant, as long as both parties reached an agreement. Their work focuses on the protection of individuals' personal information, while maintaining data transparency and users' ruled access control. Aiello et al. [35] designed IPPO, an architecture that allows users to generate and share anonymized datasets on a distributed marketplace to service providers, while monitoring the behavior of web services to discourage the most intrusive forms of tracking.

With respect to our work, these proposals build similar architectures but lack insight into decentralized access control mechanisms and or decentralized data searches.

3.2. Decentralized Access Control

DLTs have desirable features that make them a reliable alternative infrastructure for access-control systems. Their distributed nature solves the single point of failure problem and mitigates the concern for privacy leakage by eliminating third parties. Traditional access-control policies have been combined with DLTs: discretionary (DAC), to manage personal data "off-chain" (i.e., not directly stored in the DLT), through the access-control policy on the blockchain [48]; mandatory (MAC), to constrain the ability of a subject to access on a datum through smart contracts [9]; role-based (RBAC), for achieving cross-organizational authentication for user roles [49]; and attribute-based (ABAC), to grant or deny user requests based on the attributes of a user, an object and environment conditions [50]. Among DLT-based access-control mechanisms, Attribute-Based Encryption (ABE) [51] offers the best policy expressiveness without introducing many elements into the system infrastructure. ABE encrypts the data using a set of attributes that form a policy. Only those who have a secret key that meet the policy can decrypt the data. In [51], the authors designed a system using ABE-based access control and smart contracts to grant data access, with similar policies mechanism to our solution, while the authors of [52,53] proposed similar frameworks that combined DFS and blockchains to achieve fine-grained ABE-based access control. However, in any of the three previous cases, the secret attribute keys are issued directly by the data owner in the DLT or by a central authority.

3.3. Decentralized Data Search

With respect to our hypercube DHT contribution, a decentralized data search on DLT and DFS is a field that has been addressed by both scholars and developers with only a few efforts. Indeed, one of the concerns that is still open with respect to these novel technologies, is related to implementing data discovery and lookup operations in decentralized way. The Graph is one of the first protocols (actually the most used) with the aim of providing a "Decentralized Query Protocol" [32]. The Graph network consists in a Layer-2 protocol based on the use of a Service Addressable Network, i.e., a P2P network for locating nodes capable of providing a particular service such as computational work (instead of objects just as a CAN). In [54], the authors proposed a Layer-1 keyword search scheme that implements oblivious keyword search in DFS. Their protocol is based on a keyword search with authorization for maintaining privacy with retrieval requests stored as a transaction in a blockchain (i.e., Layer-1). Specifically for IPFS [20], in order to overcome the file search limitation, a generic search engine has been developed, namely "ipfs-search" [55]. This solution is rather centralized and does not escape the problem of concentration similar to the conventional web. In response to this, a decentralized solution called Siva [56] has been proposed. An inverted index of keywords is built for the published contents on IPFS and users can search through it; however, Siva is proposed as an enhancement of the IPFS public network DHT and does not feature any optimization for a keyword storage structure apart from the use of caching. Finally, a Layer-2 solution for the keyword search in DFS has been proposed in [44], where a combination of a decentralized B+Tree and HashMaps is used to index IPFS objects.

3.4. Decentralized Personal Data Management

The popularity of Internet of Things devices and smartphones and the associated generation of large amounts of data derived from their sensors [57] has resulted in an interest of individuals in the production and consumption of data via a data marketplace [11]. Making data, which are mostly personal, available for access and trade is expected to become a part of the data-driven digital economy [14]. In this context, we find a set of technologies referred to as Personal Information Management Systems, which help individuals reach the vision of Self-Sovereign Identity (SSI). SSI consists of the complete control of individuals' digital identities and their personal data through decentralization. SSI has been generically implemented as a set of technological components that are deployed in decentralized environments for the purpose of providing, requesting and obtaining qualified data in order to negotiate and/or execute electronic transactions [16].

The databox, for instance, is a PDS [8,9] that must be conceived as a concept that describes a set of storing and access-control technologies enabling users to have direct control of their data. In [11,58], the databox is a platform that provides means for individuals to manage personal data and control access by other parties wishing to use their data, supporting incentives for all parties. An undirected link to this model that puts in practice the concept of SSI is the Solid project [12]. Solid has the purpose of letting users choose where their data resides and who is allowed to access and reuse it. Semantic Web technologies are used to decouple user data from the applications that use this data. The storage itself can be conceived in a different manner, while the use of Semantic Web represents to us the core element that eases data interoperability and favors reasoning over individuals' policies. Semantic Web standards bring structure to the meaningful contents of the Web by promoting common data formats and exchange protocols, such as ontologies. The advantages consist in the fact that many ontologies are recommended by the World Wide Web Consortium (W3C) and are thus universally understood and that reasoning with the information represented using these data models is facilitated by mapping with a formal language. An example is the Open Digital Rights Language (ODRL) policy expression language. This can be used in conjunction with other standard ontologies to manage the access control to personal data in Solid [59]. Another possible approach is to program policy expression languages such as smart contracts, in order to manage control automatically [60].

4. Decentralized Personal Data Marketplace Architecture

In this paper, we are interested in describing the fundamentals of a decentralized personal data marketplace: (i) data marketplace because we intend to provide a system that enables data owners to benefit from the sharing of the data they own; the benefits can be purely economical but also linked to the participation to an ecosystem, e.g., sharing data for social good and research; on the other hand, we intend to provide an easier data access to data consumers, especially to the ones who do not have the resources to compete with Big Tech companies; (ii) personal data because we specifically focus on the type of data that is generated by individuals through their personal devices; thus, we assume that the role of data owner in the system is going to be engaged by individuals themselves or by some other entities on their behalf, with a strong emphasis to the concept of Self-Sovereign Identity [16]; and (iii) decentralized because we make use of several decentralized systems that help to more easily achieve a disintermediation in the process of transacting data.

In this section, we devise the marketplace architecture through a description of four pillar systems and their interactions. As shown in Figure 1, the different architectural components can be organized into four layers:

i   A Decentralized File Storage (DFS) is used to store personal data in an encrypted form and to create immutable universal identifiers that directly represent the content of a piece of data. This kind of system is used to take advantage of the property of high data availability that is often taken for granted in centralized file storages.

ii   Smart contracts are used to provide decentralized access control mechanisms that can be leveraged by data consumers to access data retrieved from the DFS, following a policy indicated by the data owner (e.g., access through payment).

iii A Distributed Ledger Technology (DLT) is used to enable the data indexing and validation. The ledger's untamperability property makes sure that data integrity can be validated by storing data's immutable universal identifiers, specifically in the form of hash pointers. Moreover, related pieces of data can be already linked and indexed in this layer.

iv   An hypercube-structured Distributed Hash Table (DHT) is used to provide a distributed mechanism for the search of data. This system is in charge of associating keywords to addresses or references stored in the DLT.

Decentralized Personal Data Marketplace

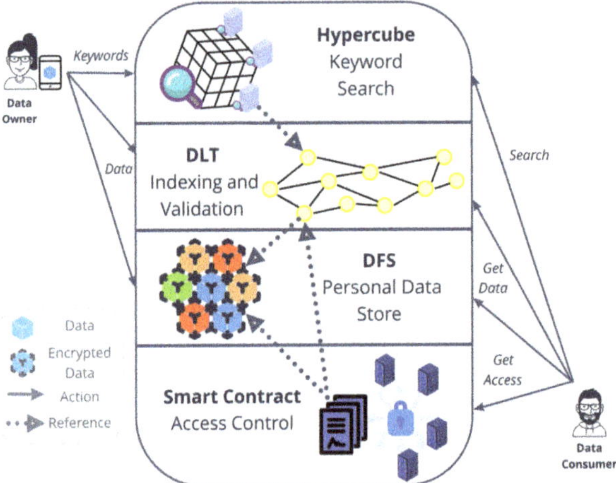

Figure 1. Decentralized data marketplace architecture.

With respect to this architecture, in the following subsections, our aim is twofold: (i) to describe in detail the hypercube DHT system and (ii) to describe the interaction between all the architectural components. More specifically, we do not go into the details of all the possible configurations of the DFS, DLT and smart contracts layers, as the discussion may become too scattered and may stray away from the issues related to the decentralized personal data market.

4.1. DFS-Based Personal Data Store

Data generated by personal devices or third-party systems on behalf of individuals are often private in nature, but incentivizing their sharing (as opposed to keeping them locked in data silos) can be beneficial in terms of economic gain and social good. However, the main challenge is often to provide access under certain conditions that data subjects find acceptable and compliant with regulations (e.g., GDPR).

A technological solution that is opposed to centralized data silos consists of the use of DFS for storing such personal data. DFS are usually built on top of a P2P network that is freely accessible and where nodes execute the same protocol to store and retrieve data. Moreover, often at the heart of such systems, we find the provision of data replication protocols that enable a high data availability. All this means that data owners holding some data in their device can easily participate in the DFS network or reach a DFS node to store and replicate data. This use, then, makes data owners confident that their data can be retrieved by any data provider that, in turn, can participate in the network or contact a DFS node. However, to be on the safer side, data owners should incentivize DFS nodes to store and replicate their data. How to do this is beyond the scope of this paper and we refer the reader to our previous work that also investigates this topic [9].

DFSs have often built in their protocols the identification of data through immutable universal identifiers that directly represent their content in order to uniquely identify contents that are disseminated in the network. An implementation of this feature would be the use of the hash digest of a piece of data with the aim of obtaining a deterministically derived identifier. Thus, any node of the network holding the same piece of data, i.e., with exact content, can use its hash to derive its immutable universal identifier. Any other node in the network can use this id to retrieve the piece of data from other nodes and to verify its integrity through the hash.

Finally, due to the fact that data can be easily replicated in the p2p network and, thus, can be easily accessed by nodes that the data owner might be not aware of, we resort to the use of encryption as a mechanism of data protection. Such a mechanism is required both by the privacy needs of data owners and, specifically, by compliance with personal data regulations. Strong and state-of-the-art cryptographic algorithms help avoid the re-identification of such pseudonymous data, i.e., encrypted personal data, when shared in the DFS network [61].

4.2. Smart Contract-Based Distributed Access Control

Smart contracts are the part of the proposed architecture where access-control logic to share encrypted personal data is performed. Through dedicated smart contracts, access to data can be purchased or can be enabled directly by the owner. Access is authorized only to consumers indicated by the policies of a data owner's contract. A policy would be for the smart contract to maintain an Access Control List (ACL) that represents the rights to access one or more pieces of data. In the rest of the paper, we focus on the application of such a policy.

According to our solution, nodes in a network that maintain a permissioned blockchain are responsible for enforcing the access rights specified in the ACLs of smart contracts. We take advantage of the high degree of trust that a blockchain provides for the data written in the ledger and, then, focus on the trust given to the nodes of this "authorization" blockchain, which must read from the ledger and follow the correct policy. If a data consumer is enlisted in the ACL, then this one is eligible to access certain data. If that is the case, then the consumer is also eligible to obtain the key used for encrypting the data in the DFS. Authorization blockchain nodes rely on ACLs to make sure that a data consumer entitled to this information can obtain such a key. For the encryption operation, we refer to a hybrid cryptographic scheme, making use of both asymmetric and symmetric keys. Generally, each piece of data is encrypted using a symmetric "content" key k, and then, this key is encrypted using an asymmetric keypair (pk_{KEM}, sk_{KEM}). This consists of a Key Encapsulation Mechanism (KEM) [62], in which the key is encapsulated and the capsule is distributed, instead of distributing the encrypted data.

4.2.1. Access Mechanism

To ensure complete protection of the individual's data, only the authorized recipient of personal data should obtain the key, and nodes on the authorization blockchain should not be able to exploit it. For this reason, we make use of a (t, n)-threshold scheme to share the capsule that contains the content keys among the blockchain nodes. In particular, the Threshold Proxy Re-Encryption (TPRE) scheme is employed:

- **Keypairs**—each actor creates a set of asymmetric keypairs, e.g., the data owner creates (pk_{DO}, sk_{DO}) while the data consumer creates (pk_{DC}, sk_{DC}).
- **Capsule**—a capsule is created by the data owner for each piece of data stored in the DFS. Recall that the content key k is used for encrypting the piece of data. Then, the result of the encryption of k results in the capsule.
- **Re-encryption key**—The re-encryption key $rk_{DO \to DC}$ is created by the data owner for each data consumer through the public key pk_{DC}.
- **Kfrags**—The data consumer divides the re-encryption key into n fragments following the (t, n)-threshold scheme. The single re-encryption key fragments are unique for each authorization blockchain node. We call these key fragments "kfrags" for simplicity.
- **Cfrags**—Each authorization blockchain node receives the same capsule and a unique kfrag. The capsule cannot be "opened" because it is encrypted with the data owner's public key. Only one kfrag (or a number less than t) cannot be used to completely re-encrypt the capsule in such a way that the data consumer can open the capsule. The authorization blockchain node only performs a re-encryption operation that takes as input the capsule and the unique kfrags, and it outputs a new capsule fragment, "cfrag". The data consumer requires at least t cfrags to reconstruct the new capsule and to decrypt it with the private key sk_{DC}.

The access mechanism is as follows: (i) the public key of the data consumer pk_{DC} is listed in the ACL (provided in detail in Section 5); (ii) the data consumer requests the release of a cfrag to at least t authorization blockchain nodes using a message signed with sk_{DC}; (iii) upon consumer request, each node checks if the signatory pk_{DC} is in the ACL through an interaction with the smart contract in the blockchain; (iv) if this is the case, then each node releases the cfrag; (v) once the data consumer obtains t cfrags, the capsule can be reconstructed and decrypted with sk_{DC}; and (vi) the decryption reveals the content key k needed to decrypt the desired piece of data stored in the DFS.

4.3. DLT Indexing and Validation

One of the main use cases of DLTs consists in data sharing due to their intrinsic property of untamperability. Once collected, in many cases, data can be stored directly on-chain, in a DLT, to validate their integrity. However, preventing the on-chain storage is a preferable solution, not only for retaining high data reads availability and better performances for data writes [9] but also because on-chain personal data are generally incompatible with data protection requirements (i.e., to guarantee personal data deletion to a data subject). Thus, our solution consists of storing personal data in a DFS and reference them in a DLT via their immutable universal identifiers, e.g., hash pointers. Moreover, due to the nature of some proposed DLTs, related pieces of data can be already linked and indexed in the ledger. That is the case of the IOTA DLT, which manages the upload of data in the form of a stream channel thanks to the Streams protocol. We refer to this DLT and this protocol to ease the description of the following parts.

Layer-2 Solution

While Layer-1 solutions in DLTs define the form of the ledger, its distribution, consensus mechanism and features, Layer-2 solutions are built on top of Layer-1 without changing its trust assumptions, i.e., the consensus mechanism or the structure [63,64]. Layer-2 protocols allow users to communicate through mediums external to the DLT network, reducing the transaction load on the underlying DLT. On top of the IOTA layer-1 DLT, we designed a Layer-2 solution using a DHT with the aim of facilitating the search for large amounts of data through specific keywords (Figure 2). In order to obtain information from a IOTA message within a stream channel, indeed, it is necessary to know the exact address of the message or of the channel, i.e., the announcement link. However, the announcement link of a stream channel does not provide any information related to the type and kind of messages. No mechanisms are provided by IOTA (and the majority of DLTs) for the discovery based on the content of certain data/streams channels that are available in the Tangle. This is the issue we deal with in this paper. In the remainder of this section, we describe how to surmount such limitations. In our system, every stream channel is indexed by a keyword set and then how such a keyword set is exploited to look for specific kinds of contents.

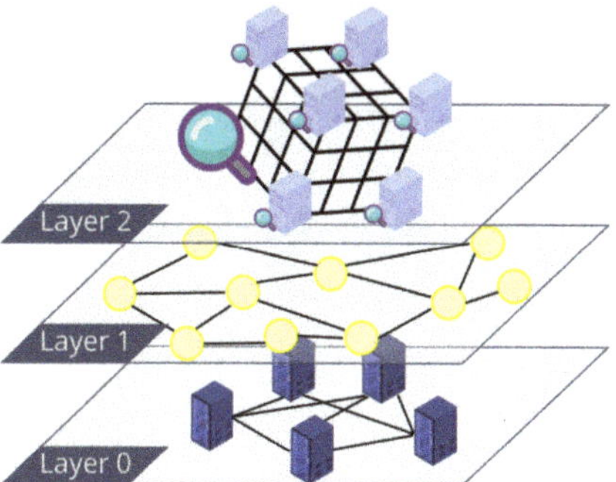

Figure 2. Layers in the context of DLTs. Layer zero consists of the DLT network, while Layer-1 is the set of software frameworks run by the network nodes (e.g., the ledger). Layer-2 solutions are the ones that leverage Layer-1 for other services, i.e., the hypercube DHT in our case.

4.4. Hypercube-Structured DHT

Considering O as the set of all stream channels in IOTA, the idea is to map each object $o \in O$ to a keyword set $K_o \subseteq W$, where W is the keyword space, i.e., the set of all keywords considered. In general, we refer to $K \subseteq W$ as a keyword set that can be associated to a data content (i.e., the metadata associated to it) or a query (i.e., we are looking to some content with a specific metadata). By using a uniform hash function $h : W \to \{0, 1, \ldots, r-1\}$, a keyword set K can be represented by the result of such a function, i.e., a string of bits u where the 1s are set in the positions given by $one(u) = \{h(k) \mid k \in K\}$. In other words, each $k \in W$ has a fixed position in the r-bit string given by $h(k)$, and that position can be associated to more than one k (i.e., hash collision). Then, every keyword set K is represented by a r-bit string where the positions are "activated", i.e., are set to 1, by all the $k \in K$.

We use these r-bit strings to identify logical nodes in a DHT network, e.g., for $r = 4$, a node id can take values such as 0100 or 1110. In particular, inspired by [24], we refer to the geometric form of the hypercube to organize the topological structure of such a DHT network. $H_r(V, E)$ is a r-dimensional hypercube, with a set of vertices V and a set of edges E connecting them. Each of the 2^r vertices represents a logical node, whilst edges are formed when two vertices differ by only one bit, e.g., 1011 and 1010 share an edge. In the network, the nodes represented by vertices that share an edge are network neighbors as well. To find out how far apart two vertices u and v are within the hypercube, the Hamming distance can be used, i.e., $Hamming(u, v) = \sum_{i=0}^{r-1}(u_i \oplus v_i)$,, where $\oplus$ is the XOR operation and u_i is the bit at the i-th position of the u string, e.g., for $u = 1011$ and $v = 1010$, we have $Hamming(u, v) = 1$.

4.4.1. Keyword-Based Complex Queries

In our system, contents can be discovered through queries that are based on the lookup of multiple keywords, associated with data. Such queries are processed by the DHT-based indexing scheme described in the previous section. The base idea is to associate a keyword set to each IOTA stream channel through the DHT. In particular each logical node locally stores an index table that associates a keyword set K_o to the announcement link of an IOTA stream channel, i.e., the reference of an object o. Then, given a keyword set K, the associated r-bit string is used to reach the logical node responsible for K through a routing mechanism, in order to obtain the set of objects $= \{o \in O \mid K_o \supseteq K\}$. For instance, with $W = \{$"Turin", "Lingotto", "Temperature", "Celsius"$\}$ and 1010 representing the keyword set $K = \{$*Turin, Temperature*$\}$, if $u \in V$ is the node that is responsible for K because the id of u is equal to 1010, then u is in charge of maintaining a list of announcement links of IOTA stream channels containing the temperature of the city of Turin. Once that node is located, the objects $= \{o \in O \mid K_o = K\}$ it stores in its index table can be returned or aggregated with other nodes' objects. These objects consist of a list of announcement links that can be used to obtain messages from IOTA.

4.4.2. Multiple Keywords Search

Our system provides two functions for making queries based on multiple keywords:

- **Pin Search**—this procedure aims at obtaining all and only the objects associated exactly with a keyword set K, i.e., $\{o \in O \mid K_o = K\}$. Upon request, the responsible node returns to the requester all the announcement links of the corresponding objects that it keeps in its table associated with K.
- **Superset Search**—this procedure is similar to the previous one, but it also searches for objects that can be described by keyword sets that include K, i.e., $\{o \in O \mid K_o \supseteq K\}$. Since the possible outcomes of this search can be quite large, a limit l is set.

For the Pin Search we need to retrieve objects only from one node, whilst for Superset Search, we need to retrieve objects from all nodes that are responsible for a Superset of K. Such nodes are contained in the sub-hypercube $SH(S, F)$ induced by the node u responsible for K, where S includes all the nodes $s \in V$ that "contain" u, i.e., $u_i = 1 \Rightarrow w_i = 1$, while F includes all the edges $e \in E$ between such nodes. Thus, during a Superset Search, the induced sub-hypercube is computed and then only nodes in such a sub-hypercube are queried using a spanning binomial tree, as described in [24] (definition 4.2). The l limit is a query parameter that indicates the maximum number of objects to return when traversing the spanning binomial tree.

4.4.3. The Query Routing Mechanism

Queries can be injected into the system by users external to the DHT to any $v \in V$ network node. Through a routing mechanism, the query reaches a node $u \in V$ that is responsible for a keyword set K. This process is described in detail in Algorithm 1.

Algorithm 1: Query Routing Mechanism

    Input: q query, K keyword set, l limit
    Data: v node string, $one(v)$, $neighbors(v)$
    Result: $\{o \in O \mid K_o \supseteq K\}$

 1 $one(u) \leftarrow \{h(k) \mid k \in K\}$
 2 **if** $one(u) \neq one(v) \wedge From(q) = \text{"User"}$ **then**
 3     $w \leftarrow \{n \mid n \in neighbors(v) \wedge Min(Hamming(n, u))\}$
 4   **return** QueryRoutingMechanism(w, q, K, l)
 5 **else**
 6   **if** $Type(q) = \text{"PinSearch"}$ **then**
 7   **return** GetObjectsFromIndexTable$(K, -1)$
 8   **else if** $one(u) \subseteq one(v)$ **then** // i.e., *SupersetSearch*
 9         $objectsList \leftarrow$ GetObjectsFromIndexTable(K, l)
10         $l \leftarrow l - Length(objectsList)$
11         $From(q) \leftarrow \text{"Node"}$
12   **while** $l > 0$ **do**
13            $c \leftarrow$ GetNextSBTreeChild(u)
14            $cList \leftarrow$ QueryRoutingMechanism(c, q, K, l)
15            $objectsList \leftarrow objectsList + cList$
16            $l \leftarrow l - Length(cList)$
17   **end**
18   **return** objectsList
19   **end**
20 **end**

---

5. *k*-DaO Use Case: Participatory Data Stewardship and Citizen-Generated Data Creation

The aim of this section is to describe a possible implementation of the above architecture through a specific use case. We first describe the scenario and then we go into the details of the technical specification. With this scenario, we find ourselves in the general context of facilitating the use of privately held data for the public interest. This is in line with the vision of the European Union's strategy on data sharing for public interest [7]. More specifically, the vision we intend to pursue with the implementation of our decentralized personal data marketplace is part of the intent to enable different stakeholders (government, businesses and citizens) to give access to and to use data transformed into non-personal form in order to create value and to make better decisions. In fact, the European Data Strategy elaborates on some points in this area, with components related to data governance and common data spaces, also by means of the Data Governance Act [14]. It enables the safe reuse of certain categories of public-sector data such as personal data.

The specific context of our scenario deals with participatory data [2] such as citizen-generated ones. Citizen-generated data, which include a range of scenarios such as participatory sensing to crowdsourced geospatial datasets, can be integrated with open data portals and, in the future, with shared data spaces. Although, to date, they are not as impactful, the aim is to increase and improve the presence of such data and to involve citizens in designing open data policy, processes and governance [65]. In most cases, citizen-generated data should be made orthogonal to the application of data protection laws and regulations, e.g., GDPR. Therefore, citizen-generated data should not contain personal data or personal data shall be appropriately anonymized or aggregated.

With this in mind, we describe the use case with the help of Figure 3. At the highest level, the flow of data is as follows: (i) citizens store and maintain their personal data in a PDS; (ii) a data aggregator undertakes the task of aggregating a specific kind of data and accesses the PDS through smart-contract access policies; (iii) the aggregator uses algorithms such as *k*-anonymity [66] to render the input personal data anonymous; and (iv) the citizen-generated anonymized aggregated dataset is published for potential data consumers. The main idea is to enable the participation of data owners in the dataset generation through a DAO. A token-based incentive for DAO members, i.e., tokenized data structures, can be used to enable participants to work together to build a curated dataset in pursuit of the instantiation of a decentralized, tokenized data marketplace [19].

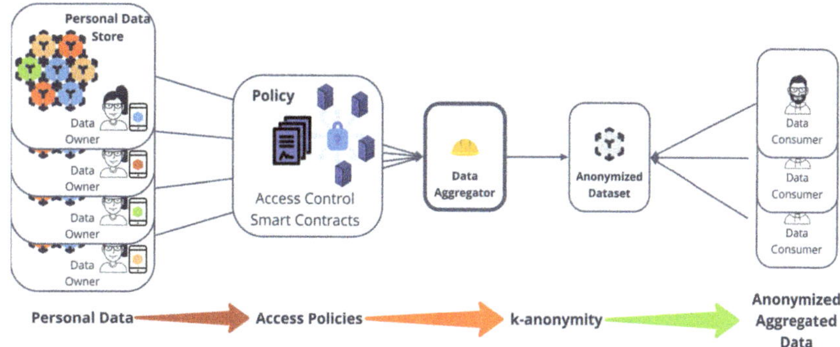

Figure 3. Citizen-generated data use case. Data owners store personal data in a PDS and set some access policies through smart contracts. A data aggregator accesses these data and produces an anonymized dataset in a participatory data stewardship framework. The anonymized aggregated dataset can then be accessed by other data consumers.

We imagine a concrete scenario of citizen-generated hiking trails or pedestrian travel routes, produced using GPS-enabled smartphones using application such as Komoot or AllTrails [67]. For this scenario, one can simply consider three kinds of personal data: (i) the user's travel trace, i.e., a set of latitude and longitude points associated with a timestamp; (ii) the user's photos taken during the travel; and (iii) the list of nearby Bluetooth devices updated with a constant interval.

5.1. Anonymizing Data by Aggregation

Figure 3 shows an overview of the interaction between the main actors. Data owners (leftmost boxes) maintain personal data in a PDS implemented using IPFS [20] as DFS. These data are travel traces, photos and Bluetooth ids recorded during the data owners' hiking sessions. Data owners also register personal data they want to share along with descriptions of what they measure, i.e., keywords in the hypercube DHT (not shown in figure; see Section 4.4). Each piece of data is then indexed in the IOTA DLT through a new stream channel for each hiking session (not shown in figure). The messages in the channel refer to data in IPFS using the CID as an immutable universal identifier. An access-control smart contract owned by the data owner (between data owners and aggregator in the figure) points to different stream channels using the associated announcement link. This smart contract is stored in a private permissioned Ethereum blockchain implemented using GoQuorum [68], i.e., the authorization blockchain. The data aggregator (at the middle of the figure) interacts with such a blockchain to request the data owners' data in line with their policies. If it manages to access the data of at least k data owners, the aggregator creates a k-DaO with the owners in the same blockchain, in order to work in a participatory data stewardship framework [2]. The anonymized dataset must meet certain requirements; otherwise, the k-DaO may decide to stop production. For instance, the data aggregator should be able to perform the data aggregation, producing a dataset that presents properties of k-anonymity and differential privacy [69]. This dataset can then be accessed by a variety of data consumers (rightmost boxes in figure) using the same data marketplace in a process where every participant to the dataset creation is rightfully rewarded.

We now make a brief digression on what it means to apply anonymization techniques in this case. The GDPR Recital 26 states that personal data becomes anonymous if it is 'reasonably likely' that no identification of a natural person can be derived [13]. This is based on the fact that the anonymization of a dataset can be defined as robust on a case-by-case basis [70]. Some techniques can provide privacy guarantees and can be used to generate efficient anonymization processes but only if their application is engineered appropriately. The k-anonymity proposal was introduced in [66], and it is considered one of the most popular approaches for syntactic protection, i.e., each release of data must be indistinguishably related to no less than a certain number (e.g., k) of individuals in the population. For instance, through a generalization approach, original values are substituted with more general values, such as the date of birth generalized by removing day and month of birth. On the other hand, we find semantic techniques, i.e., when the result of an analysis carried out on a dataset is insensitive to the insertion or deletion of a tuple in the dataset. Differential privacy [69] is the main example in this case, where a dataset is released and recipients learn properties about the population as a whole but that are probably wrong for a single individual. This can be achieved for instance by adding noise to the original dataset.

5.2. Step Zero: Search Data on the Decentralized Marketplace

The first step, or "step zero", before accessing any piece of data is the search for a specific kind of data, i.e., the data subset that a potential data consumer is interested in. This is the part where the hypercube DHT comes into play (see Section 4.4 for a detailed explanation). Figure 4 shows an example of the search of data on the decentralized marketplace. The data aggregator requests for a SuperSet Search to the hypercube, with a keyword set $K = \{$"walk", "mountain", "Tuscany"$\}$. The hypercube returns a set of aggregation links pointing to IOTA stream channels containing related data, e.g., *6bb3347...:219*. The first message of the stream channel is open to the marketplace users and includes information that points to the smart contract used for the access control. The information includes the identifier of the authorization blockchain network and the smart contract address. Each subsequent channel's message includes information to the data themselves, i.e., hash links in the form of IPFS CIDs. In particular, each message stores the CID of a IPFS directory that stores the location data, photo and Bluetooth ids in a specific timestamp, e.g., *QmW...V4b2t* is the CID of the directory and *QmW...V4b2t/1* contains a location point and timestamp. Of course, the data are encrypted; hence, the content of the IPFS data are not meaningful at this point. The next step, thus, is to gain access to the content key used for the encryption.

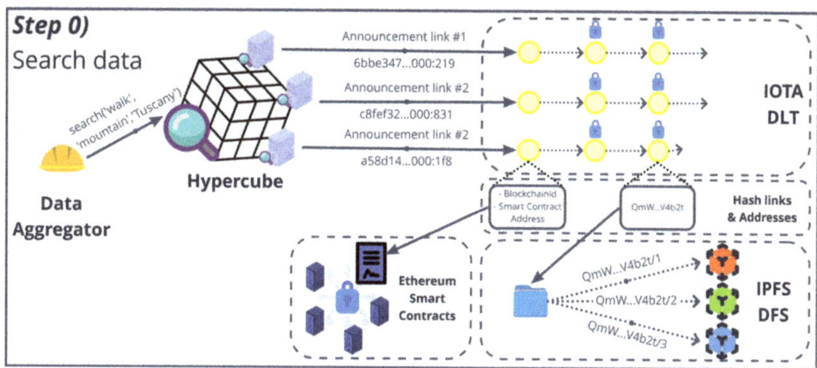

Figure 4. Example of searching data on the decentralized marketplace.

5.3. Smart Contracts Implementing the Distributed Access Control

As seen in Section 4.2, the interesting aspect of smart contracts is that an algorithm executed in a decentralized manner enables two parties, i.e., data owner and aggregator, to reach an agreement in the transaction of the data. This not only increases the disintermediation in such a process but also leaves traces to be later audited and provides incentives to all the actors to correctly behave. Figure 5 graphically shows the process of the data aggregator accessing data owner's data, while Figure 6 shows the UML Class Diagram of the smart contract implementations we discuss in this subsection.

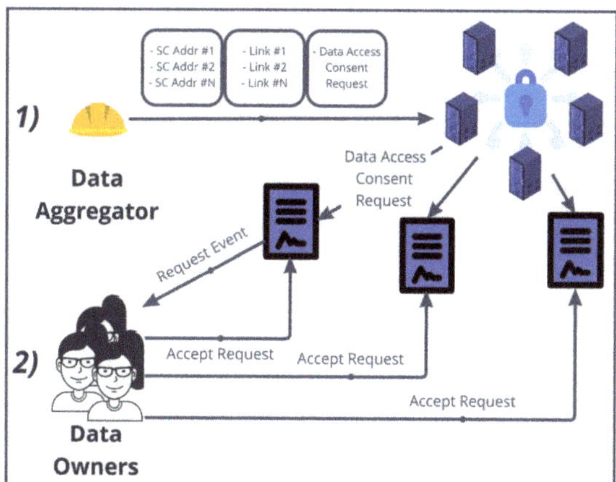

Figure 5. Example of the distributed access control where a data aggregator requests access to the data of some data owners.

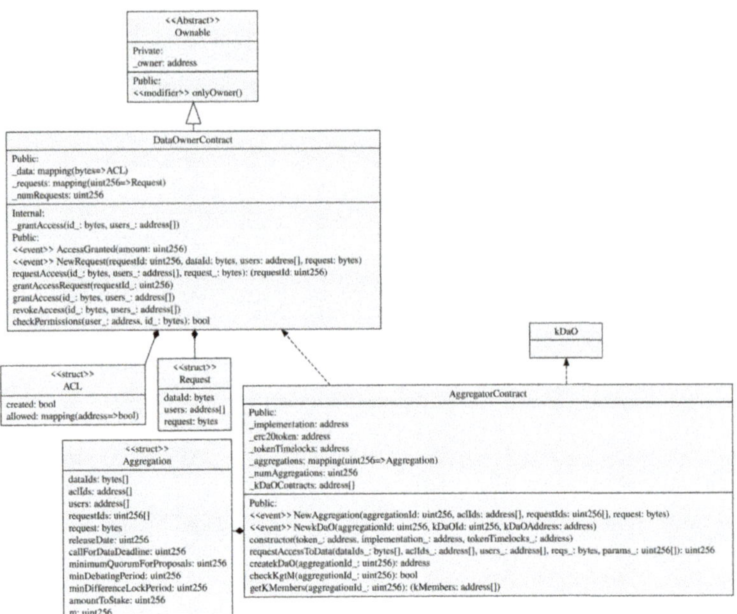

Figure 6. UML Class Diagram of *DataOwnerContract* and *AggregationContract*. Some classes, attributes and methods have been removed to render the diagram clearer.

- Each data owner has previously deployed a *DataOwnerContract* in the authorization blockchain. The data aggregator too has previously deployed an *AggregatorContract*.
- In the "step zero", the aggregator has obtained a list of *DataOwnerContract* addresses that point to IOTA stream channels through as many announcement links. Then, they produce a data access consent request in a string form for such data (the use of standardized models for the consent request, such as W3C recommended ontologies, is left as future work).
- The aggregator gives these three pieces of information as inputs to the *requestAccessTo-Data()* method in the *AggregatorContract()*, together with a series of parameters needed for the *k*-DaO (shown in the next subsection). This method implements, in only one blockchain transaction, the request to access data for each *DataOwnerContract* found as input. In particular, the method *requestAccess()* is invoked for each *DataOwnerContract*, with the associated announcement link and request as input (Figure 6 shows *id_* as parameter representing the link and an array of addresses *users* for representing the Ethereum accounts that will be granted access).
- A *NewRequest* event will reach each data owner. This one decides to consent to the access to data based on the data access consent request received through the event. If so, the data owner invokes the *grantAccessRequest()* method in the *DataOwnerContract*.
- Among the parameters set in *requestAccessToData()*, *m* was set as the minimum number of members needed to create the *k*-DaO and to start the data aggregation process. It is also the minimum number of participants required to provide "reasonable" anonymity.
- The aggregator uses the *checkKgtM()* method to check if the number *k* of data owners that granted the access to their data is greather than *m*. When this happens, the aggregator can create the *k*-DaO through the *createkDaO()* method that instantiates a new *kDaO* contract.
- The aggregator can now access all content keys for the decryption of all the data owners' data through the authorization blockchain nodes, as described in Section 4.2.1.

5.4. Smart Contracts Implementing the k-DaO

The *k*-DaO is DAO composed by the *k Data Owners* that grant access to their data to the data aggregator. Simply put, the aggregator stakes a safety deposit, and the DAO is used to start at any moment a vote to redeem this stake. The rationale behind it is to limit aggregator's malicious behavior. Not only for this but also if the creation process of the anonymized dataset involves a more complex case of curated dataset (e.g., Open-StreetMap [71]), then DAO members can make new proposals and add suggestions to vote in order to steer the development of the dataset generation. Figure 7 graphically shows the process of *k*-DaO creation and voting, while Figure 8 shows the UML Class Diagram of the smart contract implementations we discuss in this subsection.

- A *kDaO* contract is created for each aggregation process. The *AggregationContract* acts as a contract factory but implementing a proxy pattern (EIP-1167 Minimal Proxy [72]). Instead of deploying a new contract each time such as in the factory pattern, this implementation clones an already deployed contract functionalities by delegating all methods invocation to it.
- Some DAO parameters were already set up during the request data access process, such as the amount the aggregator stakes. When the *kDaO* contract is created, a transfer of an amount of ERC20 tokens [73], i.e., the *kDaOToken*, is performed automatically from the aggregator account to the *kDaO* contract. At the end of the aggregation process, the aggregator can redeem this stake if all operations have been successful.
- *k*-DaO members can call for a vote and then decide on a proposal. Any member can make a proposal using the *submitProposal()* method, and for that proposal, all members can submit a suggestion using *submitSuggestion()*. Then, all members vote on a suggestion regarding that proposal. For instance, a proposal could be to "Change anonymization technique", and some suggestions could be "Differential privacy" or "*k*-anonymity". Each proposal has their own debate period and any member can invoke *vote()* to vote for a suggestion within that time period. After the debate period, the method *executeProposal()* counts the votes, if *minQuorum* is reached, then it stores the result and possibly enacts a specific procedure.
- Indeed, any extension of the previous voting smart contract can be developed to allow for a decision taken to directly enact an operation to be executed on-chain. In this case, *submitRefundProposal()* specifically starts a vote to take the data aggregator's staked amount and to redistribute it to all members. In this case, *executeProposal()* would subdivide the staked amount to all the members if the proposal is passed.
- The *kDaOToken* is central in the DAO, as it also allows members to vote. Indeed, a member vote weight is proportional to the amount of tokens locked until a date that comes after the debate period ends. This is performed in order to avoid malicious data owners unreasonably voting to redeem the aggregator stake. A *TokenTimelockUpgreadable* is used for each token lock. This is created using the proxy pattern as well.

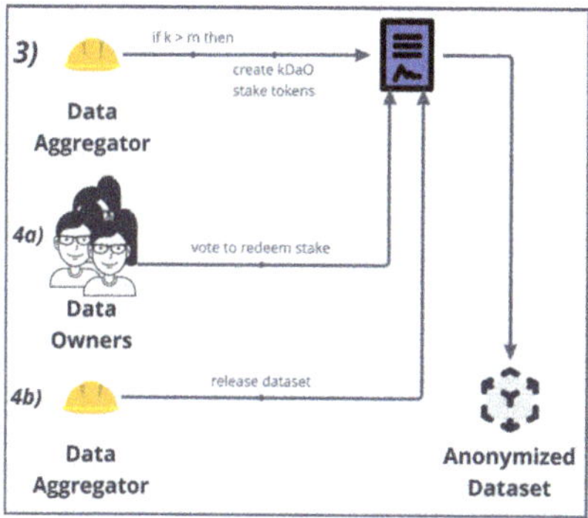

Figure 7. Example of the anonymized dataset creation and DAO voting.

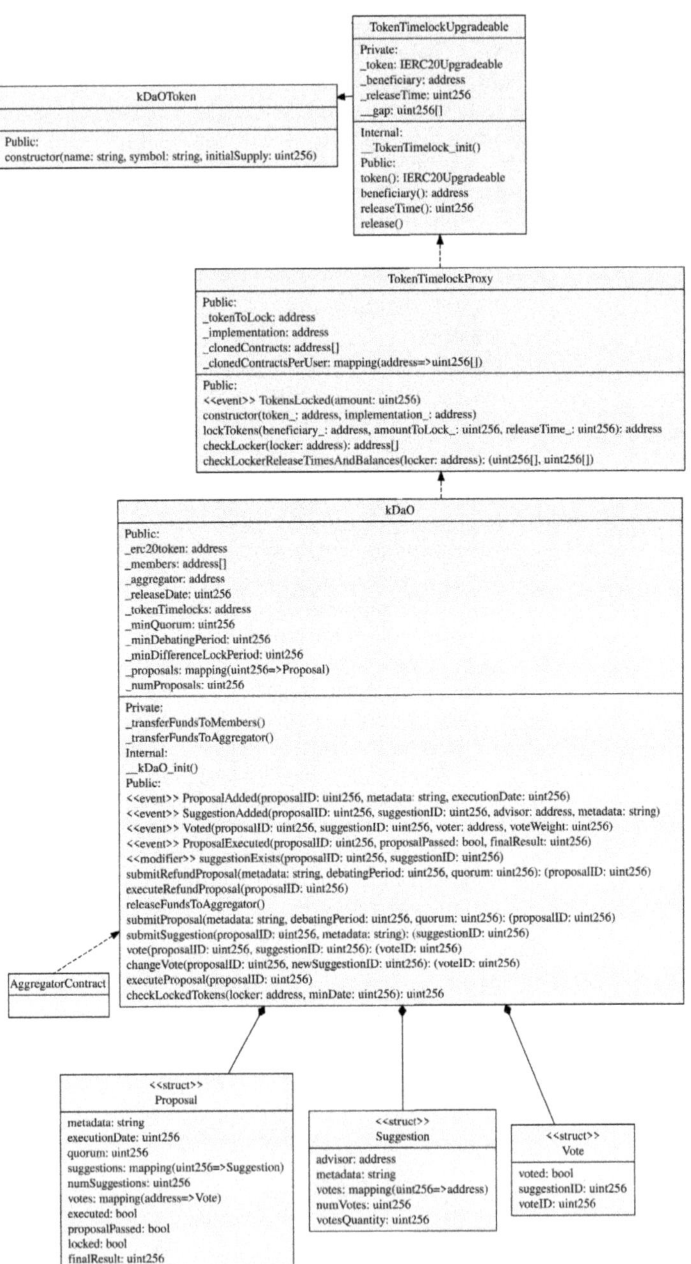

Figure 8. UML Class Diagram of *kDaO, TokenTimelockProxy, TokenTimelockUpgradeable* and *kDaOToken*. Some classes, attributes and methods have been removed to render the diagram clearer.

5.5. Anonymized Aggregated Dataset

Finally, the work of the aggregator comes to produce new data in the form of anonymized aggregated data, providing anonymity by design. Multiple configurations of aggregated data can be produced, if stated earlier. Additionally, some kind of proof can be implemented for measuring the exact quantity of data used from each subject's dataset, e.g., storing in the *kDaO* contract the root of a Merkle tree that contains all the data pieces hashes used as leaves; then, k-DAO members can validate it by requesting (off-chain) leaves to the aggregator.

For the sake of the citizen-generated data use case, the result of the whole process is stored in an open data platform. If needed, some data, such as the participants list, can be shown upon request, but it is not public, since the authorization blockchain is a private permissioned one. In other cases, the resulting dataset can be encrypted, uploaded in IPFS and then referenced in new stream channels. In this case, the dataset is treated as all the other kinds of data in the marketplace and data consumers can access to it through a *DataOwnerContract* owned by the aggregator. In this case, some kind of royalties can be transferred directly to the k-DaO members, where the payment is proportional to the contribution produced by each participant, e.g., aggregator $= 55\%$, data owner$_1 = 20\%$, data owner$_2 = 10\%$, data owner$_3 = 15\%$.

6. Performance Evaluation

Based on the above k-DaO use case, we conducted the performance evaluation in three stages: (i) in the first stage, we simulated a DHT network implementing the hypercube queries of the use case's "step zero", in order to test the average steps necessary to reach all nodes; (ii) in the second stage, we set up a local permissioned authorization blockchain to test the distributed access control in use case's steps one and two; and (iii) in the third stage, we evaluate the implementation of all smart contracts by measuring the gas usage.

In this work, we lack an analysis of the performances for storing and retrieving data from IOTA and IPFS. However, we dealt with these aspects in previous work, testing out specifically the storing of personal data such as location data and photos, i.e., testing IOTA [74] and DFS including IPFS [75]. We refer the reader to these two studies. Moreover, being separate systems, the latency in performing operations are added up one another. Meaning that a data aggregator first needs to obtain the content key from the authorization blockchain (evaluated in this work) and then operate with IOTA or IPFS.

The decentralized personal data marketplace component implementation can be found as an open source code on Github [76–79].

6.1. Hypercube DHT Simulation

We conducted a simulation assessment using PeerSim, a simulation environment developed to build P2P networks using extensible and pluggable components [80,81]. Once the hypercube-structured DHT was designed and implemented for multiple keyword search Section 4.4.2, we focused on studying the efficiency of the routing mechanism. The simulation implementation and the tests data can be found as open source code in [82]. Below are the main results obtained.

6.1.1. Tests Setup

Several tests were carried out assuming different scenarios in which the network consisted of a variable number of nodes and stored a variable number of objects. In order to evaluate Pin Search and Superset Search, tests were carried out on different sizes of the hypercube. Specifically, the number of nodes varied from 128 ($r = 7$) up to 8192 ($r = 13$). Then, for each dimension r, a different number of randomly created keyword objects, i.e., IOTA announcement links, was inserted in the DHT. The number of objects taken into consideration varies from 100, 1000 and finally 10,000.

6.1.2. Results

Given the nature of the tests, i.e., a simulated network, we considered the number of hops required for each new query as a parameter to be evaluated. A hop occurs when a query message is passed from one DHT node to the next. The query keyword sets were randomly generated, and the starting node was randomly chosen. For each type of test, 50 repetitions were performed, and then, the average results were calculated. For the Superset search, the limit value was set to $l = 10$ objects.

Pin Search

As shown in Table 1 and Figure 9 (left), the number of hops required to transmit a message from the source node to the destination node increases as the hypercube dimension increases, i.e., nodes number. The average number of hops increases from about 3.5 for 128 nodes ($r = 7$) to about 6.72 for 8192 nodes ($r = 13$). This behavior can be explained by the fact that, by increasing the hypercube dimension, the path that a message must take before reaching its destination is automatically enlarged. The number of objects in the testbed does not affect the final outcome, since the path to reach the target node only follows the rationale of the hypercube and does not depend on the number of keyword object associations stored in the DHT.

Table 1. Pin Search number of hops.

Nodes Number	Average			Standard Deviation			Confidence Interval (95%)		
	100	1000	10,000	100	1000	10,000	100	1000	10,000
128	3.64	3.2	3.5	1.33	1.32	1.12	(3.2, 4.0)	(2.8, 3.5)	(3.1, 3.8)
256	4.08	4.28	3.66	1.45	1.48	1.31	(3.6, 4.4)	(3.8, 4.6)	(3.2, 4.0)
512	4.62	4.8	4.72	1.57	1.70	1.24	(4.1, 5.0)	(4.3, 5.2)	(4.3, 5.0)
1024	5.02	4.96	4.9	1.68	1.67	1.69	(4.5, 5.4)	(4.4, 5.4)	(4.4, 5.3)
2048	5.48	6.04	5.48	1.76	1.85	1.69	(4.9, 5.9)	(5.5, 6.5)	(5.0, 5.9)
4096	6.02	6.18	5.96	1.55	1.61	1.62	(5.5, 6.4)	(5.7, 6.6)	(5.5, 6.4)
8192	6.78	7.08	6.28	1.63	1.60	1.64	(6.3, 7.2)	(6.6, 7.5)	(5.8, 6.7)

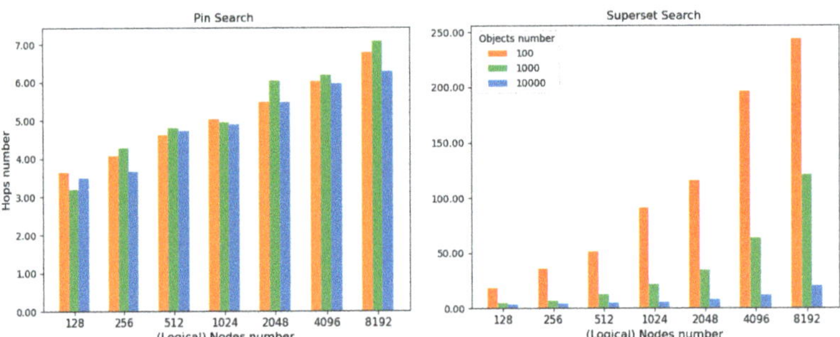

Figure 9. Number of hops on average for the Pin Search (**left**) and Superset Search (**right**).

Superset Search

The tests performed on the Superset Search present results with dissimilar values with respect to the previous case (Table 2 and Figure 9 (right)). At a first glance, in fact, those apparently anomalous values stand out, corresponding to a high number of hops between nodes, which decreases with the referenced object number. With a low number of objects referenced in the DHT, there are a high average number of hops needed to satisfy the Superset search. This phenomenon can be explained by the fact that the Superset search traverses the spanning binomial tree of the sub-hypercube induced by the node responsible for the keyword set, until it finds the number of objects indicated by the limit, i.e., $l = 10$. Hence, in a network with many nodes and few objects, the query might take longer to reach that limit because many nodes are "empty", i.e., do not reference any object. Considering the case of 4096 nodes ($r = 12$) and 10,000 objects, in a Pin search, 5.96 hops are required, on average. In a Superset search, other $11.92 - 5.96 = 5.96$ hops are needed to reach other nodes containing other results of the superset search, until the limit l is reached. If objects were uniformly distributed, the total number of nodes requested to return objects would have dropped to 4 nodes because each node would have maintained $\frac{10,000}{4096} = 2.44$ object references on average and $l = 10 (\cong 4 \times 2.44)$.

Table 2. Superset Search number of hops.

Nodes Number	Average			Standard Deviation			Confidence Interval (95%)		
	100	1000	10,000	100	1000	10,000	100	1000	10,000
128	18.28	4.54	3.52	8.44	1.54	1.19	(15.9, 20.6)	(4.1, 4.9)	(3.1, 3.8)
256	35.90	6.80	4.16	17.89	2.25	1.43	(30.9, 40.8)	(6.1, 7.4)	(3.7, 4.5)
512	51.18	12.16	4.46	37.85	3.29	1.31	(40.6, 61.6)	(11.2, 13.0)	(4.1, 4.8)
1024	91.06	21.70	5.08	72.44	6.23	1.68	(70, 111)	(19.9, 23.4)	(4.6, 5.5)
2048	115.70	34.56	7.84	98.39	13.00	1.98	(88, 142)	(30.9, 38.1)	(7.2, 8.3)
4096	196.00	63.38	11.92	186.88	25.37	2.64	(144, 247)	(56.3, 70.4)	(11.1, 12.6)
8192	243.90	120.38	20.38	253.59	68.65	6.28	(173, 314)	(101, 139)	(18.6, 22.1)

6.1.3. Discussion

The results obtained confirm what was expected due to the hypercube structure of the network: the Pin Search number of hops are of the order of the logarithm of the hypercube logical node number, i.e., $\log(n) = r$. In particular, on average, they are equal to $\frac{\log(n)}{2} = \frac{r}{2}$. For what concerns the Superset Search number of hops, on average, it is equal to $\frac{\log(n)}{2} + l$, where l is the limit of the number of nodes in the sub-hypercube to reach.

These results show the goodness of the solution in the trade-off between memory space and response time. In traditional DLTs, such as Ethereum and IOTA, searching for a datum in a transaction means traversing all the "transaction sea" in the ledger, and for this reason, the current solution is to use centralized "DLT explorers" [83]. On the other hand, in the case of sharded DLTs, the proposed solution could become a Layer-1 protocol to search the data between many shards.

Finally, while in this study we focused on DLTs as the underlying data storage, it is worth mentioning that, due to the origins of the hypercube proposal [24], DFS systems can perfectly fit with such architecture, since most of them are based on DHT already. Indeed, the implementation of the hypercube for keywords search in IPFS is a matter of future work.

6.2. Authorization Blockchain Performances

In this subsection, we present the methodology and results of the performance evaluation we carried out for the authorization blockchain. We deployed all the smart contracts in a local permissioned Ethereum blockchain, using the Consensys GoQuorum implementation [68]. ConsenSys Quorum is an open-source protocol layer with the aim of building Ethereum compatible environments for enterprises. Supporting the Ethereum protocol means the possibility to execute smart contracts compiled from Solidity. Moreover, it is composed of a suite of different technologies, among which we find GoQuorum, a fork of the Ethereum node implementation in Golang. The rationale behind this choice is to be able to implement private smart contracts and transactions for protecting personal data stored on-chain by the data owners, a feature that GoQuorum supports.

We have already tested some implementations of the authorization blockchain in [9], making a comparison between two different cryptographic methods for key distribution using two open source library implementations. In this work, we test our implementation of the TPRE Umbral protocol [23], openly available as source code [77]. This is executed by the authorization blockchain nodes and thus integrated with the GoQuorum software. The client software and the smart contracts implementation is open source too and can be found in [84].

6.2.1. Test Setup

During the test, we used the Istanbul Byzantine Fault-Tolerant (IBFT) consensus mechanism: each block requires multiple rounds of voting by the set of validators (>66%), recorded as a collection of signatures on the block [68]. During the tests, four validator nodes were deployed to create the base blockchain network. Each validator node executes the consensus mechanism with parameter values set up following the recommendations in [68], e.g., minimum inter-block validation time is set to 1 s. Moreover, these nodes also execute the TPRE service. One non-validator node is used to expose the APIs for external clients to interact with the blockchain. Several client nodes are created to interact with these APIs, which in turn disseminate transactions within the network [85]. The network was run on a server with a 10 cores Intel Xeon CPU and 8 GB of DDR4 RAM.

In the following, we evaluate this set of operations that implement the scenario shown in Section 5.3.

1. **Request Access**—this operation is executed by the data aggregator and consists of only one method invocation, i.e., the *requestAccessToData()* method in the *Aggregator-Contract*; we recall that this method requests access to data for each *DataConsumerContract* given as input.
2. **Grant Access**—this operation is executed by each data owner by invoking the *grantAccessRequest()* from their own *DataConsumerContract()*; this will store the aggregator public key pk_{DA} in the smart contract ACL.
3. **Create KFrags**—this operation includes three subsequent steps; first, the owner generates a new set of n kfrags using the data aggregator's pk_{DA} (as described in Section 4.2.1); then, the owner sends a kfrag each to the n authorization blockchain nodes; finally, the owner requests to the n nodes the creation of a cfrag using the kfrag just got (the capsule for the piece of data interested was sent in a pre-processing step, not accounted for the measuring).
4. **Get CFrags**—the last operation is executed by the data aggregator to obtain access to the content key; the aggregator first sign a challenge-response message using the secret key sk_{DA} associated to the pk_{DA}; then, the aggregator sends a Get CFrag request to k authorization blockchain nodes using the signed message; and each node validates the signature and check if pk_{DA} is in the associated ACL in the *DataConsumerContract*, and if so, each node returns a cfrag to the data aggregator.

6.2.2. Results

We recall that n is the number of validator/authorization blockchain nodes and was set to 4. We consider a round of operations the successful execution of the above described operations in order. The independent variables tested were the *threshold t*, from 1 to 4, and the *number of data owners k*, from 10 to 80 with an increase of 10 each time. We tested all the combinations of independent variables 3 times; then, we averaged the results. In each test, we initiated the round of operations 10 times for each data owner, with an interval of 3000 ms on average (value given by a Poisson Process with a mean of 3000 ms). This implies that, if overall, the set of operations lasted more than 3000 ms to be executed, probably another one was launched in parallel. This is for each data owner. The dependent metrics we measured with the tests are the *latency*, for a response to an operation, and the system *throughput*, i.e., the number of rounds of operations per second.

Round of Operations

Figure 10 shows the average response latency and standard deviation for each operation in a round. The first result that stands out is the large difference in latency between the Request Access and Grant Access operations and the Create KFrags and Get CFrags operations. This is due to the fact that the first two operations involve writing in the authorization blockchain's ledger. Thus, we can already see the impact of the blockchain in the overall system response latency.

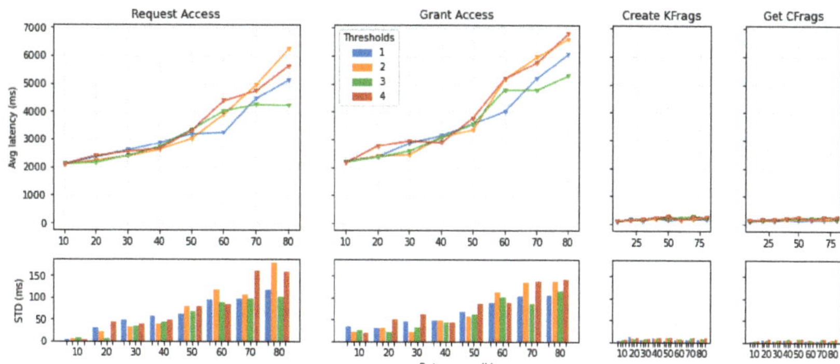

Figure 10. Average response latency and standard deviation for each operation in a round, varying the threshold from 1 to 4 and data owners from 10 to 80.

As can be seen, in general, the t value does not affect the results greatly. On the other hand, as expected, the k value that represents the number of data owners is the key factor. A slow but constant increase in the round response latency happens between 10 and 40 owners, starting from 2 s latency to 3, for both Request Access and Grant Access operations. After 40 owners, the latency increases faster per number of owners. This seems to be correlated to the fact that a new round is started on average each 3 s for each data owner. Thus, if the round takes approximately more than 3 s, as from $k = 50$ onward, many more operations start to be executed in parallel. The increase in such parallel executions seems to increase the response latency overall.

While the blockchain writing-dependent operations are in the order of the thousand milliseconds, i.e., seconds, the KFrags and CFrags operations are in the order of the hundreds and can be better analyzed using Table 3.

Table 3. Average response latency and confidence interval for the Create KFrags and Get CFrags operations in a round, varying t and k.

k	t	Create KFrags (ms)		Get CFrags (ms)	
		Average	Conf Int (95%)	Average	Conf Int (95%)
	1	75.6	(72.11, 79.09)	106.63	(104.79, 108.47)
10	2	86.58	(82.07, 91.09)	116.01	(113.49, 118.54)
	3	88.23	(82.38, 94.09)	120.17	(117.04, 123.3)
	4	100.48	(94.42, 106.53)	127.98	(124.23, 131.73)
	1	155.96	(144.22, 167.69)	128.38	(122.94, 133.82)
20	2	130.32	(122.91, 137.73)	135.27	(130.74, 139.79)
	3	144.0	(136.01, 152.0)	152.28	(146.56, 157.99)
	4	146.92	(135.99, 157.85)	163.61	(154.72, 172.49)
	1	113.11	(107.89, 118.33)	119.94	(116.93, 122.95)
30	2	146.23	(140.54, 151.92)	141.16	(137.83, 144.49)
	3	172.57	(163.51, 181.62)	167.19	(160.77, 173.62)
	4	162.65	(154.41, 170.89)	173.43	(167.14, 179.73)
	1	211.23	(200.45, 222.01)	158.86	(152.58, 165.15)
40	2	176.49	(168.25, 184.73)	166.48	(160.42, 172.53)
	3	206.08	(196.19, 215.97)	192.9	(185.59, 200.22)
	4	220.54	(210.67, 230.4)	209.77	(202.55, 216.98)

Table 3. *Cont.*

k	t	Create KFrags (ms)		Get CFrags (ms)	
		Average	Conf Int (95%)	Average	Conf Int (95%)
50	1	122.28	(117.61, 126.95)	122.32	(119.94, 124.7)
	2	189.77	(179.35, 200.2)	170.66	(163.35, 177.96)
	3	235.03	(224.69, 245.36)	215.84	(207.61, 224.08)
	4	267.82	(257.65, 277.99)	251.73	(243.17, 260.3)
60	1	172.14	(166.32, 177.95)	148.48	(144.76, 152.19)
	2	177.44	(169.55, 185.34)	172.77	(166.75, 178.8)
	3	225.4	(216.35, 234.45)	208.26	(201.29, 215.22)
	4	140.75	(135.36, 146.15)	159.98	(155.94, 164.03)
70	1	158.52	(152.33, 164.7)	141.2	(137.57, 144.83)
	2	179.65	(173.0, 186.3)	166.32	(161.58, 171.05)
	3	275.55	(264.45, 286.65)	250.54	(241.68, 259.4)
	4	230.97	(221.41, 240.53)	229.48	(221.51, 237.45)
80	1	178.65	(172.19, 185.1)	153.97	(149.92, 158.02)
	2	198.21	(190.55, 205.88)	178.61	(173.34, 183.89)
	3	204.39	(196.89, 211.89)	205.24	(198.95, 211.53)
	4	226.86	(217.05, 236.66)	231.71	(223.5, 239.92)

In both cases, we can see a direct correlation of response latency with both the t and k values. With $k = 10$, latency values for the Create Kfrag operation are around 90 ms, while those for the Get CFrag operation are around 110 ms. With $k = 80$, the values more or less double.

System Throughput

Figure 11 shows the results obtained when considering the round as a single operation, i.e., aggregating the results for each single operation. The figure thus shows the number of rounds per seconds, i.e., ops/s. The throughput results in more than 0.2 ops/s for the number of owners $k = 10$ and linearly decreases with the increase in k. With $k = 80$, we have on average a throughput of $\sim$0.07 ops/s. In this case as well, we can notice how the influence of t is almost irrelevant. As we have seen before, t influences greatly the Create Kfrag and Get CFrag operations, but these two, overall, slightly increase the round response latency with respect to the Request Access and Grant Access operations. Indeed, here too, we can see the effect of the blockchain execution in delaying the response time.

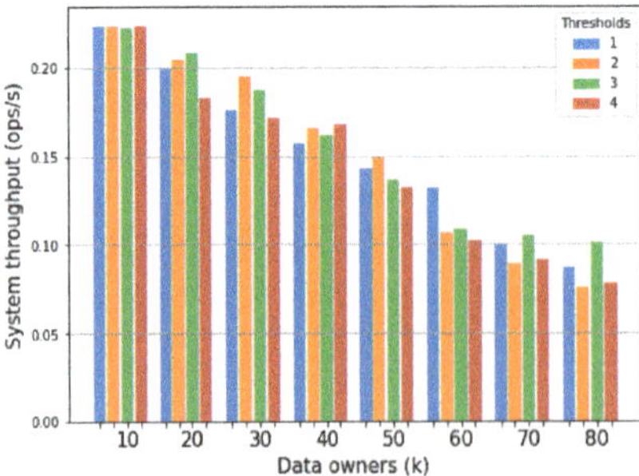

Figure 11. System throughput considering a round as a single operation, i.e., aggregating the results for each single operation, while varying t and k.

Threshold Number

Figure 12 shows the results when increasing the t value and the number of owners k for each i-th round, i.e., it shows the performances for each subsequent round instead of aggregating all rounds through their mean. In this case, the results shown confirm that the increase in t does not influence much to the overall response delay. However, this temporal point of view shows the accumulation of delay in the response time when increasing k. We can see, for instance, that up to $k = 30$ each i-th round has more or less the same average latency. When increasing k, however, the latency of rounds in the middle spikes upwards, due to the accumulation of operations to perform, and then returns to a relatively normal value in the last rounds (i.e., 9-th and 10-th).

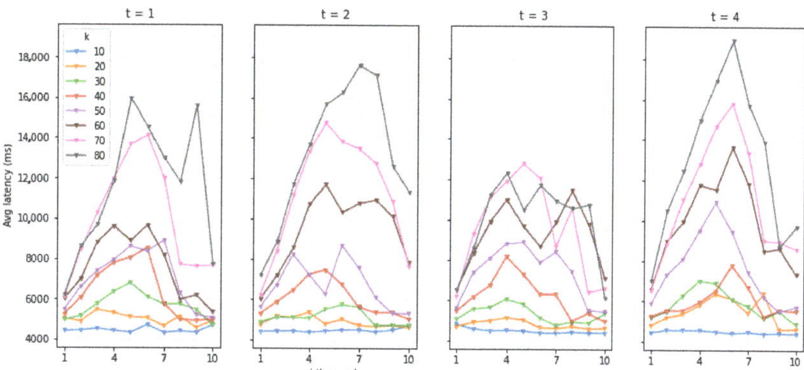

Figure 12. Average response latency when increasing the threshold t value and the number of owners k for each i-th round.

6.2.3. Discussion

Limited to the scenario we tested, it seems that a number of data owners around 30 and 40 induces the best ratio of completed rounds to response latency time. With this workload, the system can fulfill around 0.17 rounds per seconds. Overall, we can observe how the writing in the blockchain greatly impacts the whole system performance and that the number of requests related only to the TPRE operations can still scale to a larger number of data owners.

In reality, the interaction of owners with the system may be much slower, making the overall round latency increase but, at the same time, diminishing the system workload. We can imagine that the *NewRequest* event triggered by the *requestAccess()* method is shown to the data owner through a smartphone notification, thus requiring seconds, if not hours, to be read and accepted. In this context, the use of semantic web-based policy languages to express rich rules for consent and data requests could be useful in automating (and thus speeding up) this process [59]. This is left as future work.

Nonetheless, we argue that the results show the viability of our approach, especially having the possibility to tweak the authorization blockchain parameters and node hardware configuration. Moreover, the good response of the TPRE implementation gives reason to believe that, by moving this module to another blockchain that supports smart contracts but provides better latency, even improved outcomes can be achieved.

6.3. Smart Contract Gas Usage

Our focus is now on the execution of the smart contracts that we described in the use case Section 5, with regards to steps 1 to 4. In Ethereum, the *gas* is a unit that measures the amount of computational effort needed to execute operations. Thus, the higher the gas usage for a method, the more intense the computation of a blockchain node to execute the

Acknowledgments: An early version of this work appeared in [87]. This paper is an extensively revised and extended version where more than 50% is new material. We are indebted to Gabriele D'Angelo for his support in the research conducted for this work and to Cesare Giansante for his contribution on a preliminary implementation of the hypercube DHT simulation.

Conflicts of Interest: The authors declare no conflict of interest.

Abbreviations

The following abbreviations are used in this manuscript:

API	Application Programming Interface
ACL	Access Control List
CFrag	Capsule Fragment
CID	Content Identifier
DAG	Directed Acyclic Graph
DAO	Decentralized Autonomous Organization
DFS	Decentralized File Storage
DHT	Distributed Hash Table
DLT	Distributed Ledger Technology
GDPR	General Data Protection Regulation
IBFT	Istanbul Byzantine Fault-Tolerant
IPFS	InterPlanetary File System
KFrag	Key Fragment
P2P	Peer-to-Peer
PIMS	Personal Information Management System
PDS	Personal Data Store
PRE	Proxy Re-Encryption
SSI	Self-Sovereign Identity
TPRE	Threshold Proxy Re-Encryption
UML	Universal Modeling Language
W3C	World Wide Web Consortium

References

1. Cadwalladr, C.; Graham-Harrison, E. Revealed: 50 million Facebook profiles harvested for Cambridge Analytica in major data breach. *The Guardian* **2018**, *17*, 22.
2. Patel, R. *Participatory Data Stewardship*; Technical Report; Ada Lovelace Institute: London, UK, 2021.
3. Prandi, C.; Mirri, S.; Ferretti, S.; Salomoni, P. On the need of trustworthy sensing and crowdsourcing for urban accessibility in smart city. *ACM Trans. Internet Technol.* **2017**, *18*, 1–21. [CrossRef]
4. Floridi, L. The fight for digital sovereignty: What it is, and why it matters, especially for the EU. *Philos. Technol.* **2020**, *33*, 369–378. [CrossRef] [PubMed]
5. Ramachandran, G.S.; Radhakrishnan, R.; Krishnamachari, B. Towards a Decentralized Data Marketplace for Smart Cities. In Proceedings of the 2018 IEEE International Smart Cities Conference (ISC2), Kansas City, MO, USA, 16–19 September 2018; pp. 1–8. [CrossRef]
6. Zichichi, M.; Ferretti, S.; D'Angelo, G. A Framework based on Distributed Ledger Technologies for Data Management and Services in Intelligent Transportation Systems. *IEEE Access* **2020**, *8*, 100384–100402. [CrossRef]
7. High-Level Expert Group on Business-to-Government Data Sharing. *Towards a European Strategy on Business-to-Government Data Sharing for the Public Interest*; Technical Report; European Commission: Brussels, Belgium, 2021.
8. Janssen, H.; Singh, J. Personal Information Management Systems. *Internet Policy Rev.* **2022**, *11*, 1–6. [CrossRef]
9. Zichichi, M.; Ferretti, S.; D'Angelo, G.; Rodríguez-Doncel, V. Data Governance through a Multi-DLT Architecture in View of the GDPR. *Clust. Comput.* **2022**, 1–32. [CrossRef]
10. Yan, Z.; Gan, G.; Riad, K. BC-PDS: Protecting privacy and self-sovereignty through BlockChains for OpenPDS. In Proceedings of the 2017 IEEE Symposium on Service-Oriented System Engineering (SOSE), San Francisco, CA, USA, 6–9 April 2017; pp. 138–144.
11. Crabtree, A.; Lodge, T.; Colley, J.; Greenhalgh, C.; Glover, K.; Haddadi, H.; Amar, Y.; Mortier, R.; Li, Q.; Moore, J.; et al. Building accountability into the Internet of Things: The IoT Databox model. *J. Reliab. Intell. Environ.* **2018**, *4*, 39–55. doi: [CrossRef]
12. Sambra, A.V.; Mansour, E.; Hawke, S.; Zereba, M.; Greco, N.; Ghanem, A.; Zagidulin, D.; Aboulnaga, A.; Berners-Lee, T. *Solid: A Platform for Decentralized Social Applications Based on Linked Data*; Technical Report; MIT CSAIL & Qatar Computing Research Institute: Cambridge, MA, USA, 2016.
13. European Commission. *A European Strategy for Data*; European Union: Brussels, Belgium, 2020.
14. European Commission. *European Data Governance (Data Governance Act)*; European Union: Brussels, Belgium, 2020.

15. Council of European Union. *Regulation (eu) 2016/679—Directive 95/46*; European Union: Brussels, Belgium, 2016.

16. Kondova, G.; Erbguth, J. Self-sovereign identity on public blockchains and the GDPR. In Proceedings of the 35th Annual ACM Symposium on Applied Computing, Brno, Czech Republic, 30 March–3 April 2020; pp. 342–345.

17. Park, J.S.; Youn, T.Y.; Kim, H.B.; Rhee, K.H.; Shin, S.U. Smart contract-based review system for an IoT data marketplace. *Sensors* **2018**, *18*, 3577. [CrossRef]

18. Özyilmaz, K.R.; Doğan, M.; Yurdakul, A. IDMoB: IoT Data Marketplace on Blockchain. In Proceedings of the Crypto Valley Conference on Blockchain Technology (CVCBT), Zug, Switzerland, 20–22 June 2018.

19. Ramsundar, B.; Chen, R.; Vasudev, A.; Robbins, R.; Gorokh, A. Tokenized Data Markets. *arXiv* **2018**, arXiv:1806.00139.

20. Benet, J. Ipfs-content addressed, versioned, p2p file system. *arXiv* **2014**, arXiv:1407.3561.

21. Popov, S. The Tangle. 2016. Available online: https://assets.ctfassets.net/r1dr6vzfxhev/2t4uxvsIqk0EUau6g2sw0g/45eae33637ca92f85dd9f4a3a218e1ec/iota1_4_3.pdf (accessed on 24 May 2022).

22. Buterin, V. Ethereum White Paper. 2013. Available online: https://ethereum.org/en/whitepaper/ (accessed on 24 May 2022).

23. Nunez, D. *Umbral: A Threshold Proxy Re-Encryption Scheme*; University of Malaga: Malaga, Spain, 2018.

24. Joung, Y.J.; Yang, L.W.; Fang, C.T. Keyword search in dht-based peer-to-peer networks. *IEEE J. Sel. Areas Commun.* **2007**, *25*, 46–61. [CrossRef]

25. Kubach, M.; Sellung, R. On the market for self-sovereign identity: Structure and stakeholders. In *Open Identity Summit 2021*; Gesellschaft für Informatik: Bonn, Germany, 2021.

26. Zichichi, M.; Contu, M.; Ferretti, S.; D'Angelo, G. LikeStarter: A Smart-contract based Social DAO for Crowdfunding. In Proceedings of the INFOCOM 2019—IEEE Conference on Computer Communications Workshops (INFOCOM WKSHPS), Paris, France, 29 April–2 May 2019.

27. Ratnasamy, S.; Francis, P.; Handley, M.; Karp, R.; Shenker, S. A scalable content-addressable network. In Proceedings of the 2001 Conference on Applications, Technologies, Architectures, and Protocols for Computer Communications, San Diego, CA, USA, 27–31 August 2001; pp. 161–172.

28. Ferretti, S.; Ghini, V.; Panzieri, F.; Turrini, E. Seamless support of multimedia distributed applications through a cloud. In Proceedings of the 2010 IEEE 3rd International Conference on Cloud Computing, Miami, FL, USA, 5–10 July 2010; pp. 548–549.

29. Becker, M.; Bodó, B. Trust in blockchain-based systems. *Internet Policy Rev.* **2021**, *10*, 1–10. [CrossRef]

30. Pocher, N.; Zichichi, M. Towards CBDC-based Machine-to-Machine Payments in Consumer IoT. In Proceedings of the 37th ACM/SIGAPP Symposium on Applied Computing (SAC), Virtual, 25–29 April 2022; pp. 1–8.

31. Bez, M.; Fornari, G.; Vardanega, T. The scalability challenge of ethereum: An initial quantitative analysis. In Proceedings of the 2019 IEEE International Conference on Service-Oriented System Engineering (SOSE), San Francisco, CA, USA, 4–9 April 2019; pp. 167–176.

32. The Graph Protocol. 2020. Available online: https://thegraph.com/en/ (accessed on 24 May 2022).

33. De Filippi, P.; Wray, C.; Sileno, G. Smart contracts. *Internet Policy Rev.* **2021**, *10*. [CrossRef]

34. Ferretti, S.; D'Angelo, G. On the ethereum blockchain structure: A complex networks theory perspective. *Concurr. Comput. Pract. Exp.* **2020**, *32*, e5493. [CrossRef]

35. Aiello, M.; Cambiaso, E.; Canonico, R.; Maccari, L.; Mellia, M.; Pescapè, A.; Vaccari, I. IPPO: A Privacy-Aware Architecture for Decentralized Data-sharing. *arXiv* **2020**, arXiv:2001.06420.

36. Yu, L.; Zichichi, M.; Markovich, R.; Najjar, A. Intelligent Human-input-based Blockchain Oracle (IHiBO). In Proceedings of the 14th International Conference on Agents and Artificial Intelligence (ICAART), Online, 3–5 February 2022; pp. 1–12.

37. D'Angelo, G.; Ferretti, S.; Marzolla, M. A blockchain-based flight data recorder for cloud accountability. In Proceedings of the 1st Workshop on Cryptocurrencies and Blockchains for Distributed Systems, Munich, Germany, 15 June 2018; pp. 93–98.

38. Radix Knowledge Base. 2019. Available online: https://learn.radixdlt.com/ (accessed on 24 May 2022).

39. Benčić, F.M.; Žarko, I.P. Distributed ledger technology: Blockchain compared to directed acyclic graph. In Proceedings of the 2018 IEEE 38th International Conference on Distributed Computing Systems (ICDCS), Vienna, Austria, 2–6 July 2018; pp. 1569–1570.

40. Brogan, J.; Baskaran, I.; Ramachandran, N. Authenticating Health Activity Data Using Distributed Ledger Technologies. *Comput. Struct. Biotechnol. J.* **2018**, *16*, 257–266. [CrossRef]

41. IOTA Streams Specification. 2022. Available online: https://github.com/iotaledger/streams/blob/develop/specification/Streams_Specification_1_0A.pdf (accessed on 24 May 2022).

42. Ateniese, G.; Fu, K.; Green, M.; Hohenberger, S. Improved proxy re-encryption schemes with applications to secure distributed storage. *ACM Trans. Inf. Syst. Secur.* **2006**, *9*, 1–30. [CrossRef]

43. de la Vega, F.; Soriano, J.; Jimenez, M.; Lizcano, D. A peer-to-peer architecture for distributed data monetization in fog computing scenarios. *Wirel. Commun. Mob. Comput.* **2018**, *2018*, 5758741. [CrossRef]

44. Zhu, L.; Xiao, C.; Gong, X. Keyword Search in Decentralized Storage Systems. *Electronics* **2020**, *9*, 2041. [CrossRef]

45. Onik, M.M.H.; Kim, C.S.; Lee, N.Y.; Yang, J. Privacy-aware blockchain for personal data sharing and tracking. *Open Comput. Sci.* **2019**, *9*, 80–91. [CrossRef]

46. Zichichi, M.; Contu, M.; Ferretti, S.; Rodríguez-Doncel, V. Ensuring Personal Data Anonymity in Data Marketplaces through Sensing-as-a-Service and Distributed Ledger. In Proceedings of the 3rd Distributed Ledger Technology Workshop, Co-Located with ITASEC 2020, Ancona, Italy, 4 February 2020.

47. Lopez, D.; Farooq, B. A multi-layered blockchain framework for smart mobility data-markets. *Transp. Res. Part C Emerg. Technol.* **2020**, *111*, 588–615. [CrossRef]

48. Zyskind, G.; Nathan, O. Decentralizing privacy: Using blockchain to protect personal data. In Proceedings of the 2015 IEEE Security and Privacy Workshops, San Jose, CA, USA, 21–22 May 2015; pp. 180–184.

49. Cruz, J.P.; Kaji, Y.; Yanai, N. RBAC-SC: Role-based access control using smart contract. *IEEE Access* **2018**, *6*, 12240–12251. [CrossRef]

50. Maesa, D.D.F.; Mori, P.; Ricci, L. Blockchain based access control. In *IFIP International Conference on Distributed Applications and Interoperable Systems*; Springer: Cham, Switerland, 2017; pp. 206–220.

51. Zhang, Y.; He, D.; Choo, K.K.R. BaDS: Blockchain-based architecture for data sharing with ABS and CP-ABE in IoT. *Wirel. Commun. Mob. Comput.* **2018**, *2018*, 2783658. [CrossRef]

52. Wang, S.; Zhang, Y.; Zhang, Y. A blockchain-based framework for data sharing with fine-grained access control in decentralized storage systems. *IEEE Access* **2018**, *6*, 38437–38450. [CrossRef]

53. Xu, H.; He, Q.; Li, X.; Jiang, B.; Qin, K. BDSS-FA: A Blockchain-Based Data Security Sharing Platform With Fine-Grained Access Control. *IEEE Access* **2020**, *8*, 87552–87561. doi: [CrossRef]

54. Jiang, P.; Guo, F.; Liang, K.; Lai, J.; Wen, Q. Searchain: Blockchain-based private keyword search in decentralized storage. *Future Gener. Comput. Syst.* **2020**, *107*, 781–792. [CrossRef]

55. IPFS Community. Search Engine for the InterPlanetary File System. 2021. Available online: https://github.com/ipfs-search/ipfs-search (accessed on 24 May 2022).

56. Khudhur, N.; Fujita, S. Siva-The IPFS Search Engine. In Proceedings of the 2019 Seventh International Symposium on Computing and Networking (CANDAR), Nagasaki, Japan, 25–28 November 2019; pp. 150–156.

57. Serena, L.; Zichichi, M.; D'Angelo, G.; Ferretti, S. Simulation of Hybrid Edge Computing Architectures. In Proceedings of the 2021 IEEE/ACM 25th International Symposium on Distributed Simulation and Real Time Applications (DS-RT), Valencia, Spain, 27–29 September 2021; pp. 1–8.

58. Chaudhry, A.; Crowcroft, J.; Howard, H.; Madhavapeddy, A.; Mortier, R.; Haddadi, H.; McAuley, D. Personal data: Thinking inside the box. In Proceedings of the Fifth Decennial Aarhus Conference on Critical Alternatives, Aarhus, Denmark, 17–21 August 2015; pp. 29–32.

59. Esteves, B.; Pandit, H.J.; Rodríguez-Doncel, V. ODRL Profile for Expressing Consent through Granular Access Control Policies in Solid. In Proceedings of the 2021 IEEE European Symposium on Security and Privacy Workshops (EuroS PW), Vienna, Austria, 6–10 September 2021; pp. 298–306. doi: [CrossRef]

60. Davari, M.; Bertino, E. Access control model extensions to support data privacy protection based on GDPR. In Proceedings of the 2019 IEEE International Conference on Big Data (Big Data), Los Angeles, CA, USA, 9–12 December 2019; pp. 4017–4024.

61. European Union Agency for Cybersecurity. *Data Pseudonymisation: Advanced Techniques & Use Cases*; Technical Report; European Union Agency for Cybersecurity: Athens, Greece, 2021.

62. Herranz, J.; Hofheinz, D.; Kiltz, E. KEM/DEM: Necessary and Sufficient Conditions for Secure Hybrid Encryption. IACR Cryptology ePrint Archive. 2006. Available online: https://citeseerx.ist.psu.edu/viewdoc/download?doi=10.1.1.64.9369&rep=rep1&type=pdf (accessed on 24 May 2022).

63. Gudgeon, L.; Moreno-Sanchez, P.; Roos, S.; McCorry, P.; Gervais, A. SoK: Layer-two blockchain protocols. In *International Conference on Financial Cryptography and Data Security*; Springer: Cham, Switerland, 2020.

64. Yu, L.; Zichichi, M.; Markovich, R.; Najjar, A. Enhancing Trust in Trust Services: Towards an Intelligent Human-input-based Blockchain Oracle (IHiBO). In Proceedings of the 55th Hawaii International Conference on System Sciences (HICSS), Maui, HI, USA, 4–7 January 2022; pp. 1–10.

65. Corcho, O.; Jiménez, J.; Morote, C.; Simperl, E. Data.europa.eu and Citizen-Generated Data. 2022. Available online: https://data.europa.eu/sites/default/files/report/data.europa.eu_Report_Citizen-generateddataondata_europa_eu.pdf (accessed on 24 May 2022).

66. Samarati, P.; Sweeney, L. *Protecting Privacy when Disclosing Information: K-Anonymity and Its Enforcement through Generalization and Suppression*; Technical Report; SRI International: Menlo Park, CA, USA, 1998.

67. Campbell, M.J.; Dennison, P.E.; Butler, B.W.; Page, W.G. Using crowdsourced fitness tracker data to model the relationship between slope and travel rates. *Appl. Geogr.* **2019**, *106*, 93–107. [CrossRef]

68. Mazzoni, M.; Corradi, A.; Di Nicola, V. Performance evaluation of permissioned blockchains for financial applications: The ConsenSys Quorum case study. *Blockchain Res. Appl.* **2022**, *3*, 100026. [CrossRef]

69. Dwork, C. Differential privacy. In *Encyclopedia of Cryptography and Security*; Springer: New York, NY, USA, 2011; pp. 338–340.

70. Article 29 Working Party. 2014. Available online: https://ec.europa.eu/justice/article-29/documentation/opinion-recommendation/files/2014/wp216_en.pdf (accessed on 24 May 2022).

71. See, L.; Mooney, P.; Foody, G.; Bastin, L.; Comber, A.; Estima, J.; Fritz, S.; Kerle, N.; Jiang, B.; Laakso, M.; et al. Crowdsourcing, citizen science or volunteered geographic information? The current state of crowdsourced geographic information. *ISPRS Int. J. Geo-Inf.* **2016**, *5*, 55. [CrossRef]

72. Murray, P.; Nate Welch, J.M. EIP-1167: Minimal Proxy Contract. 2018. Available online: https://eips.ethereum.org/EIPS/eip-1167 (accessed on 24 May 2022).

73. Fabian Vogelsteller, V.B. EIP-20: ERC-20 Token Standard. 2015. Available online: https://eips.ethereum.org/EIPS/eip-20 (accessed on 24 May 2022).
74. Zichichi, M.; Ferretti, S.; D'Angelo, G. Are Distributed Ledger Technologies Ready for Intelligent Transportation Systems? In Proceedings of the 3rd Workshop on Cryptocurrencies and Blockchains for Distributed Systems (CryBlock 2020), London, UK, 25 September 2020; pp. 1–6.
75. Zichichi, M.; Ferretti, S.; D'Angelo, G. On the Efficiency of Decentralized File Storage for Personal Information Management Systems. In Proceedings of the 2020 IEEE Symposium on Computers and Communications (ISCC), Rennes, France, 7–10 July 2020; pp. 1–6.
76. AnaNSi-Research. Hypercube. 2022. Available online: https://github.com/AnaNSi-research/hypfs (accessed on 24 May 2022).
77. Zichichi, M. miker83z/umbral-rs. Software. 2022. [CrossRef]
78. Zichichi, M. miker83z/testingIPFS: IPFS and SIA User Client Application Tests. Software. 2021. [CrossRef]
79. AnaNSi-Research. IOTA. 2022. Available online: https://github.com/AnaNSi-research/testingIOTA (accessed on 24 May 2022).
80. Montresor, A.; Jelasity, M. PeerSim: A scalable P2P simulator. In Proceedings of the 2009 IEEE Ninth International Conference on Peer-to-Peer Computing, Seattle, WA, USA, 9–11 September 2009; pp. 99–100.
81. D'Angelo, G.; Ferretti, S. LUNES: Agent-based simulation of P2P systems. In Proceedings of the 2011 International Conference on High Performance Computing & Simulation, Istanbul, Turkey, 4–8 July 2011; pp. 593–599.
82. Giansante, C.; Zichichi, M. miker83z/Hypercube-DHT-Simulation. Software. 2022. [CrossRef]
83. Blockchain Explorer. 2020. Available online: www.blockchain.com/explorer (accessed on 24 May 2022).
84. Zichichi, M. miker83z/k-DaO. Software. 2022. [CrossRef]
85. Serena, L.; Zichichi, M.; D'Angelo, G.; Ferretti, S. Simulation of Dissemination Strategies on Temporal Networks. In Proceedings of the 2021 Annual Modeling and Simulation Conference (ANNSIM), Fairfax, VA, USA, 19–22 July 2021; pp. 1–12.
86. Wilkinson, M.D.; Dumontier, M.; Aalbersberg, I.J.; Appleton, G.; Axton, M.; Baak, A.; Blomberg, N.; Boiten, J.W.; da Silva Santos, L.B.; Bourne, P.E.; et al. The FAIR Guiding Principles for scientific data management and stewardship. *Sci. Data* **2016**, *3*, 1–9. [CrossRef]
87. Zichichi, M.; Serena, L.; Ferretti, S.; D'Angelo, G. Towards Decentralized Complex Queries over Distributed Ledgers: A Data Marketplace Use-case. In Proceedings of the 30th IEEE International Conference on Computer Communications and Networks (ICCCN), Athens, Greece, 19–22 July 2021; pp. 1–6.

Article

A Multi-Resident Number Estimation Method for Smart Homes

Andrea Masciadri [1,*], Changhong Lin [1], Sara Comai [1] and Fabio Salice [1]

Dipartimento di Elettronica, Informazione e Bioingegneria, Politecnico di Milano, 20133 Milano, Italy; changhong.lin@mail.polimi.it (C.L.); sara.comai@polimi.it (S.C.); fabio.salice@polimi.it (F.S.)
* Correspondence: andrea.masciadri@polimi.it

Abstract: Population aging requires innovative solutions to increase the quality of life and preserve autonomous and independent living at home. A need of particular significance is the identification of behavioral drifts. A relevant behavioral drift concerns sociality: older people tend to isolate themselves. There is therefore the need to find methodologies to identify if, when, and how long the person is in the company of other people (possibly, also considering the number). The challenge is to address this task in poorly sensorized apartments, with non-intrusive sensors that are typically wireless and can only provide local and simple information. The proposed method addresses technological issues, such as PIR (Passive InfraRed) blind times, topological issues, such as sensor interference due to the inability to separate detection areas, and algorithmic issues. The house is modeled as a graph to constrain transitions between adjacent rooms. Each room is associated with a set of values, for each identified person. These values decay over time and represent the probability that each person is still in the room. Because the used sensors cannot determine the number of people, the approach is based on a multi-branch inference that, over time, differentiates the movements in the apartment and estimates the number of people. The proposed algorithm has been validated with real data obtaining an accuracy of 86.8%.

Keywords: multi-person detection; sensor data; smart environment

Citation: Masciadri, A.; Lin, C.; Comai, S.; Salice, F. A Multi-Resident Number Estimation Method for Smart Homes. *Sensors* **2022**, *22*, 4823. https://doi.org/10.3390/s22134823

Academic Editors: Pietro Manzoni, Claudio Palazzi and Ombretta Gaggi

Received: 12 May 2022
Accepted: 21 June 2022
Published: 25 June 2022

Publisher's Note: MDPI stays neutral with regard to jurisdictional claims in published maps and institutional affiliations.

1. Introduction

The rapid development of the ICT sector has enabled scenarios where an ever-deeper interconnection between the physical and digital worlds is proposed (Phyigital—2007 by Chris Weil). One scenario is the home, where the interconnection between technologies and people enables the implementation of a paradigm for autonomous living, guaranteeing mutual safety (people, family members, and caregivers are in contact with each other) and allowing the identification of behavioral drifts and their subsequent compensation (behavioral drift is selectively compensated with solutions for the identified problem by containing costs and promoting personal autonomy) [1].

In recent years, applications of domiciliary technology systems interacting with the person have addressed issues such as activity recognition (e.g., [2–4]), health monitoring [5], security [6], and the prediction of future events [7].

The need to understand what is happening inside the dwelling requires that the phenomena to be tracked are observable; this assumes that there are suitable sensing systems (sensors and transducers) and that these are distributed more or less densely in the home. Considering the environmental sensing, the various proposals combine data collected from different types of sensors, such as RFIDs (Radio-Frequency IDentification), PIRs (Passive InfraRed) [8], contact sensors, pressure-sensitive mats [9], tilt sensors [10], power meters [11], inertial sensors, infrared array sensors [12], etc. In general, the types of sensor can be vision-based (e.g., cameras), wearable (generally based on inertial sensors like accelerometers, gyroscopes, etc.) or environment detection (e.g., motion or door/window sensors, temperature/humidity sensors, etc.). It is worth noting that both current regulations and perceived privacy violations make it difficult to use vision-based systems (2D and 3D cameras) over other detection techniques.

Considering wearable vs. non-wearable devices, a non-intrusive monitoring system (that is, without wearable devices) can guarantee a better trade-off between privacy and reliability because the absence of wearable devices eliminates or reduces some important critical issues, such as routine maintenance (e.g., recharging batteries) or the misuse of the device (e.g., taking it off in certain situations or forgetting to wear it). However, non-intrusive monitoring systems can provide reliable and direct measurements for many activities in a specific environment, such as room occupancy in the house, but do not achieve high levels of accuracy for many other tasks, such as calculating the number of people in a room. It is worth noting that in some scenarios, like in the case of aging, the exact calculation of the number of people in a home support system is not particularly relevant. What is often of interest is to detect whether the person is alone at home (and to inform the caregiver so that they pay more attention to the person) or whether there is a behavioral drift present, which leads the elderly to isolate themselves and progressively reduce their degree of sociality.

In these contexts, it is necessary to identify methodologies for identifying whether, when, and for how long the person is in the company of other people (possibly also considering the number).

People-counting in smart environments with distributed sensor networks has been studied in the past, but using a large number of sensors (e.g., up to 60 sensors distributed in a single apartment). Such a large number of sensors represents an important entry barrier for many households. The challenge is to tackle this task in sparsely sensorized apartments, with sensors that can be wireless (nowadays, many apartments are not equipped to host wired solutions) and can provide only local and simple information.

In this article, we propose a method that addresses technological problems (such as the blind times of the PIRs, their insensitivity in the absence of motion, and their different sensitivity depending on the distance and temperature of bodies), topological problems (such as possible sensor interference due to the inability to separate detection areas), and algorithmic problems. In the instrumented apartment, there is only one PIR per room and one on/off sensor on the front door. This is the minimum monitoring configuration: below this, one or more rooms are not 'observable'. The house is modeled as a DAG (Directed Acyclic Graph). The model is used to deal with the fragmentation of the data stream that the various sensors can generate. In particular, we need to manage unwanted transitions between rooms when there are multiple people in the house. The DAG model can constrain the transition between adjacent rooms and avoid crossing walls. However, other issues for correct detection remain. The state of each room is represented by a set of values, one for each identified person. Each value decays over time and represents the probability that the person, while not specifying who they are, is still in the room. Because the sensors used in our setting cannot identify the number of people, the approach is based on multi-branch inference that, over time, differentiates the movements in the apartment and estimates the number of people; the limitation is that the number of people must be less than the number of rooms in the dwelling [13].

The main contributions of our work are:

- The method infers, step by step, the number of people in the house. It operates on non-ubiquity: if sensors are simultaneously active in rooms that cannot interfere, there are at least as many people as the number of active sensors. If the estimation is aimed at residents only, the method derives the exact number of people in the apartment. If the estimation is aimed at guests, the method gets whether the person is alone or whether there are more people.
- The method is based on a non-intrusive and minimum-cardinality sensors network with a single PIR sensor per room and a contact sensor on the front door. The scenario is typical of real-world situations.

- The method requires little scenario information (house map, sensor position, and the observability of each PIR) to determine both adjacency between rooms and interference situations between sensors.
- Finally, the method is unsupervised (can run in different scenarios without model training). This requirement is essential in order to make it possible to monitor a wide set of apartments.

The rest of the document is organized as follows. In Section 2, previous and related work is introduced. Next, Section 3 presents the proposed approach. In Section 4, the performance is evaluated. Finally, Section 5 discusses and concludes our work.

2. Related Works

During the years, various smart environment systems have been proposed. They have been used in many scenarios, such as family houses [14], offices [15], shopping malls [16], and museums [17], and have been applied for a variety of purposes, including tracking people in buildings [18], counting people numbers [19], recognizing human behavior [20], etc. In addition, energy consumption can be monitored, and the indoor environment can be controlled automatically by using appropriate sensors and controllers [21].

The types of detection means (e.g., through sensors or transducers) that can be used in smart environments are diversified. They can be roughly divided into two main categories, wearable devices and non-wearable devices, where the latter is in turn divided into sensors–transducers and multimedia-based devices. In terms of wearable devices, Bluetooth—BLE ([22,23]), UWB, Zigbee, WiFi, and RFID technologies [24], or specialized sensors (e.g., magnetic field sensors [25]) are widely used in indoor positioning systems to track or to localize people [26]. Sensors–transducers include various types of detection systems positioned in a smart environment to detect the movement of humans [27]. A non-exhaustive list of these devices includes Passive Infrared sensors [28,29], Thermal Sensors [30], and Force-Sensing Resistors (e.g., smart floors [31]. Moreover, pressure polymer, electromechanical film (EMFi), piezoelectric sensors, load cells, or WiFi can be used [31]. Finally, multimedia-based approaches can obtain rich context from the environment with videos [32] and audio [33–35].

Various approaches for people-counting in non-intrusive monitoring environments have been proposed without multimedia-based devices. Petersen et al. [36] propose an SVM-based method (Support Vector Machine) to detect the presence of visitors in the smart homes of solitary elderly because social activity is an important factor for assessing the health status of the elderly (social, psychological, and physical are, typically, the three dimensions of the health-related quality of life). Wireless motion sensors are installed in every room, and several key features are extracted and input to a SVM classifier, which is trained to detect multi-person events. The model has been validated with a two-subjects dataset and the results demonstrate the feasibility of visitor detection. The adopted method suffers from some criticalities: time is divided into fixed slots of time, with epochs of 15 min; all possible room combinations are taken into account (i.e., $n \times (n-1)/2$) without considering that adjacent rooms could produce sensors interference; PIR blocking-time is not mentioned (blocking time could interfere with the activation order of PIRs); the time of the day is flagged a priori to consider the 'circadian rhythm'; and finally, the approach is supervised.

Müller et al. [37] implement two approaches for inferring the presence of multiple persons in a test lab equipped with 50 motion sensors (data from CASAS) and some contact sensors; only two persons are monitored in the laboratory. One approach is a simple statistical method to derive the number of people based on the raw sensor data, while the second one uses multiple hypothesis tracking (MHT) and Bayesian filtering to track the people. The first method reaches an accuracy of 90.75%; the second one reaches 83.35%. The limit of this research is that the proposed method can only distinguish whether the house has a single person or multiple persons, but does not estimate the number of people. In addition, there are dense sensors in the test smart home, and the complex installation of sensors may limit the widespread use of such a system.

The authors of [38] estimate people numbers that satisfy the house topology and sensor activation constraints; then, a Hidden Markov Model is used to refine the result. The algorithm is validated in two smart homes and obtains high accuracy results when the smart home has 0 to 3 persons, but the accuracy decreases dramatically with 4 or more persons. In this work, both simulated and real data from different scenarios were used: ARAS (limitation: small rooms, few rooms, and only partially covered), ARAS-FC (limitation: few rooms without a dataset; the authors produced some data via simulation), and House 2 (limitation: the authors produced a simulated dataset). Unfortunately, it is not clear if they took into account the limits of the PIR sensors (for example, sensitivity and blocking time) and the criticality of the map; for instance, this could create interference between the different activations.

The work in [39] proposes an unsupervised multi-resident tracking algorithm that can also provide a rough estimation of the number of active residents in the smart home. They consider two datasets. The first is the dataset TM004 from CASAS, consisting of 25 ambient sensors distributed among eight rooms and with two-bedroom apartments with two older adult residents; occasionally, their child will come and stay in their house for a couple of days. The second is named Kyoto and contains a denser grid of sensors (91 sensors installed in six rooms—hallways included) with two residents; occasionally, they received friends for a visit of a few days. The algorithm has good performance, but this decreases as the number of residents increases; moreover, the algorithm tends to generate more resident identifiers when the same resident triggers the same sensor events, and it has a higher possibility of segmentation errors when tracking residents in a location where sensors are more densely deployed.

Other papers focus on the problem of the recognition of multi-resident activities in a smart-home infrastructure [40,41] (using the CASAS dataset with 60 sensors and 2 residents), [42] (using the ARAS dataset), and they can have as a consequence the possibility of counting the number of people. However, given their main goal, the number of sensors is typically very high, and the number of residents is limited to two persons.

A recent paper [43], focuses on the recognition of some daily activities in a multi-resident family home. The recognition of daily activities is a specialized task that requires to precisely identify the type of resident. For this purpose, authors used numerous and specialized types of actuators (e.g., a sensor module for a cup and a sensor box of the fridge) to distinguish the different activities performed by the individuals and using a data-driven and knowledge-driven combination method to recognize users. The article is very interesting, but, for the obvious reasons of observability of the phenomena, it requires a conspicuous number of transducers in addition to those normally used for home monitoring.

The survey in [27] focuses on the techniques for localizing and tracking people in multi-resident environments. For the counting problem, they identify three classes of approaches: (a) binary-based techniques based on binary sensors like PIRs—they typically exploit snapshots or, possibly, the history of snapshots with spatial and temporal dependencies to understand the number of people [44]; (b) clustering-based techniques that identify multiple non-overlapping clusters containing one or more targets; and (c) statistical-based techniques based on statistical models to estimate the number of persons.

The work in [45] aims at identifying visitors by using different measures of entropy for the cases with/without visitors in a smart home equipped only with PIR sensors and a door contact sensor that is used to confirm the visits and their duration. An accuracy of 98–99% is obtained in a setting where a single occupant typically resides in the home and a visitor arrives.

Alternative approaches to estimate the number of people have been proposed. For example [31] use WiFi: the movement can be detected through the analysis of the propagation effects of radio-frequency signals. Wang et al. [46] propose a method to count people by utilizing breathing traces, reaching 86% accuracy for four people. Similarly, in [47], Fiber Bragg Grating sensors are used for the detection and number of occupants, experimented with three people. Recent techniques also include voice recording to recognize up to three persons [48].

In the literature, vision-based methods have been always considered a reliable approach to estimate the number of people because the camera can obtain rich information. Vera et al. [49] proposed a system to count people using depth cameras mounted in the zenithal position: people are detected in each camera and the tracklets that belong to the same person are determined. Even though vision-based methods are efficient and reliable [14,50,51], they are unsuitable for smart homes due to privacy reasons. Algorithms based on wearable devices are infeasible for detecting visitors who do not wear such devices. Additionally, this intrusive method may not be acceptable for those people with low compliance.

Our algorithm avoids the usage of wearable devices and cameras: it adopts a system equipped with a very low number of presence detectors to realize non-intrusive monitoring, based on architectural modules of the BRIDGe project (Behavioral dRift compensation for autonomous and InDependent livinG) [52,53].

The proposed algorithm is based on minimal data about the house structure (plans of the flat) and sensor position, such as room adjacency and possible overlapping monitored areas (sensor interference), and can update the estimated number of people dynamically. The case study is with four inhabitants (a family with two adult children) in an apartment with a living room, a kitchen (open view), three bedrooms (a double room and two single rooms), two bathrooms, and a corridor. In the apartment, there are frequent guests, especially from Friday to Sunday (typically in the evening). The maximum number of people has reached six people. The apartment is instrumented with one PIR per room (eight PIRs) and a contact sensor on the main door. The PIR has a 2 s blocking time, 2 moves sensibility (the number of moves required for the PIR sensor to report motion), and a 12 s window time (the period of time during which the number of moves must be detected for the PIR sensor to report motion).

3. Proposed Method and Algorithm

3.1. System Architecture

The use-case scenario is a classical house with typical rooms: a kitchen, a living room, and one or more bedrooms and bathrooms. In each room, a PIR sensor is installed. PIR sensors are cheap and small, but they have some shortcomings: (a) the detection area of the PIR sensor is difficult to control, so that PIR sensors in different rooms may have overlapping areas of sensing range; (b) PIR sensors can only provide a binary response to the presence or not of people regardless of the number of people; (c) the sensitivity of the PIR is not uniform (it depends on the distance, the width of the visibility area—for example, edge zone or sectors of areas—and on the speed of the subject, on the characteristics of the subject [54]); and (d) the functioning of a PIR depends on a set of motion detection parameters (e.g., the blocking time).

Moreover, a contact sensor is installed at the entrance of the smart home. The sensor sends an activation signal when the door is opened. It is worth noting that a contact sensor (e.g., door and windows perimeter monitoring) is less critical, in terms of functioning and parameters, with respect to PIRs.

Figure 1 shows an overview of the architecture and data flow of the proposed algorithm. The inputs of our algorithm are: the stream data from the PIR and contact sensors and some information stored in the database, including the house structure and the sensors settings. The stream data are processed by the Data Processor to detect the status of the sensors, which can be active or inactive. The Data Fragment Generator reads the sensor data and groups them into fragments concerning a continuous period that may represent interesting changes in the house. Then, the Event Detector detects events in the received data fragment, which may also include events coming from the door contact sensor. Based on the detected events, the status of the sensors, and the system setting, a multi-branch inference machine infers the number of people by fusing several independent inference engines that represent different possible scenarios compatible with the sequence of events. An algorithm coordinator controls all these modules and adds some functionalities that allow (a) to start the algorithm in any initial situation without any information about the number of residents and (b) to avoid accumulated errors that may occur in long-time runnings. All the modules are detailed in the next subsections.

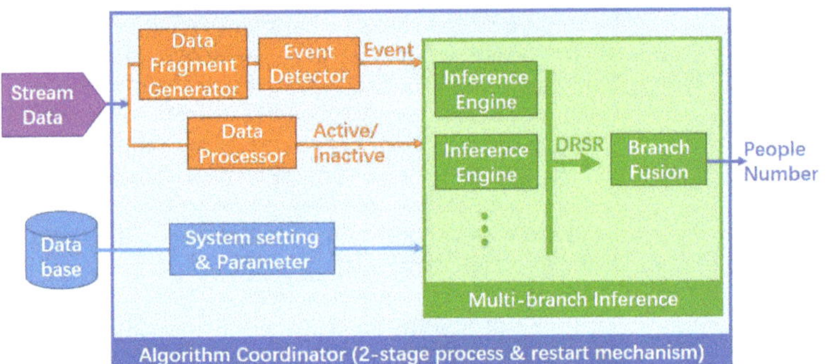

Figure 1. Architecture of the proposed algorithm.

3.2. Fragment Generation and Event Detection

3.2.1. Data Fragment Generation

Sensors produce and send data irregularly, depending on the activities that occur in the house. Typically, a series of signals are activated by the movement of a person within a small time interval. To recognize events that happen in the house, we divide stream data into semantic fragments composed of sequences of signals that occur in a certain interval as shown in Figure 2. Fragments are separated by periods that do not detect events for a given time interval.

In our Data Fragment Generator, only the active signal of PIR sensors and contact sensors installed at the entrance door are taken into consideration. Thus, data fragments represent events such as 'somebody moves from the bedroom and goes out passing through the living room'.

Figure 2. The generation of a data fragment: circles represent events that produce new data; if the time difference between two events is greater than *interval*, the new data are regarded as the beginning of a new data fragment.

3.2.2. House Event Detection

The layout of the rooms in the house is modeled as a Directed Acyclic Graph (DAG) representing their adjacency. The algorithm works also in multi-floor buildings. The detection and inference of possible events is realized by finding the transition of active signals from generated data fragments.

Besides the movement of people between adjacent rooms, there are some special events:

- **Go in:** When someone enters the house, a 'go in' event happens. The total number of people in the house increases.
- **Go out:** Similarly, a 'go out' event happens when someone goes out of the house. In this case, the total number of people decreases. Notice that the exact number of people entering/exiting the house cannot be determined, so the algorithm must take into account this aspect.

- **Overlap:** The detection area of PIR sensors in different rooms may have overlapping detection areas. An example is shown in Figure 3. Such kinds of events need to be identified to have a better inference.

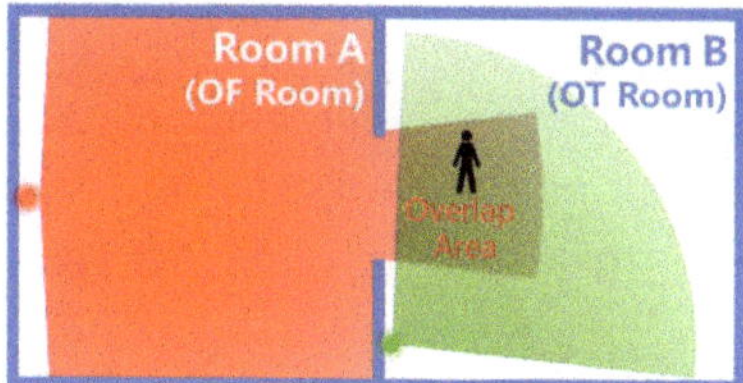

Figure 3. Overlap Example: Both sensors in Room A and B are active, but the overlap is in Room B, so this is defined as the *Overlapped True (OT)* room; instead, Room A is defined as the *Overlapped False (OF)* room.

The overlap case depends on the direction and the installation place of sensors; therefore, the possible overlap areas can be identified in advance.

If an overlapping case occurs, both sensors are active: if the difference between their timestamps is less than a predefined *overlap interval*, the detector will detect an overlap event. Figure 4 shows the typical behavior.

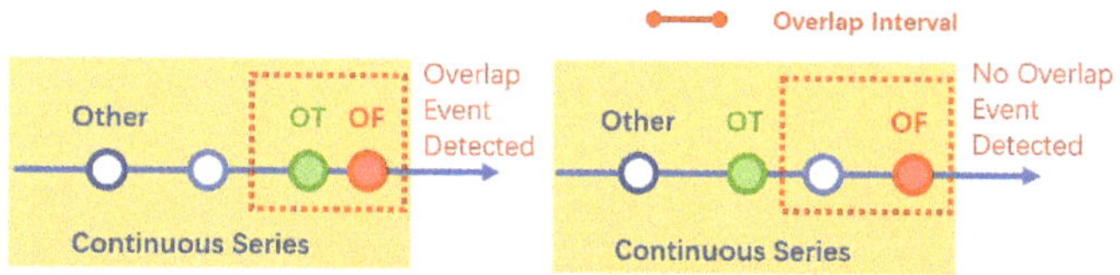

Figure 4. Overlap Event Detector: The last two signals highlighted in each data fragment are compared with the list of known overlap cases.

3.3. House Status Estimation

Decayed Room Status Representation

To represent the house status, our method considers the following facts:

- PIRs transmit a state change when they detect a change in the infrared signals they receive. After an activation, PIR sensors remain inactive for a while and do not capture other events (blocking time). It is worth noting that PIR sensors do not change state when people are motionless.
- The latest data can be considered more reliable for the representation of the current status of the house compared to previous data.

To estimate the number of occupants accurately, besides the latest data, also the previous data need to be taken into account. In each room, more than one person may be present: the status of a room is represented by a set of values, one for each estimated person, that decay over time and that represent the probability that the persons are still in the room. We call this kind of representation *Decayed Room Status Representation (DRSR$_t$)* and the status of each person j in each room i at the time instant t *Room Status Signals (RSS$_{i,t}^{j}$)*.

The value of each $RSS_{i,t}^{j}$ varies from 0 to 1: 1 means that the person has been detected, 0 that the person has left the room, an intermediate value that the person may be in the room.

Each $RSS_{i,t}^{j}$ decays over time with a given decay ratio until it reaches a lower limit, according to Equation (1), where Δt_i is the time difference from the last update of the i-th room and n_i is the number of persons estimated in the room.

$$RSS_{i,t+\Delta t_i}^{j} = max\{RSS_{i,t}^{j} - decay\_ratio \times \Delta t_i, decay\_lower\_limit\}, j = 1, 2, ..., n_i \quad (1)$$

where *decay_ratio* defines the decay speed of $RSS_{i,.}^{j}$ and *decay_lower_limit* is the limit that $RSS_{i,.}^{j}$ can reach. Such a decay mechanism can make up for the shortcomings of the PIR sensors that are insensitive to motionless people. When a person is detected in a room, if the adjacent room has not revealed an activity, we can assume that the person is still there. Therefore, the activation status can last for a certain period, until the $RSS_{i,t}^{j}$ value decays to the lower limit (*decay_lower_limit*). Thus, from the status $DRSR_t$ of the house, we can determine whether a room is occupied or not by comparing their $RSS_{i,t}^{j}$ values with a given threshold. The description of $RSS_{i,t}^{j}$ is shown in Figure 5.

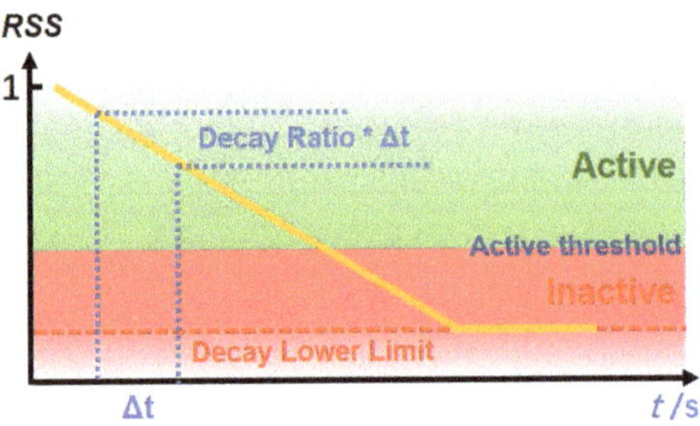

Figure 5. The graphical representation of $RSS_{i,t}^{j}$ with the active and inactive areas, the decay lower limit and the decay ratio.

The $RSS_{i,t}^{j}$ also has the following tunable parameters:

- **Additional decay**: To balance the uncertainty between the case of motionless people (but still in the room) and people that have moved to other rooms, an additional decay value is defined to be added in case of inactivity signals.
- **Active threshold**: determines the status of a room. If an $RSS_{i,t}^{j}$ is higher than the active threshold, the room status is set as occupied; when it is lower than the active threshold, the person is removed from the counting for that room.

Because the transfer of people from one room to another can be detected from a data fragment as described in Section 3.2.2, if the algorithm detects a transfer from room A to room B, while room B already has one person in it, then the number of people in the status of that room will be set to 2. For example, if the active threshold is 0.2 and the RSSs in the house are those shown in Table 1, then the total number of people in the house is estimated to be equal to 3 because three $RSS_{i,t}^{j}$ values are greater than 0.2. Notice that for Room 2, two different $RSS_{2,t}^{j}$ values are available, one for each person.

Table 1. Example of people occupation determination.

Rooms	Room 1	Room 2	Room 3	Room 4
RSS	[0.0]	[1.0, 0.3]	[0.15]	[0.73]
Status	Empty	2 people	Inactive	1 person

3.4. Inference Engine

An inference engine is an entity that infers the status of the house. It has two attributes: the status of the rooms $RSS_{i,t}$ of the house described above and a confidence score which changes with the inferring process. The confidence score represents the consistency of the state of the house. Whenever an ambiguity condition occurs, the confidence score decreases to return to 1 when the ambiguity is resolved. For example, if the inference engine finds that a transfer from one room to another is concluded successfully, the inference process finishes, and the state of the house is updated. On the contrary, if some inconsistencies are found and the process needs to continue in order to solve these problems, the confidence score is decreased. The number of inference engines depends on the events in the house and the sensors data, as different branches for all the possible cases that could be inferred are generated.

When the algorithm starts, one inference engine is initialized with the confidence score set equal to 1. All rooms are regarded as empty, i.e., all $RSS_{i,t}$ are initially empty. When the system is running, the new sensors data and the events detected by the Data Fragment Generator are fed to the inference engine to update the status of the rooms. Then, the number of people is estimated based on the $RSS_{i,t}$ values.

The Inference Engine updates status of the rooms following the main rules below:

- 'Go in' event: For the room connected to the entrance door (for example, room $i = 1$), a new $RSS_{i,t}^{j}$ with value 1 is added and the status (number of people) is increased by 1. The event is identified through the analysis of the activation sequence between the input PIR and the ON/OFF sensor;
- 'Go out' event: For the room connected to the entrance door (for example, room $i = 1$), all $RSS_{i,t}$ values and the status are set to 0; if the room is currently estimated as empty (inconsistency state), the confidence is reduced.
- 'Overlap' event: the false activation signal is ignored;
- When the system receives an active signal, the algorithm determines if a room transfer has occurred according to the topology of the house. If a transfer happens, a new $RSS_{i,t}^{j}$ of the target room is set to 1. At the same time, the $RSS_{i,t}^{j}$ with the minimum value of the room where the person comes from is deleted. In case the active signal just intercepts an activity inside the room (person movement), all $RSS_{i,t}^{j}$ for that room are incremented through the following formula (in the current settings, *arise_ratio* is equal to the *decay_ratio*)

$$RSS_{i,t+\Delta t_i}^{j} = max\{RSS_{i,t}^{j} + arise\_ratio \times \Delta t_i, 1\}, j = 1, 2, ..., n_i \qquad (2)$$

- If the PIR sensor does not activate for a certain time, but it may be possible that there are still persons in the room, the status of the room becomes inactive, and its $RSS_{i,t}^{j}$ values are decreased by the additional decay value.

Multi-Branch Inference

Because PIR and contact sensors cannot distinguish the number of people, we deal with this situation with a multi-branch inference approach to consider the context of sensors data. In some cases, the system may not be able to estimate the number of people accurately, like in the case where there are more than two candidate rooms that satisfy the room transfer condition, but after some inferences, the estimated result can finally converge to the ground-truth number. Figure 6 shows a simple example.

The maintenance of the proposed multi-branch inferring method is as follows:

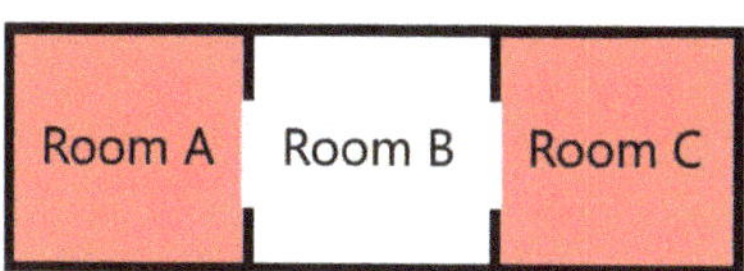

Figure 6. Example of transfer dilemma: One person is in Room A and another one is in Room C. If the PIR sensor in Room B is activated, it is hard to determine whether the person who activated it comes from Room A or Room C, until other events occur.

- **Create Branch:** When a dilemma case occurs, several new inference branches are created for every possible movement case. Every branch has a confidence attribute representing its reliability. For example, for the transfer dilemma case in Figure 6, two inference engines are created—one for a possible transfer from Room A to Room

B; another for the transfer from Room C to Room B, with independent room status values. Both continue to infer the house status simultaneously.

- **Merge Branch:** Branches that have the same status are merged. If two inference engines have the same status of the house for all the rooms, we regard them as the same inference engine, delete one of them, and sum their confidence scores.
- **Resize Confidence:** Confidence scores are scaled-up periodically according to Equation (3), where $confidence_{i,t+\Delta}$ is the new confidence for engine i, **confidence** is the vector of all confidence values of the engines.

$$confidence_{i,t+\Delta} = \frac{1}{max(\mathbf{confidence})} \times confidence_{i+t} \tag{3}$$

- **Delete Branch:** To reduce the number of inference engines, all the confidences are sorted, and the inference engines with the lowest confidence values are deleted.
- **Branch Fusion:** All inference branches have different $DRSR$ statuses and estimated people numbers. Specific methods are used to fuse the results of all the inference branches, such as voting, averaging, or weighted averaging based on branch confidence.

3.5. Algorithm

The general algorithm as well as all the steps described in the previous subsections are described in Algorithms 1–5.

Algorithm 1 Main algorithm

1: $sensorId, sensorState, sensorTimestamp \leftarrow getSensorEvent()$
2: **if** $sensorState == On$ **then**
3:     $sensorsList \leftarrow updateSensorsList(sensorID, sensorTimestamp, On)$
4:     **if** $goIn(sensorsList, sensorID) == True$ **then**
5:         **for each** $DRSR$ **in** $DRSRList$ **do**
6:             $room_{entrance} \leftarrow addNewPerson()$
7:         **end for**
8:     **else**
9:         **if** $goOut(sensorsList, sensorId) == True$ **then**
10:             **for each** $DRSR$ **in** $DRSRList$ **do**
11:                 $room_{entrance} \leftarrow zeroPerson()$
12:             **end for**
13:         **else**
14:             **if** $overlap(sensorsList, sensorId) == True$ **then**
15:                 $DRSR \leftarrow refresh(sensorsList, sensorId, time())$
16:             **else**
17:                 $DRSR \leftarrow update(sensorsList, sensorId, sensorTimestamp)$
18:             **end if**
19:         **end if**
20:     **end if**
21: **else**
22:     **if** $sensorState == Off$ **then**
23:         $sensorsList \leftarrow updateSensorsList(sensorID, sensorTimestamp, Off)$
24:         $DRSR \leftarrow update(sensorsList, sensorId, sensorTimestamp)$
25:         $DRSR \leftarrow rssDecay(sensorTimestamp)$
26:     **else**
27:         $DRSR \leftarrow refresh(sensorsList, All, time())$
28:     **end if**
29: **end if**

Algorithm 2 Function goIn

1: $def\ goIn(sensorsList, sensorId)$: *boolean*
2: **if** $(sensorId) == DOOR$ **then**
3:    **if** $time(sensorsList.entrance) - time(sensorsList.door) \leq DELTA$ **then**
4:       $returnTRUE$
5:    **else**
6:       $returnFALSE$
7:    **end if**
8: **end if**

Algorithm 3 Function goOut

1: $def\ goOut(sensorsList, sensorId)$: *boolean*
2: **if** $(sensorId) == DOOR$ **then**
3:    **if** $time(sensorsList.door) - time(sensorsList.entrance) \leq DELTA$ **then**
4:       $returnTRUE$
5:    **else**
6:       $returnFALSE$
7:    **end if**
8: **end if**

Algorithm 4 Function update

1: $def\ update(sensorsList, sensorId, time)$: $DRSR$
2: **if** $dilemma(sensorsList, time, DAG) == True$ **then**
3:    $DRSR \leftarrow addNewDRSR(sensorsList, time)$
4: **else**
5:    **if** $roomsTransfer(sensorsList, time, DAG) == True$ **then**
6:       $DRSR \leftarrow moveDRSR(sensorsList, time)$
7:    **else**
8:       $DRSR \leftarrow refresh(sensorsList, sensorId, time)$
9:    **end if**
10: **end if**

Algorithm 5 Function refresh

1: $def\ refresh(sensorsList, sensorId, time)$: $DRSR$
2: $DELTA = time - lastTime$
3: **if** $sensorId! = ALL$ **then**
4:    $DRSR \leftarrow rssArise(sensorsList, sensorId, time)$
5: **else**
6:    **if** $DELTA \geq INTERVAL$ **then**
7:       $lastTime = time$
8:       $DRSR \leftarrow rssDecay(DELTA)$
9:       $DRSR \leftarrow resize()$
10:       $DRSR \leftarrow merge()$
11:       $DRSR \leftarrow fusion()$
12:       $DRSR \leftarrow delete(numDRSR)$
13:    **end if**
14: **end if**

3.6. Algorithm Coordinator

Two further important algorithm steps are introduced: the two-stage process and the restart mechanism. The former is needed to balance the accuracy and stability of our algorithm; the latter is used to avoid an accumulated error for long-time running.

There are two challenges that our algorithm has to face: the first one is that the initial number of people and the initial status of the house are unknown because the smart home system may start at any time; the second challenge is that sensors in the smart home cannot distinguish multiple persons. For example, if two people are in the same room, they are regarded as one person, because sensors cannot see the difference with respect to the case of a single person. In this condition, the algorithm would estimate fewer people. To solve the two problems above, the algorithm works in two different stages: refresh stage and stable stage.

- **Refresh stage:** When the system starts (the initial people number is set to 0) or when the entrance door opens (some people may come in or go out), the estimated people number is uncertain. The number of people is refreshed over time, and the estimated number can converge to a correct result. This process is realized by changing the lower limit of the room status. When this is set to 0, the estimated number can increase and also decrease.
- **Stable stage:** When the estimated number remains unchanged for a specific period, the lower limit of the room status is raised to a value that is greater than the active threshold, i.e., the estimated number can only increase. In this stage, our algorithm can perform a more stable estimation because people will not disappear when the door does not open.However, the algorithm should allow the increase in the estimated number because, in the refresh stage, some motionless/sleep inhabitants may have been ignored and the corresponding room statuses are falsely decaying to 0.

A parameter called *Refresh Stage Duration* is set to switch the algorithm from the refresh stage to a stable stage. The diagram of our two-stage process is shown in Figure 7.

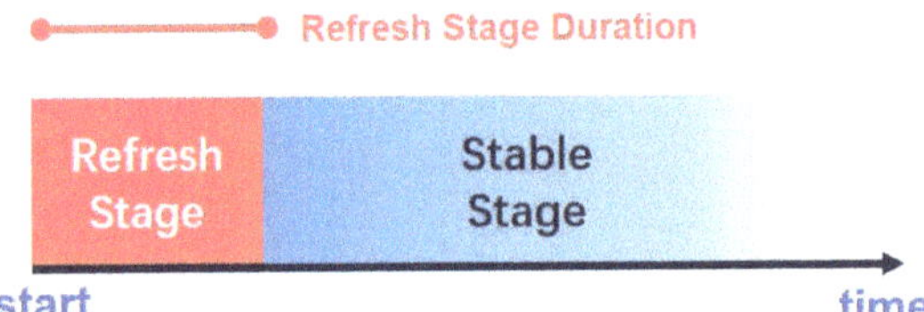

Figure 7. Diagram of the Two-stage Process: The algorithm starts, and after the Refresh Stage Duration, the algorithm switches to the stable stage.

Because smart home systems need to run for months, our proposed algorithm also needs to run for a long time. To avoid accumulated errors for the house status and people number estimation, the algorithm needs to be restarted regularly. The length of the Refresh Stage Duration has been tuned on the field, and the results are shown in the experimental section.

4. Results

Our approach has been validated in a domestic environment equipped with smart sensors using the BRIDGe platform [52]. Data have been recorded for 14 days in a house with four people (a family with two adult children). In the apartment, there are frequent guests, especially from Friday to Sunday (typically in the evening). Six is the maximum number of people at a Saturday dinner (on the other days, the typical number is less or equal to four).

4.1. House Layout and Sensor Setting

Figure 8 shows the layout of the house. It includes a kitchen (open view), a living room, two bathrooms, three bedrooms, and a corridor. The corridor connects bedrooms, bathrooms, and the living room. The entrance door of the house is in the living room.

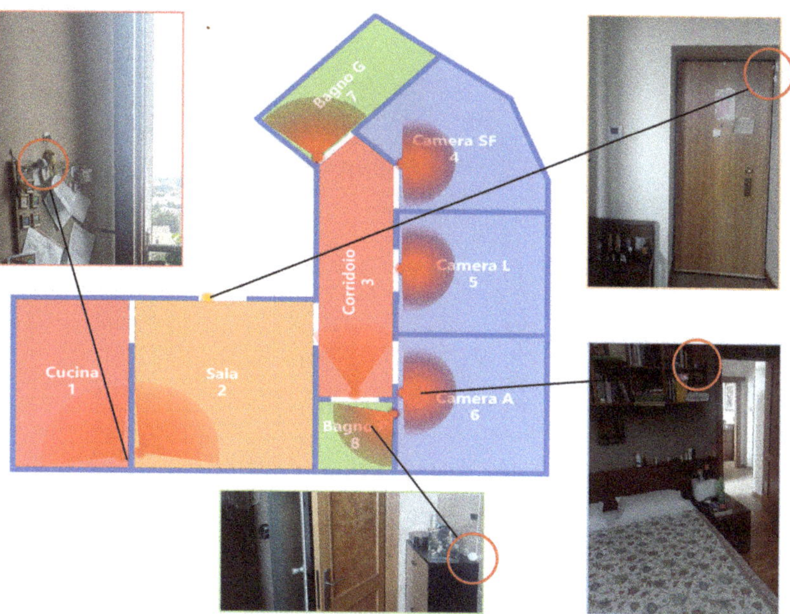

Figure 8. The Layout of Smart Environment with PIR sensors positions. The positions of the PIRs are highlighted with red circles in the pictures of the rooms. On the map represent, they are positioned in corresponding of the red circles, while the red sectors indicate the general direction of the sensors' sensing areas, not the actual sensing range.

The apartment is instrumented with one PIR per room (therefore eight PIRs) and a contact sensor on the main door. Data are collected by the FIBARO control unit (model HC2); all FIBARO sensors, Z-Wave protocol (868 MHz), are mesh networked to the FIBARO control panel. It is worth noting that the WiFi (there is a connection in the apartment) and Z-Wave connection do not interfere because they operate on different frequencies. Data transmission from the central unit to the cloud operates by events. Whenever a sensor changes state (the state—ON or OFF—and time in which the state change occurred), the record related to the sensor (SensorID, State, Time) is sent to the cloud. The PIRs are FIBARO Motion Sensors, type FGMS-001 (multi-sensor: PIR, vibration, temperature, and light), configured as follows: 2 s blocking time, 2 moves sensibility (number of moves required for the PIR sensor to report motion), and 12 s window time (period of time during which the number of moves must be detected for the PIR sensor to report motion). The PIRs are sensitive to direct sunlight. The PIRs were put in a condition not to be directly affected by the sun. There are no other particular and critical situations to take into account. The PIRS in the bathrooms are positioned far from water. By construction, they tolerate humidity; this characteristic is particularly important in Bathroom Small, where a shower is present and where it is possible to detect variations of 30% RH when taking showers in winter. Possible overlap cases exist in this smart environment as shown in Table 2. As described in Section 3.2.2, OT rooms are the correct rooms to be considered when the person is in the overlap area. For example, the first line indicates that when the person is in the corridor, there is an area where they may also be detected by the living room sensor.

Table 2. Possible Overlap Cases in Validation Dataset.

Number of Case	OT Room of Case	OF Room of Case
1	Corridor	Living Room
2	Bedroom L	Living Room
3	Bathroom Big	Corridor

There are four permanent residents, and they have private environments. In particular, there are two single rooms (Room person A and Room person L) and one master bedroom (Room persons F and S). Bathroom Big is used mainly by A, L, and F, while Bathroom Small is used mainly by S. In Bathroom Small, there is a shower used by all four family members. In the living room, there are two sofas and a television; there is no table. The table is only in the kitchen. For the ground truth, the arrival and leaving of people have been recorded manually, including in/out events, the change of people number, and the time when they were in or out. The time has been recorded manually and accurately (hours and minutes). The total people number changes when the door is opened; then, it may be distributed in different ways in the different rooms. Because the record of the contact sensor installed at the door is accurate to seconds, we used it to align the in/out time to reach second-level accuracy.

4.2. Indicators Design

To measure the performance of our approach, two kinds of indicators have been considered: the accuracy of the number of people and the stability of the number change.

4.2.1. Accuracy Design

The accuracy represents the percentage of the time with a correct estimation with respect to ground truth. As shown in Equation (4), $T_{TotalTime}$ is the total time we measured, and the unit is in seconds.

$$Accuracy = \frac{\sum_{i=0}^{n} TP_i}{T_{TotalTime}} \tag{4}$$

$$T_{TotalTime} = t - t_{begin} \tag{5}$$

$$T_{CorrectEstimate} + = [\hat{n}_t = n_t] \times (t - t_{prev}) \tag{6}$$

Accuracy is the overall accuracy of the validation dataset. TP_i is the total time that the algorithm makes a correct estimation when the ground-truth number is i. $T_{TotalTime}$ stands for the total time from which the system begins to the current moment. t and t_{begin} are the current timestamp and the moment when the system started, respectively. In the last equation, $\hat{n}_t$ is the estimated people number at the current moment, n_t is the true people number, t is the current time, and t_{prev} is the previous timestamp when the sensor data have been received.

4.2.2. Stability Indicator Design

Stability can be represented by using the notion of information entropy. The information entropy is used to measure the uncertainty of inferred numbers. We take the last 10 min of the inference results to calculate the entropy because the earlier change of numbers may possibly be caused by people's movements. The less the entropy, the better the system performance.

$$Total\_Entropy = \frac{\sum Entropy(t)}{N} \tag{7}$$

$$Entropy(t) = - \sum_{k=0}^{max(\hat{n}_t)} (p_{\hat{n}_t=k,t} ln(p_{\hat{n}_t=k,t})) \tag{8}$$

$$p_{\hat{n}_t=k,t} = \frac{\sum 1_{\hat{n}=k}}{T/F_s} \tag{9}$$

Total_Entropy is the average entropy of every moment. N is the total number of the received data from the validation dataset. $Entropy(t)$ is the calculated entropy in the 10 min before time t. $\hat{n}_t$ is the estimated people number at the current moment, $p_{\hat{n}_t=k,t}$ is the probability that the algorithm estimates the people number is equal to k in the 10 min before time t, T is the sample period, and F_s is the sample frequency; here, we set it to 1 Hz.

Notice that entropy alone is not enough to represent the stability. For example, if there are two people-counting results, such as (2,1,2,1,2,1) and (2,2,2,1,1,1), they have the same entropy value, but the former result is worse. Therefore, a measure of the frequency of changes in the estimated numbers is introduced, called *Changecost*(). The less the changes cost, the better the system performance.

$$Total\_ChangeCost = \frac{\sum ChangeCost(t)}{N} \tag{10}$$

$$ChangeCost(t) = \sum_{t=0}^{T} |\hat{n}_t - \hat{n}_{t-1}| \tag{11}$$

Total_ChangeCost is the average *ChangeCost* of the whole dataset. *ChangeCost*(t) is the calculated entropy in the 10 min before time t.

4.3. Experiment

The final selected parameters are shown in Table 3. The final accuracy result is 86.78% with about 36,000 sensors data from the eight PIRs and the contact sensors for 14 days.

Table 3. Editable Parameters in the Algorithm.

Editable Parameters	Value	Editable Parameters	Value
Decay Ratio	0.003/s	Series Interval	40 s
Additional Decay	0.2/s	Door Action Interval	60 s
Active Threshold	0.1	Max Branch Number	50
Overlap Interval	10 s	Refresh Stage Duration	300 s

Some examples of the parameter selections are reported next: the first example is the selection of the Refresh Stage Duration. We took values from 1 to 30 min and tested the data from the dataset and obtained the result in Table 4. It can be noticed that when the Refresh Stage Duration is 5 min, the algorithm can reach the highest accuracy of 91.78%, with acceptable values of the other indicators, such as *Entropy* and *ChangeCost*.

Table 4. Parameter Selection of Refresh Stage Duration.

Parameter Values	Accuracy	Entropy	ChangeCost
1 min	56.89%	0.155	0.820
3 min	88.63%	0.169	0.824
5 min	**91.78%**	**0.171**	**0.882**
10 min	89.12%	0.176	0.937
15 min	58.12%	0.213	1.375
20 min	12.17%	0.294	2.225
30 min	10.34%	0.325	2.894

The next example concerns the selection of the Max Branch Number. Similarly, several values were tested to find out the best value. The results of the tests are shown in Table 5. From the table, we can see that when the Max Branch Number is set to 30, the algorithm obtains the best result.

Table 5. Parameter Selection of Max Branch Number.

Parameter Values	Accuracy	Entropy	ChangeCost
10	49.04%	0.196	1.093
20	73.50%	0.212	1.430
30	**84.21%**	**0.177**	**1.002**
40	81.14%	0.177	0.952
50	81.14%	0.177	0.952
60	83.78%	0.173	0.913
70	81.14%	0.174	0.884

Several ablation studies of the algorithm have been undertaken to compare the performance with different methods and settings.

First of all, the validation dataset has been tested without the multi-branch inference method, i.e., only one inference engine has been used to infer the status of the smart environment. As shown in Table 6, we can see that the multi-branch method has less *Entropy* and *ChangeCost* than the single-branch method, which means that the estimated result of the former method is more stable. Moreover, the *Accuracy* of the multi-branch method is higher than the single-branch method, by over 10%. Thus, the proposed multi-branch inference method plays an important role in our algorithm.

Table 6. Comparison of multi-branch method and single-branch method.

Method	Accuracy	Entropy	ChangeCost
Multi-branch Inference	86.785%	0.152	0.757
Single-branch Inference	75.695%	0.160	0.785

In the proposed algorithm, overlap events can be detected and this message can be used as information to help the inference engine infer the house status. By using this event detector, the shortcomings of PIR sensors can be remedied. To prove this, an ablation experiment has been conducted and the result is shown in Table 7. Additionally, the detection of door actions, including the 'go in' and 'go out' events, have also been taken into account. From Table 7, we can see that without detecting the 'Overlap' event, the algorithm obtained a worse result than the proposed method in all the indicators. The former method regards the false overlapping case of the PIR sensors as real activation signals, which leads to an incorrect inference of the house status. If the 'Door Action' event detector was forbidden, the algorithm failed to make a correct estimation because the door action is important for the house status inference.

Table 7. Ablation Study on Event Detection.

Method	Accuracy	Entropy	ChangeCost
Proposed Method	86.78%	0.152	0.757
Without 'Overlap' Detector	55.71%	0.172	0.904
Without 'Door Action' Detector	fail	fail	fail

4.4. Limitations of the Method

It is worth noting that to detect the right number of people in the apartment, the following conditions must hold:

1. The number of people in the apartment is lower than the number n of rooms with a PIR sensor (or to the number of PIR sensors that cover separate areas). If this is not the case, the algorithm will identify a number of people that is at most equal to n.
2. The blocking time of the PIR sensors reduces the accuracy of our algorithm, especially when residents move around quickly and/or frequently; the lower the value, the higher the accuracy of the proposed method. It is worth noting that the parameters of the sensors strongly depend on the technology (both for connectivity and detection), on the chipset, and on the available energy. The latter aspect is the predominant factor because liveness depends on energy consumption. For example, the Tellur WiFi motion sensor and Xiaomi Aqara Zigbee have a blocking time of 60 s, the FIBARO motion sensor Z-Wave—the type used in our case study—has a blocking time that varies from 2 to 8 s, while for wired sensors, the times are extremely lower, and also with mixed detection technology (e.g., the Risco BWare DT AM microwave in K band with PIR).
3. The dynamics of the in/out events from the apartment must be lower than the dynamics of the movements of the people in the house; if the stationary condition of the number of people in the rooms is long enough, the algorithm is more likely to identify the number of people in the apartment. In fact, as the time of the people staying in the apartment increases, the certainty of the results increases (if they move among different rooms).
4. People in the apartment do not always move in pairs. If the people move in groups, the algorithm will not distinguish them from the movement of a single person.

The proposed methodology is general and without any specific needs, excluding those reported above; the rooms are those of a typical apartment (kitchen, bedroom, bathroom, etc.), and the limitations are derived from the number of the rooms and their connections. A studio apartment, for example, is an environment that does not allow, in a non-intrusive way, to draw much information about the number of people (except for special 'private' events, such as the use of the bathroom). Although it is out of the scope of this article, in some real cases we have been faced with, by increasing the PIR densities in some specific rooms, they had a 'complex' characterization. For example, in an apartment with an open-space living area gathering, where there is a kitchen, dining table, living room, etc., the area has been virtually partitioned into sub-units in order to infer where people are moving and the type of activity they are doing. In these cases (with the limits reported above), it is also possible to estimate the number of people.

5. Conclusions

In this paper, we presented a people-number estimation algorithm based on non-intrusive, sparse-distributed sensors data from a multi-resident smart environment with work on the continuous flow of data generated by the sensors. Estimating the exact number of people in a family with more than two residents is a difficult task, especially in a sparse-distributed sensor network where each room has only one binary sensor to detect the presence of a human. However, the choice for such a setting, which is basic and minimal, is affordable in practice in many situations. Moreover, having a good, even if not precise, estimation of the number of people can be sufficient in many real scenarios of older people living alone at home.

Our algorithm has several advantages: it does not need to learn any data and therefore can be applied immediately, starting at any time, and only limited information about the house settings is needed. A good accuracy has been obtained thanks to the representation of the status of the rooms and the multi-branch inference based on the context.

As future work, we plan to also test other types of sensor data, such as bed/chair sensors, to evaluate the results in motionless situations [55] where no PIR sensors are activated.

Author Contributions: Conceptualization, F.S.; Methodology, A.M.; Supervision, S.C. and F.S.; Software, C.L.; Validation, C.L. and A.M.; Writing—original draft, C.L.; resources, F.S.; data curation, F.S. and C.L.; Writing—review and editing, S.C., F.S and A.M. All authors have read and agreed to the published version of the manuscript.

Funding: This research received no external funding.

Institutional Review Board Statement: Not applicable.

Informed Consent Statement: Not applicable.

Data Availability Statement: The data presented in this study are available on request from the corresponding author. The data are not publicly available due to privacy reasons.

Conflicts of Interest: The authors declare no conflict of interest.

References

1. Alaa, M.; Zaidan, A.A.; Zaidan, B.B.; Talal, M.; Kiah, M.L.M. A review of smart home applications based on Internet of Things. *J. Netw. Comput. Appl.* **2017**, *97*, 48–65. [CrossRef]
2. Lin, B.; Cook, D.J.; Schmitter-Edgecombe, M. Using continuous sensor data to formalize a model of in-home activity patterns. *J. Ambient Intell. Smart Environ.* **2020**, *12*, 183–201. [CrossRef]
3. Fahad, L.G.; Tahir, S.F. Activity recognition in a smart home using local feature weighting and variants of nearest-neighbors classifiers. *J. Ambient Intell. Humaniz. Comput.* **2021**, *12*, 2355–2364. [CrossRef] [PubMed]
4. Bakar, U.; Ghayvat, H.; Hasanm, S.F.; Mukhopadhyay, S.C. Activity and Anomaly Detection in Smart Home: A Survey. In *Next Generation Sensors and Systems*; Mukhopadhyay, S.C., Ed.; Springer International Publishing: Cham, Switzerland, 2016; pp. 191–220.
5. Mshali, H.; Lemlouma, T.; Moloney, M.; Magoni, D. A survey on health monitoring systems for health smart homes. *Int. J. Ind. Ergon.* **2018**, *66*, 26–56. [CrossRef]
6. Dahmen, J.; Cook, D.J.; Wang, X.; Wang, H. Smart secure homes: A survey of smart home technologies that sense, assess, and respond to security threats. *J. Reliab. Intell. Environ.* **2017**, *3*, 83–98. [CrossRef]
7. Wu, S.; Rendall, J.B.; Smith, M.J.; Zhu, S.; Xu, J.; Wang, H.; Yang, Q.; Qin, P. Survey on Prediction Algorithms in Smart Homes. *IEEE Internet Things J.* **2017**, *4*, 636–644. [CrossRef]
8. Yang, D.; Sheng, W.; Zeng, R. Indoor human localization using PIR sensors and accessibility map. In Proceedings of the 2015 IEEE International Conference on Cyber Technology in Automation, Control, and Intelligent Systems (CYBER), Shenyang, China, 8–12 June 2015; pp. 577–581. [CrossRef]
9. Kasteren, T.L.; Englebienne, G.; Kröse, B.J. An Activity Monitoring System for Elderly Care Using Generative and Discriminative Models. *Pers. Ubiquitous Comput.* **2010**, *14*, 489–498. [CrossRef]
10. Chen, L.; Nugent, C.D.; Wang, H. A Knowledge-Driven Approach to Activity Recognition in Smart Homes. *IEEE Trans. Knowl. Data Eng.* **2012**, *24*, 961–974. [CrossRef]
11. Ueda, K.; Tamai, M.; Yasumoto, K. A method for recognizing living activities in homes using positioning sensor and power meters. In Proceedings of the 2015 IEEE International Conference on Pervasive Computing and Communication Workshops (PerCom Workshops), St. Louis, MO, USA, 23–27 March 2015; pp. 354–359. [CrossRef]
12. Trofimova, A.A.; Masciadri, A.; Veronese, F.; Salice, F. Indoor Human Detection Based on Thermal Array Sensor Data and Adaptive Background Estimation. *J. Comput. Commun.* **2017**, *5*, 16–28. [CrossRef]
13. Giaretta, A.; Loutfi, A. On the people counting problem in smart homes: Undirected graphs and theoretical lower-bounds. *J. Ambient. Intell. Humaniz. Comput.* **2021**, *in press.* [CrossRef]
14. Wang, L.; Gu, T.; Tao, X.; Chen, H.; Lu, J. Recognizing multi-user activities using wearable sensors in a smart home. *Pervasive Mob. Comput.* **2011**, *7*, 287–298. [CrossRef]
15. Krüger, F.; Kasparick, M.; Mundt, T.; Kirste, T. Where are My Colleagues and Why? Tracking Multiple Persons in Indoor Environments. In Proceedings of the 2014 International Conference on Intelligent Environments, Shanghai, China, 30 June–4 July 2014; pp. 190–197. [CrossRef]
16. Dogan, O.; Gurcan, O.F.; Oztaysi, B.; Gokdere, U. Analysis of Frequent Visitor Patterns in a Shopping Mall. In *Industrial Engineering in the Big Data Era, Proceedings of the Global Joint Conference on Industrial Engineering and Its Application Areas, GJCIE 2018, Nevsehir, Turkey, 21–22 June 2018*; Calisir, F., Cevikcan, E., Camgoz Akdag, H., Eds.; Springer International Publishing: Cham, Switzerland, 2019; pp. 217–227.
17. Lanir, J.; Kuflik, T.; Sheidin, J.; Yavin, N.; Leiderman, K.; Segal, M. Visualizing Museum Visitors' Behavior: Where Do They Go and What Do They Do There? *Pers. Ubiquitous Comput.* **2017**, *21*, 313–326. [CrossRef]
18. Chen, C.H.; Wang, C.C.; Yan, M.C. Robust Tracking of Multiple Persons in Real-Time Video. *Multimed. Tools Appl.* **2016**, *75*, 16683–16697. [CrossRef]

19. Adeogun, R.; Rodriguez, I.; Razzaghpour, M.; Berardinelli, G.; Christensen, P.H.; Mogensen, P.E. Indoor Occupancy Detection and Estimation using Machine Learning and Measurements from an IoT LoRa-based Monitoring System. In Proceedings of the 2019 Global IoT Summit (GIoTS), Aarhus, Denmark, 17–21 June 2019; pp. 1–5. [CrossRef]

20. Luo, X.; Guan, Q.; Tan, H.; Gao, L.; Wang, Z.; Luo, X. Simultaneous Indoor Tracking and Activity Recognition Using Pyroelectric Infrared Sensors. *Sensors* **2017**, *17*, 1738. [CrossRef] [PubMed]

21. Bamodu, O.; Xia, L.; Tang, L. An indoor environment monitoring system using low-cost sensor network. *Energy Procedia* **2017**, *141*, 660–666. Power and Energy Systems Engineering. [CrossRef]

22. Chesser, M.; Chea, L.; Ranasinghe, D.C. Field Deployable Real-Time Indoor Spatial Tracking System for Human Behavior Observations. In Proceedings of the 16th ACM Conference on Embedded Networked Sensor Systems—SenSys '18, Shenzhen, China, 4–7 November 2018; pp. 369–370. [CrossRef]

23. Oosterlinck, D.; Benoit, D.F.; Baecke, P.; Van de Weghe, N. Bluetooth tracking of humans in an indoor environment: An application to shopping mall visits. *Appl. Geogr.* **2017**, *78*, 55–65. [CrossRef]

24. Belmonte Fernández, O.; Puertas-Cabedo, A.; Torres-Sospedra, J.; Montoliu-Colás, R.; Trilles Oliver, S. An Indoor Positioning System Based on Wearables for Ambient-Assisted Living. *Sensors* **2016**, *17*, 36. [CrossRef]

25. Gozick, B.; Subbu, K.P.; Dantu, R.; Maeshiro, T. Magnetic Maps for Indoor Navigation. *IEEE Trans. Instrum. Meas.* **2011**, *60*, 3883–3891. [CrossRef]

26. Liu, H.; Darabi, H.; Banerjee, P.; Liu, J. Survey of Wireless Indoor Positioning Techniques and Systems. *IEEE Trans. Syst. Man Cybern. Part C Appl. Rev.* **2007**, *37*, 1067–1080. [CrossRef]

27. Ngamakeur, K.; Yongchareon, S.; Yu, J.; Rehman, S.U. A Survey on Device-Free Indoor Localization and Tracking in the Multi-Resident Environment. *ACM Comput. Surv.* **2020**, *53*, 1–29. [CrossRef]

28. Yang, D.; Xu, B.; Rao, K.; Sheng, W. Passive Infrared (PIR)-Based Indoor Position Tracking for Smart Homes Using Accessibility Maps and A-Star Algorithm. *Sensors* **2018**, *18*, 332. [CrossRef] [PubMed]

29. Suzuuchi, S.; Kudo, M. Location-associated indoor behavior analysis of multiple persons. In Proceedings of the 2016 23rd International Conference on Pattern Recognition (ICPR), Cancun, Mexico, 4–8 December 2016; pp. 2079–2084. [CrossRef]

30. Singh, S.; Aksanli, B. Non-Intrusive Presence Detection and Position Tracking for Multiple People Using Low-Resolution Thermal Sensors. *J. Sens. Actuator Netw.* **2019**, *8*, 40. [CrossRef]

31. Al-Naimi, I.; Wong, C.B. Indoor human detection and tracking using advanced smart floor. In Proceedings of the 2017 8th International Conference on Information and Communication Systems (ICICS), Irbid, Jordan, 4–6 April 2017; pp. 34–39. [CrossRef]

32. Nielsen, C.; Nielsen, J.; Dehghanian, V. Fusion of security camera and RSS fingerprinting for indoor multi-person tracking. In Proceedings of the 2016 International Conference on Indoor Positioning and Indoor Navigation (IPIN), Alcala de Henares, Spain, 4–7 October 2016; pp. 1–7. [CrossRef]

33. Yun, S.S.; Nguyen, Q.; Choi, J. Distributed sensor networks for multiple human recognition in indoor environments. In Proceedings of the 2016 13th International Conference on Ubiquitous Robots and Ambient Intelligence (URAI), Xi'an, China, 19–22 August 2016; pp. 753–756. [CrossRef]

34. Chen, X.; Chen, Y.; Cao, S.; Zhang, L.; Zhang, X.; Chen, X. Acoustic Indoor Localization System Integrating TDMA+FDMA Transmission Scheme and Positioning Correction Technique. *Sensors* **2019**, *19*, 2353. [CrossRef] [PubMed]

35. Cho, H.S.; Ko, S.S.; Kim, H.G. A robust audio identification for enhancing audio-based indoor localization. In Proceedings of the 2016 IEEE International Conference on Multimedia Expo Workshops (ICMEW), Seattle, WA, USA, 11–15 July 2016; pp. 1–6. [CrossRef]

36. Petersen, J.; Larimer, N.; Kaye, J.A.; Pavel, M.; Hayes, T.L. SVM to detect the presence of visitors in a smart home environment. In Proceedings of the Annual International Conference of the IEEE Engineering in Medicine and Biology Society, San Diego, CA, USA, 28 August–1 September 2012; pp. 5850–5853. [CrossRef]

37. Müller, S.M.; Steen, E.E.; Hein, A. Inferring Multi-person Presence in Home Sensor Networks. In *Ambient Assisted Living*; Wichert, R., Klausing, H., Eds.; Springer International Publishing: Cham, Switzerland, 2016; pp. 47–56.

38. Renoux, J.; Köckemann, U.; Loutfi, A. Online Guest Detection in a Smart Home using Pervasive Sensors and Probabilistic Reasoning. In Proceedings of the European Conference on Ambient Intelligence, Larnaca, Cyprus, 12–14 November 2018.

39. Wang, T.; Cook, J.D. sMRT: Multi-Resident Tracking in Smart Homes with Sensor Vectorization. *IEEE Trans. Pattern Anal. Mach. Intell.* **2020**, *43*, 2809–2821. [CrossRef]

40. Riboni, D.; Murru, F. Unsupervised Recognition of Multi-Resident Activities in Smart-Homes. *IEEE Access* **2020**, *8*, 201985–201994. [CrossRef]

41. Chen, D.; Yongchareon, S.; Lai, E.M.K.; Yu, J.; Sheng, Q.Z. Hybrid Fuzzy C-means CPD-based Segmentation for Improving Sensor-based Multi-resident Activity Recognition. *IEEE Internet Things J.* **2021**, *8*, 11193–11207. [CrossRef]

42. Jethanandani, M.; Sharma, A.; Perumal, T.; Chang, J.R. Multi-label classification based ensemble learning for human activity recognition in smart home. *Internet Things* **2020**, *12*, 100324. [CrossRef]

43. Li, Q.; Huangfu, W.; Farha, F.; Zhu, T.; Yang, S.; Chen, L.; Ning, H. Multi-resident type recognition based on ambient sensors activity. *Future Gener. Comput. Syst.* **2020**, *112*, 108–115. [CrossRef]

44. Li, T.; Wang, Y.; Song, L.; Tan, H. On Target Counting by Sequential Snapshots of Binary Proximity Sensors. In *Wireless Sensor Networks, Proceedings of the 12th European Conference, EWSN 2015, Porto, Portugal, 9–11 February 2015*; Abdelzaher, T., Pereira, N., Tovar, E., Eds.; Springer International Publishing: Cham, Switzerland, 2015; pp. 19–34.
45. Howedi, A.; Lotfi, A.; Pourabdollah, A. An Entropy-Based Approach for Anomaly Detection in Activities of Daily Living in the Presence of a Visitor. *Entropy* **2020**, *22*, 845. [CrossRef]
46. Wang, F.; Zhang, F.; Wu, C.; Wang, B.; Liu, K.J.R. Respiration Tracking for People Counting and Recognition. *IEEE Internet Things J.* **2020**, *7*, 5233–5245. [CrossRef]
47. Vanus, J.; Nedoma, J.; Fajkus, M.; Martinek, R. Design of a New Method for Detection of Occupancy in the Smart Home Using an FBG Sensor. *Sensors* **2020**, *20*, 398. [CrossRef] [PubMed]
48. Tran, S.N.; Zhang, Q. Towards Multi-resident Activity Monitoring with Smarter Safer Home Platform. In *Smart Assisted Living: Toward an Open Smart-Home Infrastructure*; Chen, F., García-Betances, R.I., Chen, L., Cabrera-Umpiérrez, M.F., Nugent, C., Eds.; Springer International Publishing: Cham, Switzerland, 2020; pp. 249–267.
49. Vera, P.; Monjaraz, S.; Salas, J. Counting Pedestrians with a Zenithal Arrangement of Depth Cameras. *Mach. Vis. Appl.* **2016**, *27*, 303–315. [CrossRef]
50. Wang, L.; Gu, T.; Tao, X.; Lu, J. Sensor-Based Human Activity Recognition in a Multi-user Scenario. In *Ambient Intelligence, Proceedings of the European Conference, AmI 2009, Salzburg, Austria, 18–21 November 2009*; Tscheligi, M., de Ruyter, B., Markopoulus, P., Wichert, R., Mirlacher, T., Meschterjakov, A., Reitberger, W., Eds.; Springer: Berlin/Heidelberg, Germany, 2009; pp. 78–87.
51. Komai, K.; Fujimoto, M.; Arakawa, Y.; Suwa, H.; Kashimoto, Y.; Yasumoto, K. Beacon-based multi-person activity monitoring system for day care center. In Proceedings of the 2016 IEEE International Conference on Pervasive Computing and Communication Workshops (PerCom Workshops), Sydney, NSW, Australia, 14–18 March 2016; pp. 1–6. [CrossRef]
52. Mangano, S.; Saidinejad, H.; Veronese, F.; Comai, S.; Matteucci, M.; Salice, F. Bridge: Mutual Reassurance for Autonomous and Independent Living. *IEEE Intell. Syst.* **2015**, *30*, 31–38. [CrossRef]
53. Veronese, F.; Comai, S.; Matteucci, M.; Salice, F. Method, Design and Implementation of a Multiuser Indoor Localization System with Concurrent Fault Detection. In Proceedings of the 11th International Conference on Mobile and Ubiquitous Systems: Computing, Networking and Services, London, UK, 2–5 December 2014; pp. 100–109. [CrossRef]
54. Veronese, F.; Comai, S.; Mangano, S.; Matteucci, M.; Salice, F. PIR Probability Model for a Cost/Reliability Tradeoff Unobtrusive Indoor Monitoring System. In Proceedings of the International Conference on Smart Objects and Technologies for Social Good, Venice, Italy, 30 November– 1 December 2017; pp. 61–69.
55. Rosato, D.; Masciadri, A.; Comai, S.; Salice, F. Non-Invasive Monitoring System to Detect Sitting People. In Proceedings of the 4th EAI International Conference on Smart Objects and Technologies for Social Good—Goodtechs '18, Bologna, Italy, 28–30 November 2018; pp. 261–264. [CrossRef]

Article

Sensoring the Neck: Classifying Movements and Actions with a Neck-Mounted Wearable Device †

Jonathan Lacanlale, Paruyr Isayan, Katya Mkrtchyan and Ani Nahapetian *

Computer Science Department, California State University, Northridge (CSUN), Northridge, CA 91330, USA;
jonathan.lacanlale.608@my.csun.edu (J.L.); paruyr.isayan.536@my.csun.edu (P.I.);
katya.mkrtchyan@csun.edu (K.M.)

* Correspondence: ani@csun.edu
† This is an extended version of conference paper Lacanlale, J.; Isayan, P.; Mkrtchyan, K.; Nahapetian, A. Look Ma, No Hands: A Wearable Neck-Mounted Interface. In Proceedings of the Conference on Information Technology for Social Good, Roma, Italy, 9–11 September 2021.

Abstract: Sensor technology that captures information from a user's neck region can enable a range of new possibilities, including less intrusive mobile software interfaces. In this work, we investigate the feasibility of using a single inexpensive flex sensor mounted at the neck to capture information about head gestures, about mouth movements, and about the presence of audible speech. Different sensor sizes and various sensor positions on the neck are experimentally evaluated. With data collected from experiments performed on the finalized prototype, a classification accuracy of 91% in differentiating common head gestures, a classification accuracy of 63% in differentiating mouth movements, and a classification accuracy of 83% in speech detection are achieved.

Keywords: wearable computing; interaction design; neck-mounted interface; flex sensor; machine learning (ML)

Citation: Lacanlale, J.; Isayan, P.; Mkrtchyan, K.; Nahapetian, A. Sensoring the Neck: Classifying Movements and Actions with a Neck-Mounted Wearable Device. *Sensors* **2022**, *22*, 4313. https://doi.org/10.3390/s22124313

Academic Editors: Claudio Palazzi, Ombretta Gaggi, Pietro Manzoni and Mehmet Rasit Yuce

Received: 6 March 2022
Accepted: 30 May 2022
Published: 7 June 2022

Publisher's Note: MDPI stays neutral with regard to jurisdictional claims in published maps and institutional affiliations.

1. Introduction

The ever-increasing prevalence of mobile phones, wearable devices, and smart speakers has spurred intense exploration into user interfaces. These new user interfaces need to address the challenges posed by the ubiquitous interaction paradigm, while having available the possibilities that these varied smart technologies provide.

Arenas for exploration of mobile user interfaces include improving gesture-based interfaces to enable interaction in limit mobility settings or by decreasing the social disruption that is caused by repeated disruptive interactions. Interfaces have been developed that use the movement of the hands, arms, eyes, and feet.

Touch gesture controls still dominate mobile system interfaces because of the ubiquity of touch screens [1]. However, the dominant tap, scroll, and pinch gestures have been linked to repetitive strain injuries on smart phones [2,3]. In addition, they have their limitations on wearable devices because of the limited screen size and, in turn, the available interface surface. The gestures on smartwatch screens need to be done with greater precision and with more constriction of the hand muscles, since the smartwatch screens are significantly smaller than the smartphone screens.

Voice user interfaces (VUIs) that are used for smart speakers have been another arena for improvement, with voiceless speech being explored for situations where there is background noise and for microinteractions.

In this work, we examine the benefits that sensoring the neck can provide within the breadth of mobile user interfaces. We explore and develop a new user interface for mobile systems, independent of limb motions. For example, in place of a scroll down, the head can be tilted forward. In place of a tap, the head can be turned to one side, all with only an inexpensive sensor affixed to the neck or shirt collar.

We sensor the neck with an inexpensive and nonintrusive flex sensor and show the range of interfaces that are possible with the incorporation of this simple wearable technology into our lives. Our efforts provide a proof of concept that common actions, such as head tilts, mouth movements, and even speech, can be classified through the interpretation of the bend angle received from the neck. We explore the size of the flex sensor and the positioning of the sensor on the neck and use our classification results to tailor the prototype.

Applications for neck interfaces include use in assistive devices where limb motion is limited, in gaming and augmented reality systems for more immersive experiences, and in wearable and vehicular systems where hand and/or voice use is restricted or inconvenient. Neck interactions expand a user's bandwidth for information transference, in conjunction with or in place of the typically saturated visual and the audial channels.

A neck-mounted prototype was designed and developed, as detailed in Section 3. The system design considered comfort and the range of motion in the neck and upper body. The form factor and the positioning of the system was finalized to enable the embedding in clothing, such as in a shirt collar. A range of sensor types, sizes, and positions were considered and evaluated.

The prototype's head gesture and position classification accuracy was evaluated for five different classes of common head tilt positions. These experimental evaluations are detailed in Section 4. Head tilt classification is important because it enables user interface input with simple and subtle head gestures.

The encouraging results from the head gesture classification motivated us to explore more possibilities, including using the prototype for mouth movement and speech classification. The experimental evaluations of mouth movements and speech classification are detailed in Section 5. By also incorporating speech and/or mouth movement detection, head gestures for software interactions can be differentiated from head gestures that arise during regular conversation.

The main contributions of this work are (1) the development of a neck-mounted prototype, with an evaluation of sensor types, sizes, and positions; (2) the evaluation of the prototype's head-position classification accuracy; (3) mouth movement detection; and (4) speech detection and classification.

2. Related Work

Interfaces that sense hand and arm gestures are widespread [4], including those that rely on motion sensors [5–8], changes in Bluetooth received signal strength [9], and light sensors [10,11]. Interfaces that leverage the movement of the legs and the feet have also been explored [12,13]. Computer vision-based approaches using the camera to capture head and body motions [14,15], facial expressions [16], and eye movement [17] also exist.

Detection of throat activity has been explored using different enabling technologies. Acoustic sensors have been used for muscle movement recognition [18], speech recognition, ref. [19] and actions related to eating [20–22]. Prior research has been done on e-textiles used in the neck region for detecting posture [23] and swallowing [24], but those efforts have relied on capacitive methods that have limitations in daily interactions. Researchers have explored sensoring the neck with piezoelectric sensors for monitoring eating [25] and medication adherence [26].

In addition to the neck-mounted sensors systems, there has been an exploration of actuation at the neck region using vibrotactile stimulation for accomplishing haptic perception [27–29].

The use of video image processing for speech recognition has been applied to lip reading [30–32]. More recently, as part of the silent or unvoiced speech recognition research efforts, mobile phone and wearable cameras have been used for speech classification from mouth movements. Researchers have used bespoke wearable hardware for detecting mouth and chin movements [33], or leveraged smart phone cameras [34].

Electromyography (EMG) has also been used for speech and/or silent speech classification. Researchers have used EMG sensors on the fingers placed on the face for mouth movement classifications [35]. EMG sensoring of the face for speech detection has also been carried out [36].

Tongue movement has been monitored for human–computer interfaces, including using a magnetometer to track a magnet in the mouth [37], using capacitive touch sensors mounted on a retainer in the mouth [38], using EMG from the face muscles around the mouth [39], and using EMG coupled with electroencephalography (EEG) as sensed from behind the ear [40]. Detecting tooth clicks has also been explored including a teeth-based interface that senses tooth clicks using microphones placed behind the ears [41].

Head position classification has been carried out with motion sensors on the head [42], pairing ultrasound transmitters and ultrasonic sensors mounted on the body [43] and barometric pressure sensing inside the ear [44].

This work is an expansion on our previously published conference paper [45] that classified head gestures using on a single neck-mounted bend sensor. In this expanded work, we look not only at head gesture classification using our neck-mounted sensor interface, but also at mouth movement classification, speech detection, and speech classification.

3. Prototype

A neck-mounted wearable prototype was developed and used for classifying neck movement, mouth movement, and speech. The prototype consists of a sensor affixed to the neck which is connected to a microcontroller. The data collected from the sensor is wirelessly transferred via Bluetooth by the microcontroller to the user's paired smart phone. On the smart phone, the time-series data is in real time filtered, classified, and then used as input to a software application. Figure 1 provides an overview of the wearable system and its components interactions.

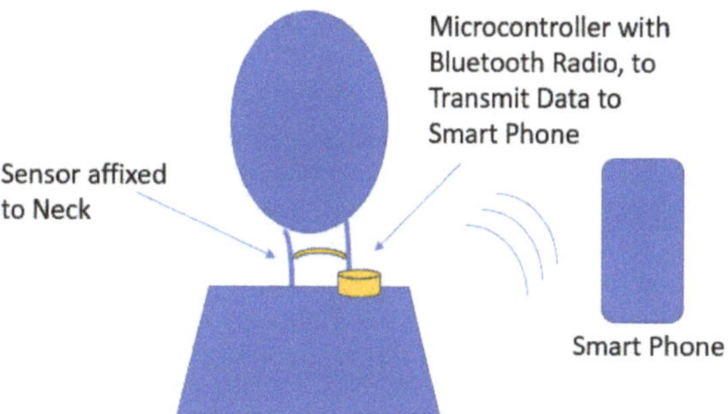

Figure 1. Prototype system's component overview, with sensor placed on neck and wearable hardware placed on collar for communicating data to a smartphone for processing and for interfacing with the application.

E-textile and flex sensors were investigated as potential candidates for the prototype. E-textiles can be used as capacitive sensors or as resistive sensors. With the capacitive method, the e-textile worked well as a proximity sensor to detect when the sensor was near human skin. However, once the sensor was in contact with or in close proximity of the skin, the sensor data became saturated and did not provide valuable features or respond to movements. Using the e-textile sensor as a resistive sensor was more successful in displaying features when actively bending or pulling the material.

The flex sensor proved to be the most appropriate for sensoring the neck. The flex sensor acts as a flexible potentiometer, whose resistance increases as the bend angle increases. Unlike the e-textile, which did not return to a static level after deformation and was prone to noise, the flex sensor performed reliably under bending and returned to a stable level when straight.

A variety of positions for the sensor around the neck, chin, and side of face were explored with the neck being the most practical in terms of data collection and ease of wear.

The hardware of the final prototype consists of an inexpensive (approximately USD 10) flex sensor, whose change in resistance signaled change in the bend of the sensor. The flex sensor was placed against the neck by weaving it under a small piece of paper that was taped to the neck. An Arduino microcontroller collected and wirelessly transmitted the data from the sensor to a smart phone for processing and display. Both an Arduino Nano and an Arduino Mega 2560 were used in the experiments.

A simple moving average (SMA) filter was used to smooth the measured resistance signal. SMA filters replace the current data value with the unweighted mean of the k previous points in the data stream, in effect smoothing the data by flattening the impact of noise and artifact that is outside the bigger trend of the data. As the window size is decreased, the smoothness of the data is decreased. In this application, a window size that is too small can result in artifact and/or noise in the time-series data being improperly classified as a neck movement event. As the window size is increased, the impact of noise and artifact is also decreased, but the likelihood that relevant information is filtered out is increased. In this application, with a window size that is too large, there is the risk of delaying the recognition of neck movement events or even missing the events altogether. A window size of k = 40 was selected, which roughly maps to one second of data.

4. Head Tilt Detection

In a series of experiments, two types of flex sensors in a variety of positions on the neck are evaluated to determine the feasibility of differentiating and classifying head tilt and positioning.

In the experiments conducted, both a short sensor in three different positions and a long sensor were considered. Each sensor placement and sensor received 10 experiments per head-tilt with a time duration of 30 s. The tilts were held static for the entire 30 s. For each experiment, approximately 1100 data points were collected.

4.1. Flex Sensor Types and Placement

Two types of flex sensors are considered: a short sensor and a long sensor. With the short sensor, three different placements are considered: a low placement, a center placement, and a high placement. The low placement is at the bottom of the neck, closest to the collar, as shown in Figure 2a. The center placement is directly over the larynx, at the middle of the neck, as shown in Figure 2b. The high placement is the top of the throat, closest to the chin, as shown in Figure 2c. The long sensor spans the three positions along the neck, from the base of the neck to under the chin, as shown in Figure 3.

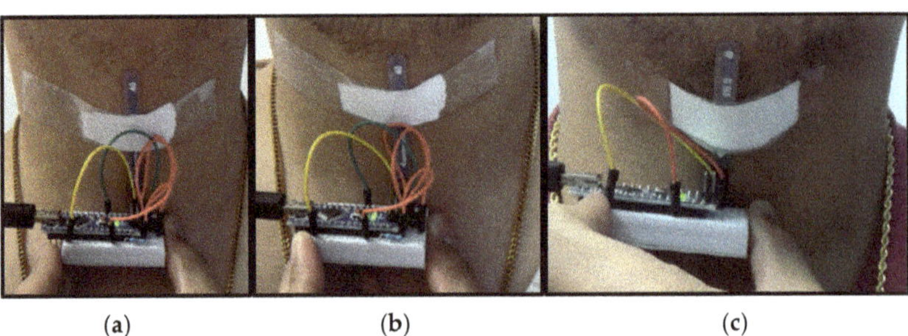

(a)   (b)   (c)

Figure 2. (a) Low, (b) center, and (c) high placement of the short flex sensor along the center line of the neck.

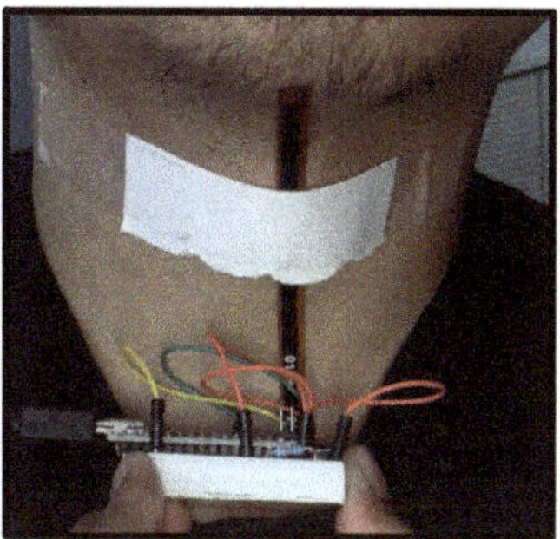

Figure 3. The placement of the long flex sensor along the center line of the neck.

4.2. Data Visualization

We visualize here some of the data collected across various placements of the sensors and for different head tilts. Figures 4–6, respectively, display the collected resistance data over a 30-s time frame across the first three classes of head tilts, namely down, forward/no tilt, and up, for each placement of the short sensor, namely low, center, and high placement. Figure 7 displays the collected resistance data over a 30-s time frame for the long sensor, across the first three classes of head tilts, namely down, forward, and up. The data represented has been filtered using a moving average filter.

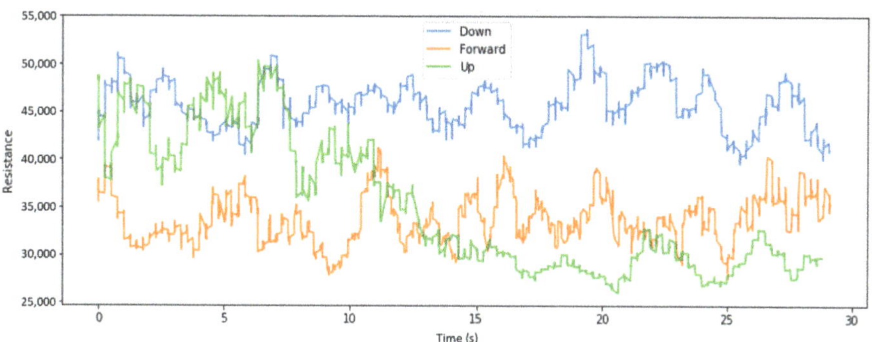

Figure 4. With low placement of short sensor, head tilt filtered data.

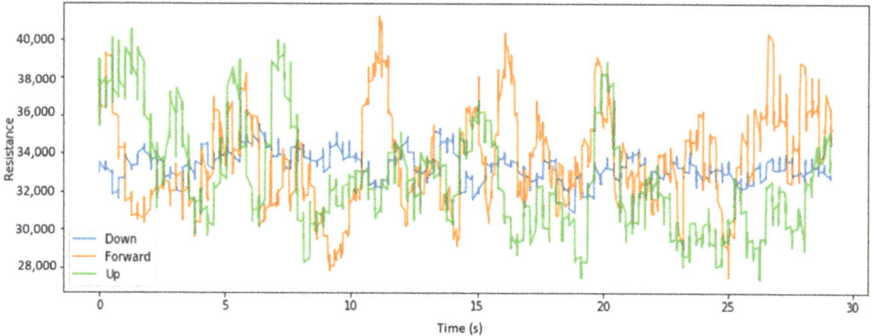

Figure 5. With center placement of short sensor, head tilt filtered data.

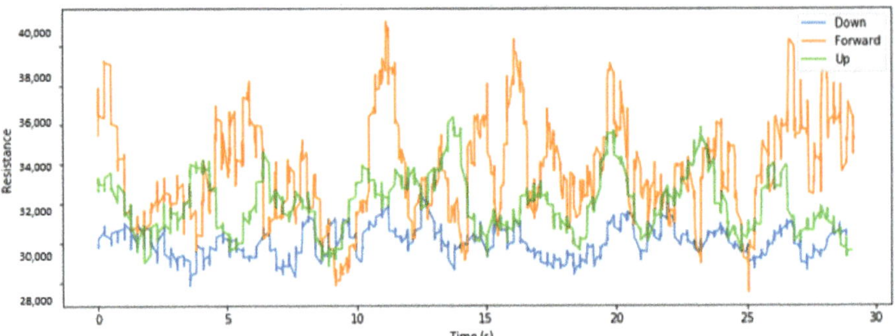

Figure 6. With high placement of short sensor, head tilt filtered data.

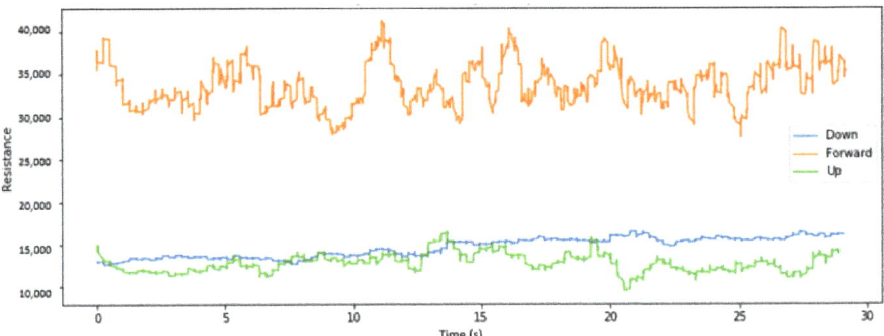

Figure 7. With long sensor, head tilt filtered data.

The short, low sensor placement and the long sensor (Figures 4 and 7, respectively) show the clearest distinction between the three classes. Therefore, the short, low sensor placement and the long sensor were further evaluated using all five classes of head tilts, namely down, forward, up, right, left. The collected resistance data over a 30-s time frame are shown in Figures 8 and 9, respectively.

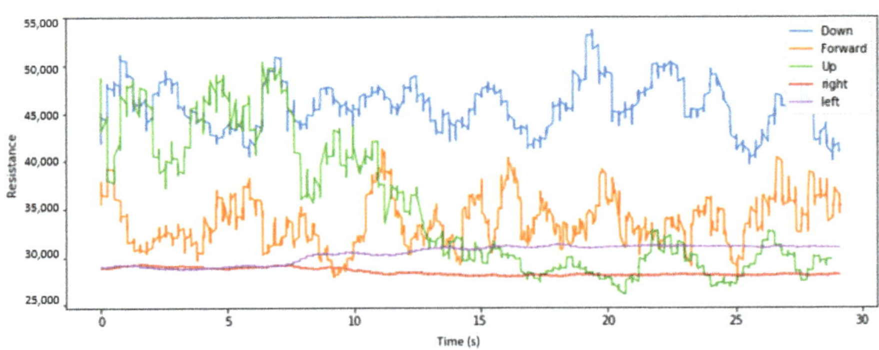

Figure 8. With low placement of short sensor, head tilt filtered data, with right and left tilts added.

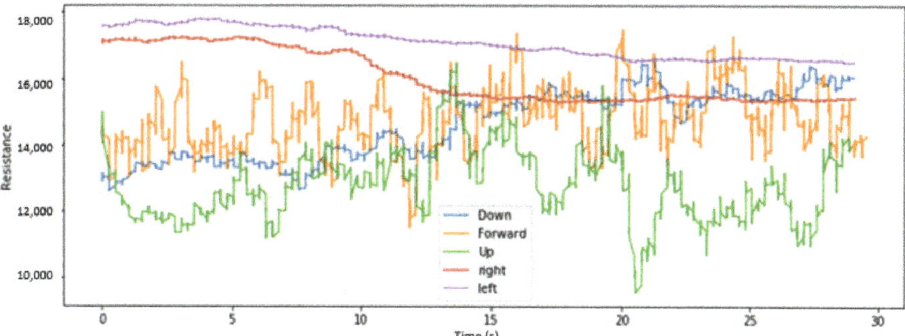

Figure 9. With long sensor, head tilt filtered data, with right and left tilts added.

4.3. Head Tilt Detection Machine Learning Results

We evaluated the accuracy of classifying a three-class dictionary of head tilts. We then went on to evaluate the accuracy of classifying an expanded five-class dictionary of head tilts. The classification results are presented in this subsection.

Three different classical machine learning (ML) classifiers were considered, specifically logistic regression, SVM, and random forest. The labeled dataset was partitioned into a train and held-out test set with an 80:20 ratio. To ensure the consistency of the models, a k-fold cross-validation was performed. A fivefold cross-validation of the train set was performed, with a random fourth of the examples in the training fold being used for validation during hyper-parameter tuning. For all the classical ML models, the Scikit-learn library in Python was used.

All four configurations, i.e., the long sensor and the three (low, center, and high) placements of the short sensor, were evaluated using the three head tilts (down, forward/not tilt, and up).

Table 1 displays our fivefold accuracy based on the model and placements of the sensors. In all cases, Logistic Regression was not sufficient in classifying the three-class dictionary. The short and low sensor placement and the long sensor had the best results. In both cases, random forest is the best performing model with test accuracies reaching ~83.4% and ~96% for the short, low placement and the long sensor, respectively.

Table 1. Fivefold training, cross-validation, and held-out test accuracy of classical ML models with different feature sets. The bold font denotes the cases with the highest accuracy for that model. These results are for the three-class dictionary.

Model		Short Sensor Low Placement	Short Sensor Center Placement	Short Sensor High Placement	Long Sensor
Logistic Regression	*Train*	0.744	0.379	0.629	0.603
	Validate	0.74	0.379	0.622	0.602
	Test	0.76	0.349	0.589	0.608
SVM	*Train*	0.825	0.594	0.648	0.891
	Validate	0.809	0.547	0.612	0.881
	Test	0.824	0.555	0.575	0.891
Random Forest	*Train*	0.955	0.918	0.854	**0.989**
	Validate	0.821	0.665	0.694	**0.945**
	Test	0.834	0.669	0.671	**0.960**

To the best performing results, two additional classes were added. The two additional classes are the user's head facing right and the user's head facing left.

Table 2 shows the performance of the short sensor with low placement and the long sensor when classifying against this five-class dictionary. As with previous results, random forest had the best performance with a test accuracy of ~83% for the short sensor and ~91% for the long sensor.

Table 2. Fivefold training, cross-validation, and held-out test accuracy of classical ML models with different feature sets. The bold font denotes the cases with the highest accuracy for that model. These results are for the five-class dictionary that includes facing right and facing left.

Model		Short Sensor Low Placement	Long Sensor
Logistic Regression	*Train*	0.734	0.337
	Validate	0.733	0.338
	Test	0.755	0.363
SVM	*Train*	0.756	0.869
	Validate	0.741	0.812
	Test	0.76	0.818
Random Forest	*Train*	0.956	**0.977**
	Validate	0.824	**0.915**
	Test	0.828	**0.91**

Table 3 shows the confusion matrix for the short sensor with low placement with the random forest classifier. The largest source of misclassifications are from the up data points, with only 65 out of 157 labels predicted correctly.

Table 3. Five-class confusion matrix for the short sensor with low placement. Rows represent actual class and columns represent predicted class.

	Random Forest	Predicated				
		Down	Forward	Up	Right	Left
	Down	259	0	10	0	0
	Forward	1	285	40	1	3
Actual	Up	25	47	65	15	5
	Right	0	0	8	185	18
	Left	0	0	3	29	194

Table 4 shows the confusion matrix for the long sensor using the random forest classifier. With the long sensor, only 17 out of 182 up data points are mislabeled. The largest confusion is between left and right tilts.

From the confusion matrix the neck gesture language can be created. The most frequent or the most important gestures can be assigned to the head tilts that achieve the highest classification accuracy, both in terms of sensitivity and specificity. For example, the following mapping of neck gestures would be appropriate for the social media app Instagram. While on their feeds, users would tilt their heads forward to signal scrolling and would turn their heads to the side, either right or left, to 'like' an image.

Table 4. Five-class confusion matrix for the long sensor. Rows represent actual class and columns represent predicted class.

Random Forest		Predicated				
		Down	Forward	Up	Right	Left
	Down	202	3	11	0	0
	Forward	0	494	2	0	0
Actual	Up	17	0	182	0	0
	Right	0	0	0	204	49
	Left	0	0	0	36	166

5. Speech and Mouth Movement Detection

In this section, we explore a larger range of opportunities that the neck-mounted sensor can provide in addition to the head gesture detection detailed in Section 4. Section 5.1 addresses speech detection using the prototype, by differentiating speech from static breathing. Section 5.2 address mouth movement classification, namely the determination of how many times the mouth has been opened and closed. Section 5.3 tackles the challenging task of speech classification using only the detection of movement in the neck.

Speech and mouth movement detection provide contextual information that can be used to trigger or to mute the head tilt interface. For instance, if the system detects that the user is talking, then the user's head tilts are not relayed to application software.

5.1. Speech Detection

Figure 10 shows an example sensor reading from static breathing and from talking, specifically saying 'hello', on the same graph. The visualization demonstrates that the presence of speech can potentially be differentiated from static breathing using only the data collected from the flex sensor on the neck-mounted prototype.

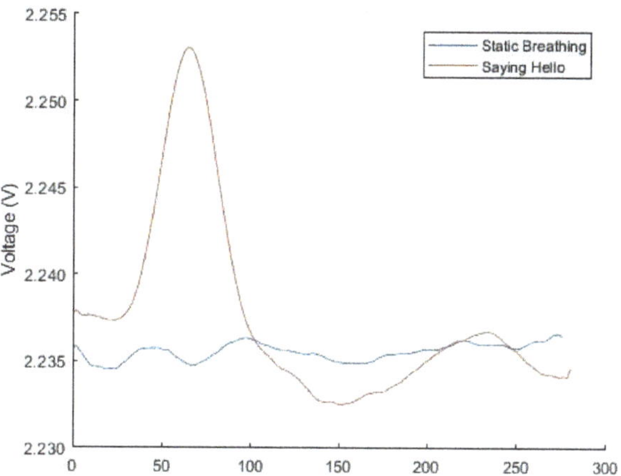

Figure 10. Sensor readings from static breathing and saying 'hello'.

Using the neck-mounted prototype, an experiment was conducted to see if static breathing can indeed be differentiated from speech. Three-second-long samples with the prototype's flex sensor were collected of both static breathing and of saying 'hello'. A total of 60 samples, 30 of each class, were collected. The samples were classified using K-nearest neighbors (k-NN) with dynamic time warping (DTW), with k set to 3.

Dynamic time warping measures the similarity between two time-series signals, which may vary in speed and in length. It calculates the minimal distance between the signals allowing for warping of the time axis, with similar signals having lower cost than dissimilar signals.

Each test signal is compared against all the training signals, and the DTW cost between the test signal and each training signals is calculated. The DTW cost of the k nearest neighbors, i.e., most similar training signals, is then used to classify the signal.

Table 5 shows the confusion matrix for the classification results. The overall accuracy of the classification was 83.3% with 3 of the 30 talking samples misclassified as breathing.

Table 5. Two-class confusion matrix for static breathing and talking. Rows represent actual class and columns represent predicted class.

		Predicated	
		Static Breathing	**Talking**
Actual	Static Breathing	23	7
	Talking	3	27

5.2. Mouth Movement Classification

In another experiment, the classification of mouth movements without the generation of any sound was examined. The mouth was opened and closed without sound being generated. It was a four-class dictionary, with static breathing (no mouth movement), opening and closing of the mouth once, opening and closing of the mouth twice, and opening and closing of the mouth three times.

Three-second-long samples with the prototype's flex sensor were collected with a total of 60 samples, 15 of each class. The samples were classified using K-nearest neighbors (k-NN) with dynamic time warping, with k set to 3.

Table 6 shows the confusion matrix for the classification results. The overall accuracy of the classification was 67.5%. The classification of static breathing resulted in most of the misclassifications. By considering sample's peak-to-valley amplitude, this misclassification can be decreased.

Table 6. Four-class confusion matrix for mouth movements. Rows represent actual class and columns represent predicted class.

		Predicated			
		Breathing	**One Cycle**	**Two Cycles**	**Three Cycles**
	Breathing	2	3	3	12
Actual	One cycle	0	19	1	0
	Two cycles	0	7	13	0
	Three cycles	0	0	0	20

5.3. Speech Classification

The final experiments explored speech classification. Two different experiments of speech classification were carried with each having a set of four different sentences or phrases being spoken with the prototype affixed to the neck and the bend sensor capturing the neck activity.

For each of the two experiments, three-second-long samples with the prototype's flex sensor were collected. For the first experiment with sentences, a total of 40 samples were collected, 10 of each class. The sentences used in the experiments were "I am a user who is talking right now"; "This is me talking with a sensor attached"; "Who am I talking to at this very moment?"; and "Can you recognize what I am saying while attached to a sensor?" For the second experiment with famous idioms, a total of 80 samples were collected, 20 of each class. The idioms used in the experiment were "a blessing in disguise"; "cut somebody some slack"; "better late than never"; and "a dime a dozen." The samples were classified using K-nearest neighbors (k-NN) with dynamic time warping, with k set to 3.

Tables 7 and 8 show the confusion matrices for the classification results for the two experiments, respectively. The overall accuracy of the classification was 62.5% and 32.5%, respectively.

Table 7. Four-class confusion matrix for spoken sentences. Rows represent actual class and columns represent predicted class.

		Predicated			
		"I Am … "	"This Is … "	"Who … "	"Can You … "
Actual	"I am a user who is talking right now."	0	9	1	0
	"This is me talking with a sensor attached."	0	10	0	0
	"Who am I talking to at this very moment?"	0	4	6	0
	"Can you recognize what I am saying while attached to a sensor?"	0	0	1	9

Table 8. Four-class confusion matrix for spoken phrases. Rows represent actual class and columns represent predicted class.

		Predicated			
		"A Blessing in Disguise"	"Cut Somebody Some Slack"	"Better Late than Never"	"A Dime a Dozen"
Actual	"A blessing in disguise"	0	0	14	6
	"Cut somebody some slack"	0	2	1	17
	"Better late than never"	0	0	19	1
	"A dime a dozen"	0	0	15	5

6. Discussion

The experiments with sensor data captured from the neck-mounted prototype show that the short sensor with low placement on the neck and the long sensor had the best results. For a three-class dictionary of head tilts, random forest is the best performing model with test accuracy of ~83.4% for the short sensor with low placement and ~96% for the long sensor. For a five-class dictionary of head tilts, random forest again had the best performance with a test accuracy of ~83% for the short sensor with low placement and ~91% for the long sensor.

Movements farther from the neck were also successfully detected and classified. Sensor data captured from the neck was able to differentiate speaking from static breathing, with ~83% accuracy. The presence and the number of mouth movements was classified with ~68% accuracy. Speech classification was more challenging, achieving up to 62.5% accuracy in differentiating spoken sentences from a four-class dictionary.

7. Conclusions

In this work, we show that subtle neck tilts, mouth movements, and speech can be detected and classified using an inexpensive flex sensor placed at the neck, and thus can prove to be enabling technology for use in software interfaces.

A flex sensor incorporated into a shirt collar or as part of a necklace opens new possibilities for software interaction. The accuracy of the classification of head tilts and their socially undisruptive nature makes head tilting a good option for signally software micro-interactions. For example, a tilt of the head can dismiss a smartwatch notification.

As head gestures can be made during the course of natural speech, the detection of speech and mouth movements allows for the interface to be tailored to times when a person is not speaking and thus improve the interface with greater context awareness.

Author Contributions: Conceptualization, A.N.; methodology, J.L., P.I., K.M. and A.N.; software, J.L. and P.I.; validation, J.L. and P.I.; investigation, J.L., P.I., K.M. and A.N.; writing—original draft preparation, J.L., P.I. and A.N.; writing—review and editing, J.L., A.N. and K.M.; visualization, J.L., P.I. and A.N.; supervision, A.N. and K.M.; project administration, A.N. All authors have read and agreed to the published version of the manuscript.

Funding: This research was funded by CSUN Research, Scholarship, and Creative Activity (RSCA) 2021–2022, PI: Ani Nahapetian.

Institutional Review Board Statement: Not applicable.

Informed Consent Statement: Informed consent was obtained from all subjects involved in the study.

Conflicts of Interest: The authors declare no conflict of interest.

References

1. Orphanides, A.K.; Nam, C.S. Touchscreen interfaces in context: A systematic review of research into touchscreens across settings, populations, and implementations. *Appl. Ergon.* **2017**, *61*, 116–143. [CrossRef] [PubMed]

2. Jonsson, P.; Johnson, P.W.; Hagberg, M.; Forsman, M. Thumb joint movement and muscular activity during mobile phone texting—A methodological study. *J. Electromyogr. Kinesiol.* **2011**, *21*, 363–370. [CrossRef]

3. Lee, M.; Hong, Y.; Lee, S.; Won, J.; Yang, J.; Park, S.; Chang, K.-T.; Hong, Y. The effects of smartphone use on upper extremity muscle activity and pain threshold. *J. Phys. Ther. Sci.* **2015**, *27*, 1743–1745. [CrossRef] [PubMed]

4. Dannenberg, R.B.; Amon, D. A gesture based user interface prototyping system. In Proceedings of the 2nd Annual ACM SIGGRAPH Symposium on User Interface Software and Technology (UIST '89), Williamsburg, VA, USA, 13–15 November 1989; Association for Computing Machinery: New York, NY, USA, 1989; pp. 127–132. [CrossRef]

5. McGuckin, S.; Chowdhury, S.; Mackenzie, L. Tap 'n' shake: Gesture-based smartwatch-smartphone communications system. In Proceedings of the 28th Australian Conference on Computer-Human Interaction (OzCHI '16), Launceston, TAS, Australia, 29 November–2 December 2016; Association for Computing Machinery: New York, NY, USA, 2016; pp. 442–446. [CrossRef]

6. Deponti, D.; Maggiorini, D.; Palazzi, C.E. Smartphone's physiatric serious game. In Proceedings of the 2011 IEEE 1st International Conference on Serious Games and Applications for Health (SEGAH '11), Braga, Portugal, 9–11 November 2011; pp. 1–8.

7. Deponti, D.; Maggiorini, D.; Palazzi, C.E. DroidGlove: An android-based application for wrist rehabilitation. In Proceedings of the 2009 IEEE International Conference on Ultra Modern Telecommunications (ICUMT 2009), St. Petersburg, Russia, 12–14 October 2009; pp. 1–7. [CrossRef]

8. Moazen, D.; Sajjadi, S.A.; Nahapetian, A. AirDraw: Leveraging smart watch motion sensors for mobile human computer interactions. In Proceedings of the 2016 13th IEEE Annual Consumer Communications & Networking Conference (CCNC), Las Vegas, NV, USA, 9–12 January 2016; pp. 442–446. [CrossRef]

9. Vance, E.; Nahapetian, A. Bluetooth-based context modeling. In Proceedings of the 4th ACM MobiHoc Workshop on Experiences with the Design and Implementation of Smart Objects (SMARTOBJECTS '18), Los Angeles, CA, USA, 25 June 2018; Association for Computing Machinery: New York, NY, USA, 2018; pp. 1–6. [CrossRef]

10. Holmes, A.; Desai, S.; Nahapetian, A. LuxLeak: Capturing computing activity using smart device ambient light sensors. In Proceedings of the 2nd Workshop on Experiences in the Design and Implementation of Smart Objects (SmartObjects '16), New York City, NY, USA, 3–7 October 2016; Association for Computing Machinery: New York, NY, USA, 2016; pp. 47–52. [CrossRef]

11. Papisyan, A.; Nahapetian, A. LightVest: A wearable body position monitor using ambient and infrared light. In Proceedings of the 9th International Conference on Body Area Networks (BodyNets '14), London, UK, 29 September–1 October 2014; ICST (Institute for Computer Sciences, Social-Informatics and Telecommunications Engineering): Brussels, Belgium, 2014; pp. 186–192. [CrossRef]

12. Velloso, E.; Schmidt, D.; Alexander, J.; Gellersen, H.; Bulling, A. The Feet in Human–Computer Interaction. *ACM Comput. Surv.* **2015**, *48*, 1–35. [CrossRef]

13. Scott, J.; Dearman, D.; Yatani, K.; Truong, K.N. Sensing foot gestures from the pocket. In Proceedings of the 23nd Annual ACM Symposium on User Interface Software and Technology (UIST '10), New York City, NY, USA, 3–6 October 2010; Association for Computing Machinery: New York, NY, USA, 2010; pp. 199–208. [CrossRef]

14. Ohn-Bar, E.; Tran, C.; Trivedi, M. Hand gesture-based visual user interface for infotainment. In Proceedings of the 4th International Conference on Automotive User Interfaces and Interactive Vehicular Applications (AutomotiveUI '12), Portsmouth, NH, USA, 17–19 October 2012; Association for Computing Machinery: New York, NY, USA, 2012; pp. 111–115. [CrossRef]

15. Davis, J.W.; Vaks, S. A perceptual user interface for recognizing head gesture acknowledgements. In Proceedings of the 2001 Workshop on Perceptive user Interfaces (PUI '01), Orlando, FL, USA, 15–16 November 2001; Association for Computing Machinery: New York, NY, USA, 2001; pp. 1–7. [CrossRef]

16. Li, H.; Trutoiu, L.C.; Olszewski, K.; Wei, L.; Trutna, T.; Hsieh, P.-L.; Nicholls, A.; Ma, C. Facial performance sensing head-mounted display. *ACM Trans. Graph.* **2015**, *34*, 1–9. [CrossRef]

17. McNamara, A.; Kabeerdoss, C.; Egan, C. Mobile User Interfaces based on User Attention. In Proceedings of the 2015 Workshop on Future Mobile User Interfaces (FutureMobileUI '15), Florence, Italy, 19–22 May 2015; Association for Computing Machinery: New York, NY, USA, 2015; pp. 1–3. [CrossRef]

18. Yatani, K.; Truong, K.N. BodyScope: A wearable acoustic sensor for activity recognition. In Proceedings of the 2012 ACM Conference on Ubiquitous Computing (UbiComp '12), Pittsburgh, Pennsylvania, 5–8 September 2012; Association for Computing Machinery: New York, NY, USA, 2012; pp. 341–350. [CrossRef]

19. Bi, Y.; Xu, W.; Guan, N.; Wei, Y.; Yi, W. Pervasive eating habits monitoring and recognition through a wearable acoustic sensor. In Proceedings of the 8th International Conference on Pervasive Computing Technologies for Healthcare (PervasiveHealth '14), Oldenburg, Germany, 20–23 May 2014; ICST (Institute for Computer Sciences, Social-Informatics and Telecommunications Engineering): Brussels, Belgium, 2014; pp. 174–177. [CrossRef]

20. Erzin, E. Improving Throat Microphone Speech Recognition by Joint Analysis of Throat and Acoustic Microphone Recordings. *IEEE Trans. Audio Speech, Lang. Process.* **2009**, *17*, 1316–1324. [CrossRef]

21. Turan, M.T.; Erzin, E. Empirical Mode Decomposition of Throat Microphone Recordings for Intake Classification. In Proceedings of the 2nd International Workshop on Multimedia for Personal Health and Health Care (MMHealth '17), Mountain View, CA, USA, 23 October 2017; Association for Computing Machinery: New York, NY, USA, 2017; pp. 45–52. [CrossRef]

22. Cohen, E.; Stogin, W.; Kalantarian, H.; Pfammatter, A.F.; Spring, B.; Alshurafa, N. SmartNecklace: Designing a wearable multi-sensor system for smart eating detection. In Proceedings of the 11th EAI International Conference on Body Area Networks (BodyNets '16), Turin, Italy, 15–16 December 2016; ICST (Institute for Computer Sciences, Social-Informatics and Telecommunications Engineering): Brussels, Belgium, 2016; pp. 33–37.

23. Hirsch, M.; Cheng, J.; Reiss, A.; Sundholm, M.; Lukowicz, P.; Amft, O. Hands-free gesture control with a capacitive textile neckband. In Proceedings of the 2014 ACM International Symposium on Wearable Computers (ISWC '14), Seattle, WA, USA, 13–17 September 2014; Association for Computing Machinery: New York, NY, USA, 2014; pp. 55–58. [CrossRef]

24. Cheng, J.; Zhou, B.; Kunze, K.; Rheinländer, C.C.; Wille, S.; Wehn, N.; Weppner, J.; Lukowicz, P. Activity recognition and nutrition monitoring in every day situations with a textile capacitive neckband. In Proceedings of the 2013 ACM Conference on Pervasive and Ubiquitous Computing Adjunct Publication (UbiComp '13 Adjunct), Zurich, Switzerland, 8–12 September 2013; Association for Computing Machinery: New York, NY, USA, 2013; pp. 155–158. [CrossRef]

25. Kalantarian, H.; Alshurafa, N.; Sarrafzadeh, M. A Wearable Nutrition Monitoring System. In Proceedings of the 2014 11th International Conference on Wearable and Implantable Body Sensor Networks (BSN '14), Zurich, Switzerland, 16–19 June 2014; IEEE Computer Society: Washington, DC, USA, 2014; pp. 75–80. [CrossRef]

26. Kalantarian, H.; Motamed, B.; Alshurafa, N.; Sarrafzadeh, M. A wearable sensor system for medication adherence prediction. *Artif. Intell. Med.* **2016**, *69*, 43–52. [CrossRef] [PubMed]

27. Morrow, K.; Wilbern, D.; Taghavi, R.; Ziat, M. The effects of duration and frequency on the perception of vibrotactile stimulation on the neck. In Proceedings of the 2016 IEEE Haptics Symposium (HAPTICS '16), Philadelphia, PA, USA, 8–11 April 2016; pp. 41–46.

28. Yamazaki, Y.; Hasegawa, S.; Mitake, H.; Shirai, A. Neck strap haptics: An algorithm for non-visible VR information using haptic perception on the neck. In Proceedings of the ACM SIGGRAPH 2019 Posters (SIGGRAPH '19), Los Angeles, CA, USA, 28 July 2019; Association for Computing Machinery: New York, NY, USA, 2019. Article 60, pp. 1–2. [CrossRef]

29. Yamazaki, Y.; Mitake, H.; Hasegawa, S. Tension-based wearable vibroacoustic device for music appreciation. In Proceedings of the International Conference on Human Haptic Sensing and Touch Enabled Computer Applications (EuroHaptics '16), London, UK, 4–7 July 2016; Springer: Berlin/Heidelberg, Germany, 2016; pp. 273–283.

30. Ephrat, A.; Peleg, S. Vid2speech: Speech reconstruction from silent video. In Proceedings of the 2017 IEEE International Conference on Acoustics, Speech and Signal Processing (ICASSP '17), New Orleans, LA, USA, 5–9 March 2017; pp. 5095–5099.

31. Ephrat, A.; Halperin, T.; Peleg, S. Improved speech reconstruction from silent video. In Proceedings of the IEEE International Conference on Computer Vision, Venice, Italy, 22–29 October 2017; pp. 455–462.

32. Kumar, Y.; Jain, R.; Salik, M.; Shah, R.; Zimmermann, R.; Yin, Y. MyLipper: A Personalized System for Speech Reconstruction using Multi-view Visual Feeds. In Proceedings of the 2018 IEEE International Symposium on Multimedia (ISM), Taichung, Taiwan, 10–12 December 2018; pp. 159–166. [CrossRef]

33. Kimura, N.; Hayashi, K.; Rekimoto, J. TieLent: A Casual Neck-Mounted Mouth Capturing Device for Silent Speech Interaction. In Proceedings of the International Conference on Advanced Visual Interfaces, Salerno, Italy, 28 September–2 October 2020; Association for Computing Machinery: New York, NY, USA, 2020; pp. 1–8. [CrossRef]

34. Sun, K.; Yu, C.; Shi, W.; Liu, L.; Shi, Y. Lip-Interact: Improving Mobile Device Interaction with Silent Speech Commands. In Proceedings of the 31st Annual ACM Symposium on User Interface Software and Technology (UIST '18), Berlin, Germany, 14–17 October 2018; Association for Computing Machinery: New York, NY, USA, 2018; pp. 581–593. [CrossRef]

35. Manabe, H.; Hiraiwa, A.; Sugimura, T. Unvoiced speech recognition using EMG—Mime speech recognition. In *CHI '03 Extended Abstracts on Human Factors in Computing Systems (CHI EA '03)*; Association for Computing Machinery: New York, NY, USA, 2003; pp. 794–795. [CrossRef]

36. Maier-Hein, L.; Metze, F.; Schultz, T.; Waibel, A. Session independent non-audible speech recognition using surface electromyography. In Proceedings of the IEEE Workshop on Automatic Speech Recognition and Understanding, Cancun, Mexico, 27 November–1 December 2005; pp. 331–336. [CrossRef]

37. Sahni, H.; Bedri, A.; Reyes, G.; Thukral, P.; Guo, Z.; Starner, T.; Ghovanloo, M. The tongue and ear interface: A wearable system for silent speech recognition. In Proceedings of the 2014 ACM International Symposium on Wearable Computers (ISWC '14), Seattle, WA, USA, 13–17 September 2014; Association for Computing Machinery: New York, NY, USA, 2014; pp. 47–54. [CrossRef]
38. Li, R.; Wu, J.; Starner, T. TongueBoard: An Oral Interface for Subtle Input. In Proceedings of the 10th Augmented Human International Conference 2019 (AH2019), Reims, France, 11–12 March 2019; Association for Computing Machinery: New York, NY, USA, 2019; pp. 1–9. [CrossRef]
39. Zhang, Q.; Gollakota, S.; Taskar, B.; Rao, R.P. Non-intrusive tongue machine interface. In Proceedings of the SIGCHI Conference on Human Factors in Computing Systems (CHI '14), Toronto, ON, Canada, 26 April–1 May 2014; Association for Computing Machinery: New York, NY, USA, 2014; pp. 2555–2558. [CrossRef]
40. Nguyen, P.; Bui, N.; Nguyen, A.; Truong, H.; Suresh, A.; Whitlock, M.; Pham, D.; Dinh, T.; Vu, T. TYTH-Typing on Your Teeth: Tongue-Teeth Localization for Human-Computer Interface. In Proceedings of the 16th Annual International Conference on Mobile Systems, Applications, and Services (MobiSys '18), Munich, Germany, 10–15 June 2018; Association for Computing Machinery: New York, NY, USA, 2018; pp. 269–282. [CrossRef]
41. Ashbrook, D.; Tejada, C.; Mehta, D.; Jiminez, A.; Muralitharam, G.; Gajendra, S.; Tallents, R. Bitey: An exploration of tooth click gestures for hands-free user interface control. In Proceedings of the 18th International Conference on Human-Computer Interaction with Mobile Devices and Services (MobileHCI '16), Florence, Italy, 6–9 September 2016; Association for Computing Machinery: New York, NY, USA, 2016; pp. 158–169. [CrossRef]
42. Crossan, A.; McGill, M.; Brewster, S.; Murray-Smith, R. Head tilting for interaction in mobile contexts. In Proceedings of the 11th International Conference on Human-Computer Interaction with Mobile Devices and Services (MobileHCI '09), Bonn, Germany, 15–18 September 2009; Association for Computing Machinery: New York, NY, USA, 2009; pp. 1–10. [CrossRef]
43. LoPresti, E.; Brienza, D.M.; Angelo, J.; Gilbertson, L.; Sakai, J. Neck range of motion and use of computer head controls. In Proceedings of the Fourth International ACM Conference on Assistive Technologies (Assets '00), Arlington, VA, USA, 13–15 November 2000; Association for Computing Machinery: New York, NY, USA, 2000; pp. 121–128. [CrossRef]
44. Ando, T.; Kubo, Y.; Shizuki, B.; Takahashi, S. CanalSense: Face-Related Movement Recognition System based on Sensing Air Pressure in Ear Canals. In Proceedings of the 30th Annual ACM Symposium on User Interface Software and Technology (UIST '17), Quebec City, QC, Canada, 22–25 October 2017; Association for Computing Machinery: New York, NY, USA, 2017; pp. 679–689. [CrossRef]
45. Lacanlale, J.; Isayan, P.; Mkrtchyan, K.; Nahapetian, A. Look Ma, No Hands: A Wearable Neck-Mounted Interface. In Proceedings of the Conference on Information Technology for Social Good (GoodIT '21), Rome, Italy, 9–11 September 2021; Association for Computing Machinery: New York, NY, USA, 2021; pp. 13–18. [CrossRef]

sensors

MDPI

Article

Novel Perspectives for the Management of Multilingual and Multialphabetic Heritages through Automatic Knowledge Extraction: The DigitalMaktaba Approach [†]

Sonia Bergamaschi [1], Stefania De Nardis [2], Riccardo Martoglia [1,*], Federico Ruozzi [1], Luca Sala [2], Matteo Vanzini [2] and Riccardo Amerigo Vigliermo [1]

[1] University of Modena and Reggio Emilia, 41125 Modena, Italy; sonia.bergamaschi@unimore.it (S.B.); federico.ruozzi@unimore.it (F.R.); r.a.vigliermo@unimore.it (R.A.V.)

[2] mim.fscire, 40125 Bologna, Italy; denardis@fscire.it (S.D.N.); salaluca.info@gmail.com (L.S.); vanzinimatteo.info@gmail.com (M.V.)

* Correspondence: riccardo.martoglia@unimore.it

† This paper is an extended version of our paper published in ACM International Conference on Information Technology for Social Good (GoodIT), Rome, Italy, 9–11 September 2021.

Abstract: The linguistic and social impact of multiculturalism can no longer be neglected in any sector, creating the urgent need of creating systems and procedures for managing and sharing cultural heritages in both supranational and multi-literate contexts. In order to achieve this goal, text sensing appears to be one of the most crucial research areas. The long-term objective of the *DigitalMaktaba* project, born from interdisciplinary collaboration between computer scientists, historians, librarians, engineers and linguists, is to establish procedures for the creation, management and cataloguing of archival heritage in non-Latin alphabets. In this paper, we discuss the currently ongoing design of an innovative workflow and tool in the area of text sensing, for the automatic extraction of knowledge and cataloguing of documents written in non-Latin languages (Arabic, Persian and Azerbaijani). The current prototype leverages different OCR, text processing and information extraction techniques in order to provide both a highly accurate extracted text and rich metadata content (including automatically identified cataloguing metadata), overcoming typical limitations of current state of the art approaches. The initial tests provide promising results. The paper includes a discussion of future steps (e.g., AI-based techniques further leveraging the extracted data/metadata and making the system learn from user feedback) and of the many foreseen advantages of this research, both from a technical and a broader cultural-preservation and sharing point of view.

Keywords: digital libraries; minority languages; humanistic informatics; computer archiving; intercultural communication

Citation: Bergamaschi, S.; De Nardis, S.; Martoglia, R.; Ruozzi, F.; Sala, L.; Vanzini, M.; Vigliermo, R.A. Novel Perspectives for the Management of Multilingual and Multialphabetic Heritages through Automatic Knowledge Extraction: The DigitalMaktaba Approach. *Sensors* **2022**, *22*, 3995. https://doi.org/10.3390/s22113995

Academic Editors: Pietro Manzoni, Claudio Palazzi and Ombretta Gaggi

Received: 8 April 2022
Accepted: 23 May 2022
Published: 25 May 2022

1. Introduction

Since 1700, when the difficulty of establishing a stable system of norms arose, Europe has been studying the management and cataloguing of documentary heritages. Organic codes were devised for catalog compilation in several countries between the 1800s and 1900s, and worldwide agreements were established to create a common system of descriptive cards. The need to manage multimedia content today imposes new and urgent demands: creating systems and procedures for managing and sharing cultural heritages in both supranational and multi-literate contexts. This is the challenging scenario of the recently started *DigitalMaktaba* (in Arabic, the word *maktaba* is derived from the root k-t-b which originates the words: *kitāb* ("book"), *kutub* ("books"), *kātib* ("writer"), *kuttāb* ("writers", also "Koranic school") and so on. The prefix *ma-* indicates the place where something is found or carried out; therefore, *maktaba* literally means: the "place where books are found", "library") project, born from the collaboration between computer scientists, historians, librarians, engineers and linguists gathered together from the mim.fscire start-up, the University of Modena and Reggio Emilia (UniMoRe) and the Fondazione per le Scienze Religiose (FSCIRE), leader institution of the RESILIENCE European research infrastructure on Religious Studies (ESFRI Roadmap, 2021). The intersection of the knowledge of religious studies, digital humanities, corpus linguistics, educational studies and engineering and computer science guarantees a broad reflection on various aspects related to the design theme: technological, ethical, cultural, social, economic, political and religious. This synergy between academic and extra-academic science-sector skills and varied professional experience is fundamental to effectively address the challenges that a technologically advanced, multicultural and historically rich community, such as the European one, poses in the field of the conservation and enhancement of one's own cultural heritage. The long-term objective is to establish procedures for the creation, management and cataloguing of librarian and archival heritage in non-Latin alphabets. In particular, the project test case is the large collection of digital books made internally available by the "Giorgio La Pira" library in Palermo, which is a hub of FSCIRE foundation, dedicated to history and doctrines of Islam. Documents such as these pose a number of non-trivial issues in their computer-assisted management, especially optical character recognition (OCR) and knowledge extraction, since their texts are presented in several non-Latin alphabets (in particular, Arabic, Persian and Azerbaijani) and, for each alphabet, in multiple characters, also in a single work (see Figure 1 for a sample).

Figure 1. A sample frontispiece with multiple Arab characters.

DigitalMaktaba focuses on innovative solutions in the context of digital libraries, providing several techniques to support and automate many of the tasks (OCR, linguistic-resource linking, metadata extraction, and so on) related to the text sensing/knowledge extraction and cataloguing of the documents in a multi-lingual context. Even if the text sensing/OCR/machine-learning research area is in general very active concerning Latin script documents [1,2], up to now only few projects (e.g., [3–5]) have been proposed in the state of the art research for the curation of new and innovative digital libraries in the considered Arabic-script languages; furthermore, most of them require consistent manual work and none of them returns rich information and metadata beyond the extracted text. However, we deem that the linguistic and social impact of multiculturalism can no longer be neglected in any sector. Until a few years ago, only few highly specialized libraries possessed texts in non-Latin alphabets; now, even the smallest ones must adapt acquisitions to the needs of culturally heterogeneous users and are often unable to do so due to the difficulty of managing this data. Hence, the urgency of a global sharing of multicultural heritages.

The present work extends our previous paper [6] in several directions and discusses the currently ongoing design of an innovative workflow and tool for the automatic extraction of knowledge and cataloguing of documents written in non-Latin languages, and in particular for the Arabic, Persian and Azerbaijani languages. The Material and Methods section (Section 3) presents an overview of the tool that is being developed, whose information-extraction pipeline (Section 3.1) smartly combines the output of several techniques that are described in detail, with special emphasis on the text-sensing aspect:

- *Text extraction*, leveraging and combining the output of the best-performing OCR libraries in order to extract text information in a more accurate and uniform way than current proposals;
- *Metadata* enrichment, enabling to automatically capture useful metadata: (a) *syntactic metadata*, including text-regions information (improved w.r.t. the state of the art approaches by means of a newly proposed text-region numbering and merging approach), identified language(s) and character(s), text size and position on page, and the self-assessed quality of extraction through an ad-hoc metric; (b) *linguistic metadata*, including links to external linguistic resources providing useful information such as word definitions for further (semantic) processing; and (c) *cataloguing metadata*, through a novel approach for automatic title and author identification in a frontispiece.

Besides information extraction, which is our current focus, we also take a look (Section 3.2), for the first time, at the data-management foundations enabling convenient and efficient access to the stored data and simple data exchange. Results (discussed in Section 4) include a look at the user interface and overall functionalities of the current prototype incorporating the above-described techniques, and several preliminary evaluation tests. The tests, performed on a subset of our use case dataset provide promising results on the effectiveness of the text, text-regions and cataloguing-metadata extraction, also w.r.t. the state of the art techniques. Generally speaking, the tool already overcomes typical limitations of current proposals, including uneven performance/limited support for different languages/characters, difficulties in automating batch extraction and very limited additional metadata availability.

The discussed techniques and their rich metadata output will be the groundwork for the complete semi-automated cataloguing system we are aiming to obtain, whose future steps, including intelligent and AI-based techniques providing even greater assistance to the librarian and incremental learning with system use, are discussed in Section 4.3. The paper is complemented by a detailed discussion of the state of the art research (Section 2). Finally, Section 5 concludes the paper by detailing some of the many foreseen advantages of this research, both from a technical and broader cultural point of view. In short, we hope this research will ultimately help in preserving and conserving culture, a crucial task, especially in this particular and interesting scenario, and to facilitate the future consultation and sharing of knowledge, thus encouraging the inclusiveness of the European community and beyond.

2. Related Works

In this section, we discuss related works by specifically focusing on projects that have been proposed for the curation of digital libraries in Arabic-script languages (Section 2.1). We also specifically examine what is available on the text sensing/extraction front, always in Arabic script (Section 2.2). We conclude the section by comparing the features of the DigitalMaktaba proposal to existing state of the art techniques, specifically identifying the innovative aspects (Section 2.3).

2.1. Projects and Proposals for the Curation of Digital Libraries in Arabic-Script Languages

From an academic point of view, even though the information-retrieval and text-extraction/sensing fields on Arabic scripts have made huge strides in the last decades, there have been not many projects aimed at exploiting them for the curation of new and innovative digital libraries. In 2009 the Alexandria library announced the creation of the Arabic Digital Library as a part of the DAR project (Digital Assests Repository), with text-extraction tools for Arabic-language characters implemented with a high accuracy, despite being designed only for extracting short information in the text [7]. In addition, worth mentioning here are more recent projects concerning the digitization and the building of Arabic and Persian texts corpora. The first example is represented by the Open Islamicate Text Initiative (OpenITI) [3], which is a multi-institutional effort to construct the first machine-actionable scholarly corpus of premodern Islamicate texts. Led by researchers at the Aga Khan University International (AKU), University of Vienna/Leipzig University (LU), and the Roshan Institute for Persian Studies at the University of Maryland, OpenITI contains almost exclusively Arabic texts, which were put together into a corpus within the OpenArabic project, which was developed first at Tufts University in the frame of the Perseus Project [8] and then at Leipzig University. The main goal of OpenArabic is to build a machine-actionable corpus of premodern texts in Arabic collected from open-access online libraries such as Shamela [9] and the Shiaonline library [10]. From this important partnership, two other interesting projects have been developed: KITAB [4] at the AKU and the Persian Digital library (PDL) at the Roshan Institute for Persian Studies [5]. The first one provides a toolbox and a forum for discussions about Arabic texts and its main goal is to research relationships between Arabic texts and discover the inter-textual system laying underneath the Arabic rich textual tradition. The PDL project is part of the larger Open ITI project and is focused primarily on the construction of a scholarly verified and machine-actionable corpus. PDL has already created an open-access corpus of more than 60,000 Persian poems collected from the Ganjoor site [11] and then integrated with a lemmatizer [12] and a digital version of the Steingass persian dictionary [13]. Another similar project is Arabic Collections Online (ACO), another multi-institutional project between NYU, Princeton, Cornell, and the American University of Cairo and Beirut in collaboration with the UAE National Archives and the Qatar National Library (QNL). It provides a publicly available digital library of Arabic language content. ACO currently provides digital access to 17,262 volumes across 10,148 subjects drawn from rich Arabic collections of distinguished research libraries [14]. It aims to digitize, preserve, and provide free open access to a wide variety of Arabic language books in subjects such as literature, philosophy, law, religion, and more. Although of a different kind, we would like to mention a few other important projects focusing on the digitization of Arabic and Persian manuscripts that involve handwritten-text recognition (HTR), such as The British Library projects [15,16] with the partnership of the Qatar National Library (Qatar Digital Library) [17] and the Iran Heritage foundation [18].

DigitalMaktaba has a number of significant differences and innovative aspects w.r.t. all the above mentioned approaches; these will be discussed in Section 2.3.

2.2. Text Sensing/Extraction/OCR in Arabic-Script Languages

Talking more specifically about text sensing and OCR, one of the areas where the first steps in DigitalMaktaba are being performed, we can distinguish between research projects and publicly available tools. From a research perspective, Arabic-script OCR is not an easy topic, since many issues have to be dealt with, including character skewing, the noisy structure of the titles and the presence of diacritical marks (vowels) mixing with diacritical dots. Studies on hidden Markov models (HMM) such as al-Muhtasib [19] have given good results on character variation. Obaid [20] proposed a segmentation-free approach for the recognition of *naskh*, derived from the verb *nasakha* "to transcribe, to copy, (to abrogate)", one of the most popoluar forms of Arabic script: now, more Qurans are written in *naskh* than in all other scripts combined. Popular for writing books because of its legibility and adapted for printing, it is still the most common font in printed Arabic. The model is extensible, robust, and adaptive to character variation and to text degradation. The use of symbolic AI combined with algorithms (such as the C4.5 algorithm) has shown high tolerance to noisy documents with a high training speed [21]. In more recent times, contour-based systems for character recognition have been proposed. As shown in the study of Mohammad [22], the systems demonstrate robustness to noise resulting in high average recognition accuracy. Other works have targeted the difficulties posed by Arabic or Persian manuscripts when disentangling overlapped characters that cause diacritic points to nudge forward (right to left) their original position, creating recognition errors or failure. Many attempts have been made to provide useful algorithms able to recognize the slanting and overlapping script typical of the Arabic handwritten script (in particular *nasta'līq*) [23]. Different typologies of neural networks (NN) have been indagated, such as the simple artificial neural network (ANN) [24], bidimensional long-short memory (BLSTM) and recurrent neural network (RNN), sometimes implemented with some HMM [25]. In addition, different ML techniques have been implemented, such as K-means or K-nearest neighbour (KNN), in order to cluster diacritical dots and segment different characters in a proper way. Persian-manuscript recognition has also been an active field of studies. Early in 1997, Dehgan and Faez extracted images utilizing Zernike moments, pseudo-Zernike and Legendre moments [26]. By using an ART2 neural network they obtained very good results. Mowlaei developed a recognition system of Persian digits and characters by using Haar wavelet to extract features and then insert them into an NN [27]. A different approach is represented by fuzzy logic, particularly indicated in ambiguous contexts. Linguistic fuzzy models have demonstrated robustness to Persian script manuscripts variations [28]. More recently, RNN and Deep NN has been introduced along with new segmentation techniques [29] or architectures such as DensNet and Xception [30].

While the above works are certainly interesting, they often do not offer publicly available OCR tools. Therefore, we will now focus specifically on publicly available OCR libraries supporting the required languages. Among the free and open source ones, there are systems such as Tesseract (Available online: https://github.com/tesseract-ocr/tesseract (accessed on 4 February 2022)), EasyOCR (available online: https://github.com/JaidedAI/EasyOCR (accessed on 4 February 2022)), GoogleDocs (available online: https://docs.google.com (accessed on 4 February 2022)) and Capture2Text (available online: http://capture2text.sourceforge.net/ (accessed on 4 February 2022)). While certainly a good starting point, these systems have a number of drawbacks that will be discussed in Section 2.3. Regarding metadata extraction, the most notable multilingual resources supporting the considered languages are the Open Multilingual WordNet thesauri (available online: http://compling.hss.ntu.edu.sg/omw/ (accessed on 9 February 2022)), including Arabic and Persian WordNet (see Section 3.1 for more details).

2.3. Comparison and Discussion of Innovative Aspects of DigitalMaktaba w.r.t. State of the Art Techniques

Let us now consider the specific contributions of our proposal w.r.t. state of the art techniques, discussing their innovative aspects.

Overall workflow and tool aim and context. As seen in Section 2.1, not so many projects have been proposed in this context; in any case, all the projects that we have mentioned target only a part of the languages considered in DigitalMaktaba and aim at the pure digitization of a (smaller) library of books, often with consistent manual work. To give just a brief example, the Italian National Librarian System (SBN) does not provide the opportunity to insert metadata in non-Latin alphabets, thus relying heavily on ineffective transliteration systems, which seems to be in contrast to the adjustments that other countries are preparing and to the standards dictated by the International Standard Bibliographic Description (ISBD). Instead, DigitalMaktaba includes:

- *Multiple languages*: the presented innovative workflow and tool works in the Arabic, Persian and Azerbaijani languages, which have not been considered together in other works;
- *A larger size*: the project is aimed at the creation of a very large digital library (300,000+ books, much more than other projects, which are aimed at thousands of books at most), thanks also to the innovative automation features which are not present in the discussed other projects (see below);
- *Non-Latin alphabet metadata*: the automatically extracted metadata, besides being very rich (see points below), contribute to the creation of a digital library integrated with SBN without relying on transliteration (in contrast to the cited state of the art projects).

Text-extraction approach. Regarding text extraction, we have seen in Section 2.2 that, even if some approaches are available in the literature for the considered languages, they are very specific since they do not target all the languages involved in DigitalMaktaba, and, most importantly, they are not publicly available and therefore impossible to be experimentally compared. Concerning the discussed publicly available libraries, considered alone they do not always offer consistent and high-quality results on all the required languages; moreover, many require manual work (batch process is not always possible). Therefore, the novel combined approach we propose in DigitalMaktaba exceeds the scope of the best-performing free libraries and combines/enriches their features (see Section 3.1) in order to obtain a completely automatic system producing high-quality outputs:

- *Better effectiveness*: thanks to the proposed text and text-regions extraction approaches, the achieved effectiveness is better than the state of the art techniques (see tests in Section 4);
- *Across all considered languages*: in contrast to the discussed OCR libraries, all the considered languages are automatically identified without manual work and supported without uneven performance issues.

Metadata-extraction approach. Automatic metadata extraction is a unique feature w.r.t. the approaches discussed in Sections 2.1 and 2.2, which are aimed at pure text extraction and (possibly) manual metadata entering. Instead, DigitalMaktaba offers:

- *Rich metadata*: syntactic, linguistic and cataloguing metadata are extracted and stored for each new processed document;
- *Automated extraction*: the metadata extraction is a fully automated process, including the identification of title and authors, which in state of the art projects (Section 2.1) is performed as a time-consuming manual activity.

Proposed tool: batch automation, data management and UI. Further innovative aspects are the following (we are not aware of similar features in the discussed works):

- *Flexible data management*: the system can, at any time, access the data stored in a standard DBMS and convert the bibliographic information toward most desired outputs;
- *Less manual work*: the extraction pipeline and associated UI features help users in performing less manual work, fully automating batch text and metadata extraction (for instance, the discussed OCR libraries do not support this when working with multiple languages).

3. Materials and Methods

The information-extraction/text-sensing process we propose is depicted in Figure 2 and is divided into three steps, for which we will now give an overview: document preprocessing, text extraction and metadata extraction. Even if the current phase of the project is particularly focused on title pages elaboration, the described approach is sufficiently general for any kind of documents/page; in particular, it is devised so as to provide, for each processed page, information about the identified text regions, the contained text in the best-possible quality and a number of associated metadata.

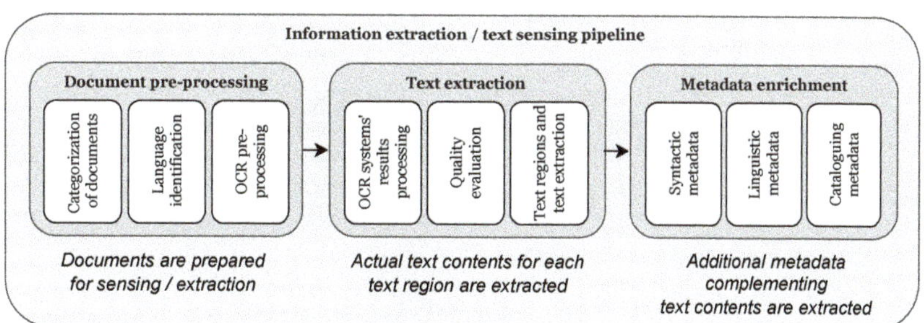

Figure 2. The information-extraction/text-sensing pipeline of the proposed approach: first, documents are pre-processed in order to identify their language and submit them to the available OCR engines (left box); then, text extraction is performed, where OCR raw data is processed and evaluated and the text-region extraction, renumbering, merging and fusion approaches are performed (center box); finally, syntactic, linguistic and cataloguing metadata is sensed (right box). Detailed descriptions of the different phases are available in Section 3.1.

See Section 3.1 for more details on the processing steps.

Document preprocessing. In the first step the documents are classified into *digitized* or *non-digitized* ones. As non-digitized documents do not provide editable text, OCR approaches must be used in order to extract image content (for most-complex case, on which we focus in this paper; for digitized documents, the text is directly extracted and processing goes on to the subsequent steps). To enable effective OCR processing, but also successive metadata extraction, it is necessary to detect in advance the language(s) of the text; in contrast to many state-of-the-art systems, this process is completely automated, then the document is processed by means of several OCR engines, returning a preliminary output which will be processed and merged in the subsequent step.

Text extraction. In the second step, the raw output of the OCR engines is analyzed, elaborated and smartly merged in order to extract: (a) for each document page, the different text regions present in it (for instance, a large central text region containing the document title, and so on, a feature that is crucial for automatic cataloguing); (b) for each text region, the contained text with the best-possible quality. The above points require to solve a number of technical issues, including *identification and linking of the different text regions* among the output of the different systems (for (a)) and definition/exploitation of a quality-evaluation metric enabling the *choice/merge of the best text output* (for (b)) (see Section 3.1).

Metadata enrichment. Eventually, the output is enriched with additional metadata information, going beyond typical state-of-the-art tools: (i) *syntactic metadata*, i.e., text-regions information, identified language(s) and character(s), text size and position on page, and self-assessed quality of extraction; (ii) *linguistic metadata*, i.e., links to external linguistic resources; and (iii) *cataloguing metadata*, i.e., automatically extracted author and title information (see Section 3.1 for details on their extraction).

3.1. Text-Sensing Aspects: Information Extraction

We will now provide more details of the techniques used in the information-extraction steps and the implementation choices behind them. In order to better understand their rationale, we will first of all discuss the preliminary exploratory analyses that were performed on the OCR systems identified in the state-of-the-art approaches and are to be exploited in the processing.

Analysis and selection of OCR libraries. Evaluating the best state-of-the-art libraries (and their strengths/weaknesses) on which to base document processing was crucial to define pipeline implementation. In order to reach this aim, we selected a subset of 100 sample documents from the La Pira digital archive, chosen so as to be representative of different languages and characters involved (we will also exploit this subset in the preliminary evaluation of the system compared to state of the art, discussed in Section 4.2). Then, we manually applied several available OCR libraries (including the ones cited in the related work discussion) to test their features and quality. The first filter that allowed us to discard some libraries was the supported language: we eliminated from the choice the libraries that do not support the languages of our interest. Furthermore, we decided to focus on open-source systems, which allowed us to discard many other items from the list. Other tools were discarded as academic projects still under development or carried out at an amateur level that did not seem to suit our purpose. In the end, we selected three libraries: *GoogleDocs* (and in particular its OCR features when importing documents), *EasyOCR* and *Tesseract*.

For testing text-extraction effectiveness, we defined an ad-hoc evaluation framework by taking into account: (a) the quality of the output (oq, range [0–2]) as quantified by linguistic experts; (b) the quality of the input (iq, range [0–2], taking into account the document scan quality/resolution). Since typical OCR evaluations (including accuracy) are not suited to the above requirements, we defined two ad-hoc *quality metrics* that, from two different points of view, depend on the quality of the documents and also have a strong dependence on expert feedback:

- *qdiff* (range [−2, 2]), expressing whether the output is in line (0), superior (positive values) or inferior (negative values) to the input quality;
- *qscore* (range [1, 5]), expressing the quality of the result of the OCR system given the input quality.

The specific definitions are the following:

$$qscore = \begin{cases} 5 - ((2 - oq) * (iq + 1)) & \text{if } oq \neq 0, \\ 1 & \text{otherwise,} \end{cases} \tag{1}$$

$$qdiff = oq - iq \tag{2}$$

While Equation (2) is quite straightforward, the idea behind Equation (1) is to subtract from the best score a penalization that is the more pronounced the higher the input quality and the lower the obtained output quality. The performed tests (whose numerical results will be summarized in Section 4.2 in comparison with our proposal) highlighted several critical issues in available OCR libraries, with each one having its strengths and weaknesses. On one hand, Tesseract and EasyOCR are capable of extracting a few portions of text with medium quality, and they are among the few to return some metadata (limited to the position of the text in the original image, even if not very precise for Tesseract); on the other hand, they require manual specification of the language before processing. GoogleDocs provides automatic language identification and better output quality; however, at the same time, its output is devoid of metadata. The overall processing pipeline combines such libraries in a new and more comprehensive approach, satisfying our goals for a rich and high-quality output without the need of manual intervention.

Combined approach, language identification. On the basis of the technical strengths and weaknesses of the various libraries, specific choices were made to make them work together in an automated way. As to language identification, the documents are first processed with GoogleDocs, whose output is used to obtain the language via GoogleTranslate, then this information is passed to EasyOCR and Tesseract for further processing.

Text (and text-regions) extraction. As to text region identification and text extraction, we devised a way to exploit both GoogleDocs (in many cases superior) OCR quality and the other libraries richer output (including text position on the page): for each page, (i) the page is processed in EasyOCR, Tesseract and GoogleDocs in parallel; (ii) from the libraries providing approximate text-region metadata (specifically, EasyOCR, since Tesseract metadata are not sufficiently precise), text-region information is extracted; (iii) text regions are renumbered and merged by means of ad-hoc techniques; (iv) each of the identified regions is "linked" to the text output from the different libraries (including those not supporting region identification, i.e., GoogleDocs); and (v) the best output for each of the regions is selected.

Point (iii) will be described in detail in the following subsection. In order to perform point (iv), the text for each of the regions (from EasyOCR) is compared to (parts of) the raw text obtained from GoogleDocs and Tesseract by means of the edit distance metric: in this way, each sentence (or group of sentences) can be associated to the belonging region. As to point (v), automatic quality evaluation is performed by means of a simple metric *wcount* defined with the aid of the external linguistic resources. Furthermore, *wcount* simply corresponds to the count of existing words present within the results of the considered multilingual corpora (Open Multilingual WordNet and others as described in the linguistic metadata description); for each region, the output having the higher *wcount* is selected.

Text (and text-regions) extraction: text-region renumbering and merging. Text regions (which we will now call boxes, for simplicity) are crucial to the subsequent processing and user-interaction steps; in particular, it is essential to: (a) have them numbered in a way that reflects the logical flow of information; (b) avoid excessive fragmentation (e.g., multiple boxes for information that has to be considered as one piece of text). Unfortunately, due both to the specific complexities given by the considered languages and the often suboptimal quality of the available document images, even the raw output of OCR libraries most suited to box metadata extraction (in our case, EasyOCR) does not meet the above requirements, for instance, fragmenting text into too many boxes and not correctly ordering them following the Arabic right-to-left convention. For this reason, we devised a text-region renumbering and merging phase that proceeds following these steps (see also Abbreviation part for an overview of the used abbreviations):

- **Horizontal grouping:** first of all, the boxes of a page are grouped into horizontal groups following a horizontal grouping criterion (see Figure 3): two boxes will belong to the same group if the vertical distance *v-dist* between their medians w.r.t. the height of the tallest box does not exceed a given threshold $th_{v\text{-}dist}$;

- **Merging:** inside each group of boxes are boxes satisfying the merging criteria (i.e., relative height difference below a given threshold $th_{h\text{-}diff}$ and horizontal distance relative to page width below a given threshold $th_{h\text{-}dist}$, as depicted in Figure 3);

- **(Re)numbering:** the resulting boxes are renumbered from top to bottom (different groups) and from right to left (inside each group): Figure 3 shows the resulting numbers on the top right corner of each box.

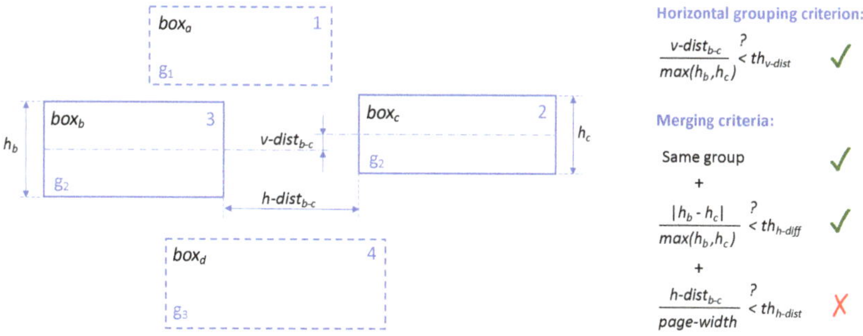

Figure 3. Visual example for horizontal box grouping and merging criteria: box_b and box_c are grouped into the same horizontal group (g_2) but are not merged. Groups are shown on the bottom left corner of each box. The resulting box numbering is shown on the top right corner of each box.

All thresholds are expressed as a ratio between 0 and 1. As we will see from the tests (Section 4.2), this process enables us to obtain text-region information that is much closer to the desired one (the tests will also discuss how we derive the three best-performing threshold values).

Metadata-enrichment overview. As discussed in the pipeline overview, besides text, different metadata are added to the output; let us now detail the different processing phases where each of them comes from. The language of the document is the one extracted from the pre-processing phase; text-region information comes from the text-region extraction/renumbering/merging described in the previous section; text size is not directly available from the OCR libraries output; however, it is extracted by analyzing font sizes in GoogleDocs raw output; quality metadata is the *wcount* metric corresponding to the best selected output; linguistic metadata is extracted for the document words by searching each of them in the multilingual corpora (this provides additional information including word definitions and synonyms, see next section for more details). Linguistic metadata also enable various methods for the extraction of further cataloguing metadata (automatic identification of title and authors, see next sections) and will as well support future automatic cataloguing tasks (e.g., automatic identification of document topics and categories through semantic processing on linguistic metadata).

Metadata enrichment: linguistic metadata. Searching for linguistic information in the languages covered by the project (Arabic, Persian, and Azerbaijani) is certainly a complex task, as we also underlined in the related works: for instance, the Arabic language has many more words and variations, including vocalized and unvocalized, than languages deriving from Latin such as English. To date, there are no open-source linguistic resources providing a coverage level at least comparable to those available for the English language. In order to partially overcome this issue, we decided not to base our tool on a single resource but to exploit a pool of them. After evaluating them in terms of linguistic features and size (i.e., number of words), we designed linguistic-metadata extraction techniques jointly exploiting:

- *Open Multilingual WordNet* (http://compling.hss.ntu.edu.sg/omw/ (accessed on 10 March 2022)) providing word lookup for the three different languages (only unvocalized words, in the case of Arabic) and access to extended semantic information (including word definitions). Arabic and Persian/Azerbaijani Wordnet contain 17,785 and 17,560 synsets, respectively;
- *Arramooz* (https://github.com/linuxscout/arramooz (accessed on 10 March 2022)) an open-source Arabic dictionary for morphological analysis providing unvocalized word lookup and definitions for the Arabic language. It contains 50,000 words;
- *Tashaphyne* (https://pypi.org/project/Tashaphyne (accessed on 10 March 2022)), a light stemmer that is used as a devocalizer in order to extend the Arabic coverage of the two previously described resources.

The joint exploitation of the above resources enables us to enhance the overall linguistic coverage: for instance, for the Arabic language, among 458 "test" terms (229 in unvocalized and 229 in vocalized form), we were able to obtain an overall coverage of 73%, compared, for instance, to less than 10% and 36% for Wordnet and Arramooz used alone.

Metadata enrichment: cataloguing metadata (title/author identification). One of the most important but time-consuming activities for cataloguing a new document is to manually insert (or select among the OCR output text) its title, authors and other information. Currently available tools (including those discussed in Section 2) do not propose ways to automate/support this process. The tool we propose aims to exploit the extracted metadata (including box size and position, and linguistic metadata) and text in order to automatically suggest to the librarian the text regions that most likely contain specific fields. At the time of writing, we have designed and tested some preliminary but promising (see Section 4) strategies for identifying the text regions (boxes) containing title and authors in a frontispiece:

- *DIM method*: boxes are sorted on vertical dimension, then the first box in the ranking is suggested as title, the second one as author(s) (following the intuition that the largest texts on a frontispiece are typically the title and the authors' names, in this order);

- *RES method*: external linguistic resources (including a list of names and surnames in the different languages) are searched and boxes are sorted on the basis of the percentage of found words and names (for identifying title and author, respectively);
- *WGH method*: method combining the contributions of the previous methods (a linear combination ranking fusion technique [31] is exploited in order to produce final rankings for both title and authors).

In the future, we plan to extend these methods and combine them with machine-learning techniques in order to learn from system usage (see also next section).

3.2. Data Management

While we are currently most focused on the information extraction techniques which will be key to the effectiveness of our proposal, work is also already undergoing on some of the subsequent steps that will lead to a complete and usable cataloguing tool. In this section, we will specifically discuss the data-management foundations, while a look at the user interface and functionalities we are currently considering on our preliminary prototype implementation will be given in Section 4.1.

The extracted data and metadata are stored on a DBMS in order to guarantee good efficiency levels for both data insertion/update and querying in a typical usage scenario. Currently, our database design is focused on relational DBMSs; in the future, we will also consider extending this design to possibly exploit specific big-data-management techniques and tools for even larger workloads.

Figure 4 shows the entity-relationship schema for our database. The database is designed to easily store and retrieve the document data and metadata whose extraction we described in Section 3.1, along with the definitive catalogue data that the user will insert while using the system:

- The *document* entity is at the heart of the schema and stores the catalogued documents data/metadata, including title, author, language and the path of the folder containing the actual document file(s);
- The details about the scanned images of a document are stored in the *Image* entity (each image corresponds to a specific document page), including file location and image dimensions;
- The *category* entity stores information about the specific category (name and description) to which each document belongs. Categories are currently manually selected by the user from a 3-level hierarchy (modelled through a self-association) containing more than 560 entries; however, in the future we plan to provide "smart" techniques based on AI to assist the process by means of relevant suggestions;
- The entities on the left part of the image store useful metadata about both the OCR results and the feedback coming from system use: for each image, the *Box* entity stores the text regions associated to an image (position, dimensions, text, and automatic quality evaluation as described in previous section), while *Box_info* stores the semantic information about the box content (i.e., the fact that the box contains author, title, and other information is stored in the *association_type* property).

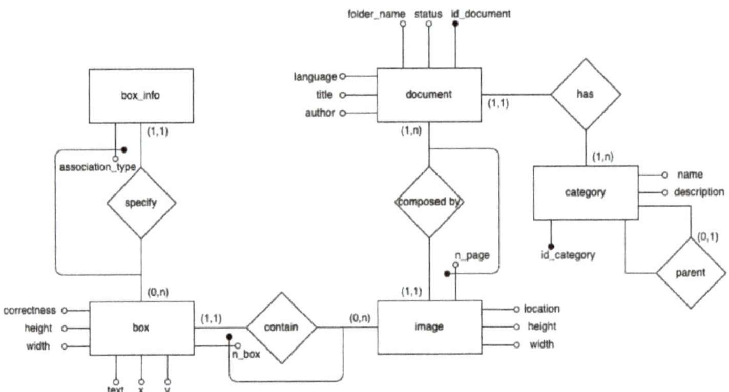

Figure 4. Data management: entity-relationship diagram for database conceptual schema.

The database is designed to support not only standard cataloguing needs but also to store the data that will be key to provide future smart assitance to the user: in particular, the information contained in *Box* and *Box_info* will enable machine-learning techniques that will be able to enhance the system effectiveness through use (e.g., for title/author recognition).

The database is implemented in PostgreSQL; several kinds of indexes enable its fast querying, in particular GIN (generalized inverted indexes) supporting title and author full-text search and b+trees for category lookup. Further advanced search techniques (including fuzzy approximate search) will be developed in order to make searches more efficient and effective w.r.t. typical cataloguing needs.

4. Results and Discussion

In this section, we will consider what we have achieved so far both in terms of the resulting cataloguing-tool prototype we are implementing (whose current user interface and functionalities are described in Section 4.1) and of the experimental evaluation of the presented techniques (Section 4.2).

4.1. Prototype: User Interface and Functionalities.

The techniques described in the previous sections have been incorporated in a preliminary application prototype that we are designing. The cataloguing tool is implemented in Python and exploits the Flask framework in order to provide a user-friendly, even if in an initial design phase, user interface. Among the already enabled functionalities are:

- Batch document preprocessing: this allows to preprocess an entire folder of PDF documents. The current implementation exploits multithreading for faster processing, using one OCR thread per page. The UI informs users on the documents being processed and allows them to proceed to document cataloguing for the ones that are ready (see Figure 5);
- Catalogue an already processed document: this guides the user on a series of steps where (s)he is provided with a graphical user interface in order to finalize document properties input (title, author, category, etc.). For instance, the UI for title selection (Figure 6) shows the document frontispiece (on the left) and automatically selects the text region(s) (whose extracted text content is shown on the right) that are most likely to contain the title. The user is able to select different text regions to modify/merge their text (text is concatenated from right to left based on the order in which they are clicked). The tool also visually helps users by showing at a glance the automatically discovered links to the available linguistic resources: the found words are shown in green color and, when the "clip" icon is clicked, a popup displays which words are found in

which linguistic resource and, for each of them, related information such as vocalized versions and definitions (lower part of Figure 6). See Section 3.1 for a description of the involved extraction process for text, text regions (including automatic region merge and sorting), and linguistic and cataloguing metadata (including automatic author and title identification);

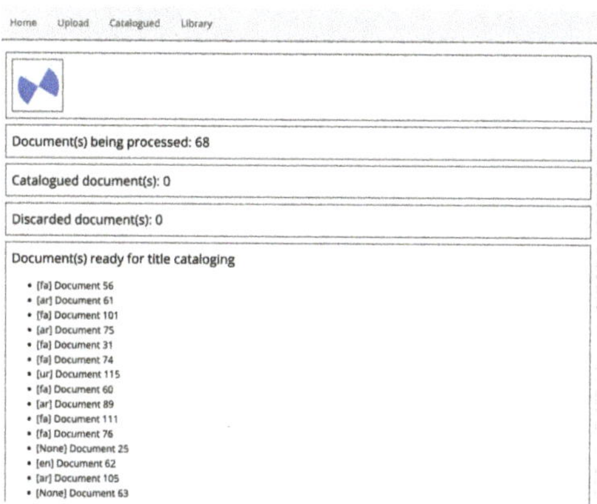

Figure 5. Preliminary resulting cataloguing tool prototype UI: Document preprocessing.

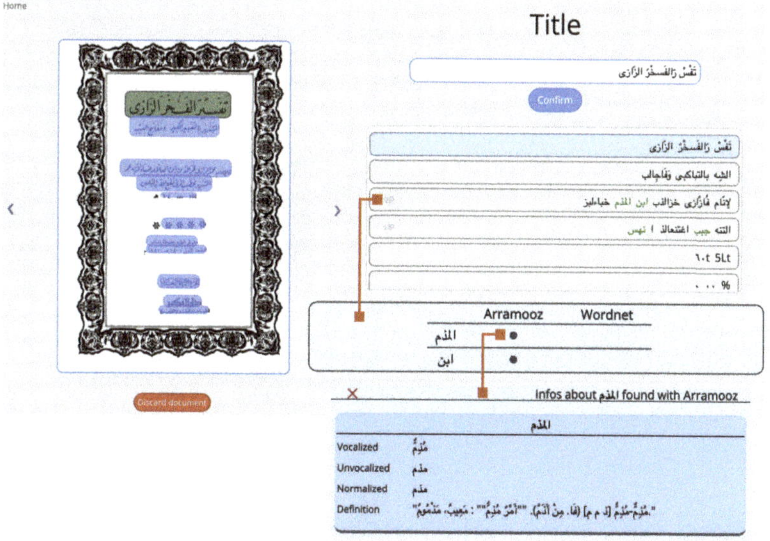

Figure 6. Preliminary resulting cataloguing tool prototype UI: document cataloguing (title window), showing the actual extraction of the title (highlighted in the green box, on the frontispiece displayed on the left). From top to bottom on the left side, the title is hinted before confirming and also selected in blue under the confirm button. Arramoz and Wordnet are the employed linguistic resources, in this case activated (in green) on the word *ibn* "son" and *al-mudhimm* "the one who reprehends, the censor".

- Various search functions on the catalogue database, including author and title full-text search and category search (see Section 3.2);
- see a summary of the currently catalogued documents data (Figure 7);

Document id	Title	Author name	Author surname	Language	Macro category	Category	Sub category	Edit
78	خاء	ماء	محمد	fa	Theology / Kalām	māturīdī		title author category
74	حرب الردة	سقر	طنطي	ar	Šarī'a and Islamic Jurisprudence	Šāfi'ī school		title author category
73	تفسير الرازي	ونقاجاني	الله بالتباكي	ar	Philosophy and Sciences	History of Sciences and Arts		title author category

Figure 7. Preliminary resulting cataloguing tool prototype UI: catalogued documents summary. After the semi automatic selection of title, author name (and eventually surname), all selected data are put together in the cataloguing interface in relation to a specific topic and field which represents a category (or sub-category) of the library. Title, author name and surname are shown in the original language (arabic script) as shown here in the figure.

- Other miscellaneous functions: modify already entered data, delete documents from database, restore deleted documents, view catalogued documents.

4.2. Experimental Evaluation

In this section, we report on the tests we carried out to perform an initial evaluation of the effectiveness of the approaches we propose (even if the complete tool discussed in the previous section is still in a very early implementation phase). In particular, we will discuss the evaluation of the effectiveness of the text-region renumbering and merging, OCR/text extraction, and of the title and author identification techniques (all described in Section 3.1). All tests are performed on a subset of 100 sample documents from the project library which, thanks to their variety, are representative of the complete collection (both in terms of image quality and linguistic contents). In the future, as the development of the tool continues and as it will be employed for actual librarian and cataloguing work and its database populated, we aim to be able to extend the scope to larger document sets.

Effectiveness of text-region extraction. In this first batch of tests, we aimed to evaluate the effectiveness of the text-region renumbering and merging described in Section 3.1. Effectiveness is evaluated on two metrics w.r.t. a gold standard manually determined by experts: *average percentage error*—the percentage of boxes in each document having a wrong number, averaged on the whole document set, and *percentage of documents with box sort error*—the percentage of documents having at least one error in the numbering of their boxes.

Let us first consider text region renumbering. The first test (the left part of Figure 8) shows the effect of moving the vertical distance $th_{v\text{-}dist}$ threshold: as expected, there is a trade off between very low threshold values (which tend to make too-selective horizontal groups) and higher ones (which tend to produce too-inclusive groups).

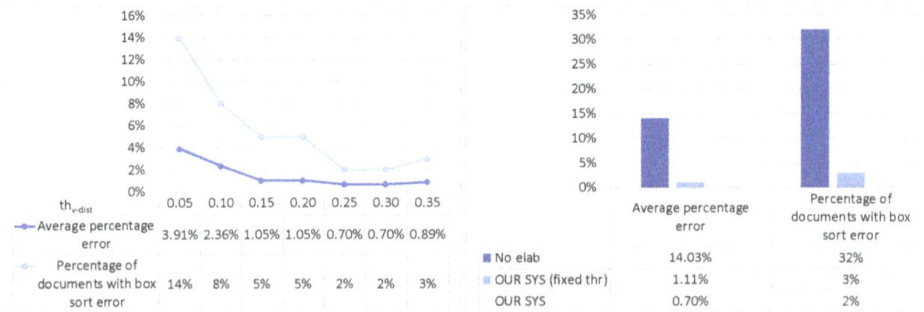

Figure 8. Text-region renumbering tests: effect of vertical-distance threshold $th_{v\text{-}dist}$ (**left**) and error comparison between different approaches (**right**).

Note that all threshold "tunings" were performed on a separate tuning dataset of the same size of the main dataset, in order to keep such phases separate from the final evaluation. The trade off is at $th_{v\text{-}dist} = 0.25$–$0.30$. This setting enables a very low percentage error of 0.7% for the first metric and 2% for the second one. The right part of Figure 8 compares the final effectiveness achieved by the approach as described in this paper and two baselines (no renumbering, i.e., taking the box region numbers as provided by EasyOCR, and adopting a fixed threshold expressed in pixels instead of the relative one described in this paper): as we can see, the considered metrics drop from 14.03% to 0.7% and from 32% to 2%, respectively.

As to text-region merging, we performed similar tests in order to analyze the effect of moving the two thresholds (horizontal distance $th_{h\text{-}dist}$ and height difference $th_{h\text{-}diff}$) and evaluating the overall effectiveness of the approach. Being that the two thresholds are practically independent, we first evaluated the effect of moving the first with the second one set at a default value, then we moved the second one with the first set at the value suggested by the first test. Figures 9 and 10 (left part) show that we a have good trade offs at $th_{h\text{-}dist} = 0.1$ and $th_{h\text{-}diff} = 0.5$. The evaluation of the effectiveness of the merging approach (with the above threshold values) on the main dataset is shown in the right part of Figure 10: the two metrics are basically confirmed at 2% and 0.1% for our approach (as opposed to 15% and 2.54%, respectively, when no box merging is performed).

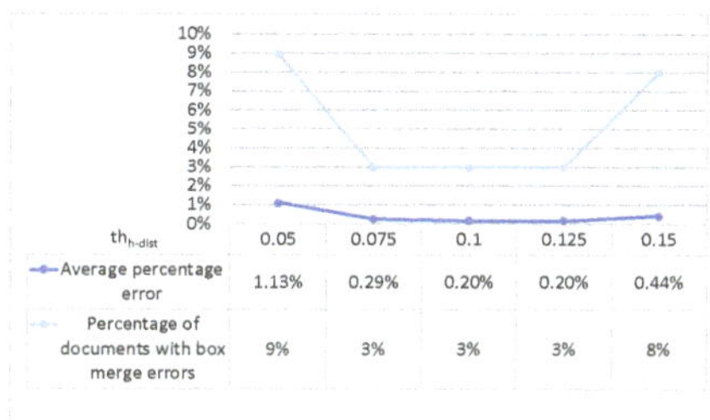

Figure 9. Text-region merging: effect of horizontal-distance threshold $th_{h\text{-}dist}$.

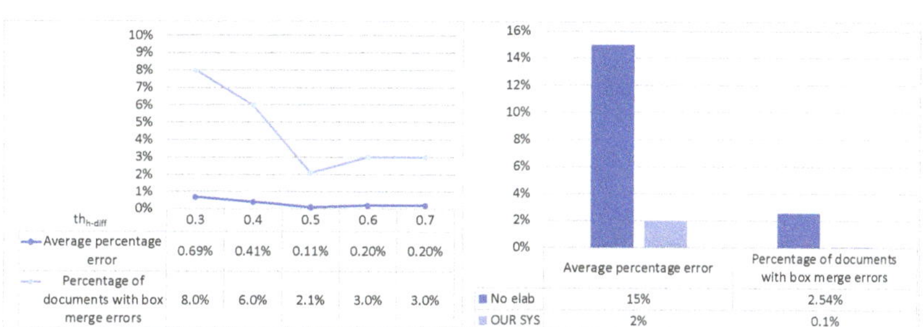

Figure 10. Text-region merging: effect of height-difference threshold $th_{h\text{-}diff}$ (**left**) and error comparison with and without merging (**right**).

Effectiveness of OCR/text extraction. We will now discuss the results of the evaluation of the OCR libraries, as described in Section 3.1, and the effectiveness that our tool is able to reach. The metrics *qdiff* and *qscore* (as defined in Section 3.1) were used. In particular, the input quality *oq* was defined in a range from 0 (a low-quality scan of a page that contains a lot of noise, or poorly defined or damaged writing) to 2 (a well-defined, high-quality scan); as to output quality, it was evaluated by linguistic experts on a range from 0 (completely wrong results) to 2 (completely correct results).

Figure 11 shows the average performance of each system in terms of *qdiff* (left part of figure) and *qscore* (right part of figure). Starting our analysis from state-of-the-art systems, as we can see from the *qdiff* metric, GoogleDocs generally performs better than EasyOCR (with a score near 0, confirming an output that is typically in line with the quality of the processed input), while the worst-performing library is Tesseract, with a *qdiff* near -1. This is also confirmed by the *qscore* values. As to our approach, we can see that its scores are slightly better than GoogleDocs, which is the best-performing system. In particular, this is the first evidence that the best output selection strategy, based on the count of words found in existing multilingual corpora, is working well (indeed, in this regard, we verified that, on the sample of documents considered, our approach correctly classifies the best output in 95% of cases).

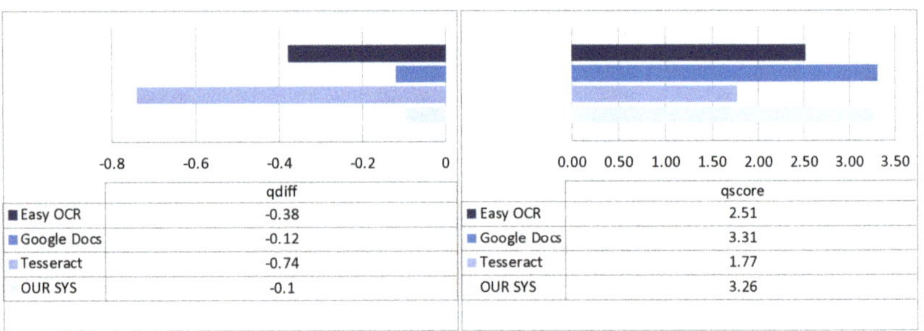

	qdiff
■ Easy OCR	-0.38
■ Google Docs	-0.12
Tesseract	-0.74
OUR SYS	-0.1

	qscore
■ Easy OCR	2.51
■ Google Docs	3.31
Tesseract	1.77
OUR SYS	3.26

Figure 11. Text extraction: qdiff (**left**) and qscore (**right**) overall results.

Moreover, we were also interested in analyzing the comparative performances w.r.t. the specific languages of the sample documents (results shown in Figure 12). As we can see, considering state-of-the-art libraries and *qdiff* (left part of figure), we see that some languages are more difficult to deal with than others (e.g., Azerbaijani), while only for Persian do some systems provide an output quality exceeding the input quality of the documents (positive *qdiff* scores). In particular, the best-performing system for Azerbaijani is EasyOCR, with GoogleDocs being very close (also looking at *qscore* on the right), then Tesseract.

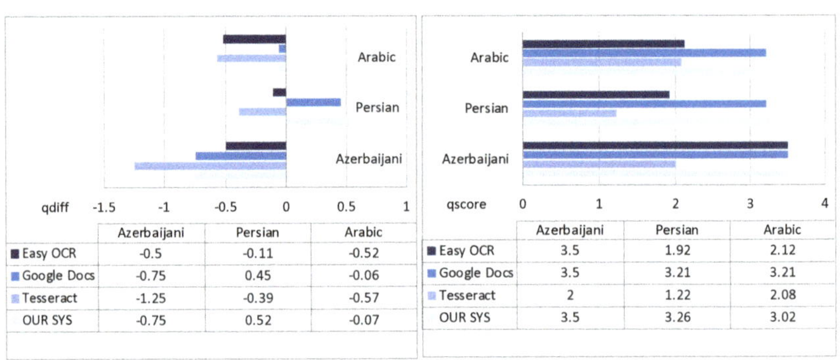

qdiff	Azerbaijani	Persian	Arabic
■ Easy OCR	-0.5	-0.11	-0.52
■ Google Docs	-0.75	0.45	-0.06
Tesseract	-1.25	-0.39	-0.57
OUR SYS	-0.75	0.52	-0.07

qscore	Azerbaijani	Persian	Arabic
■ Easy OCR	3.5	1.92	2.12
■ Google Docs	3.5	3.21	3.21
Tesseract	2	1.22	2.08
OUR SYS	3.5	3.26	3.02

Figure 12. Text extraction: qdiff (**left**) and qscore (**right**) results per language.

GoogleDocs appears as the general best choice (in particular for Persian and Arabic, with good results also for Azerbaijani); anyway, by going beyond the average values shown in the graphs and analyzing the performance on the single document cases, we note that there are indeed some cases (especially for Arabic and Persian) where GoogleDocs is not always able to return a better output than others.

Effectiveness of title/author identification. Figure 13 shows the accuracy (% of correct guesses) we currently achieve on the considered dataset for the three methods discussed in Section 3.1. The DIM and WGH method achieve the best accuracy for both title (65% for both) and author (40% and 41%, respectively) and their performance is quite close. We have to remember that the quality of the scanned images is generally quite low (this is to reflect the actual digital data that is available to cataloguers) and this, in some cases, prevents the full exploitation of the external linguistic resources' potential on the extracted text. While the overall figures can certainly be improved, in any case, they represent a promising result since none of the systems available in the state of the art projects aims at automating this task, thus requiring the completely manual insertion/selection of both titles and authors. In the future, we will consider further improvements to the methods so that the system will adapt to the quality of the documents and only provide the suggestions for which it is most confident.

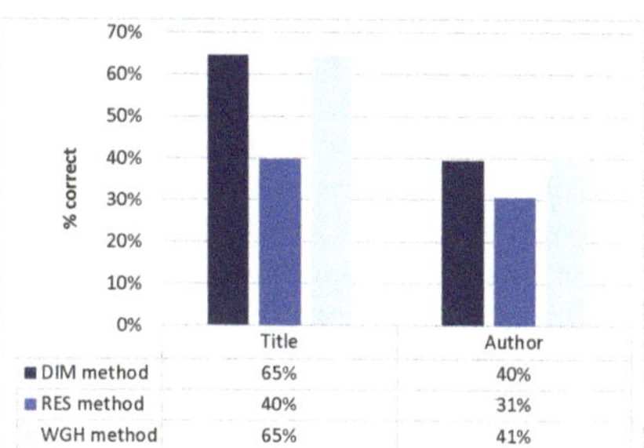

Figure 13. Title/author identification: correctness of the best guess for the different methods.

4.3. Future Work

Generally speaking, besides the specific improvements to the presented methods discussed in the relevant sections, many are the steps we envision that will lead to the creation of the final complete supervised intelligent cataloguing tool, for which we will also exploit our past expertise in semantic [32] and machine-learning techniques in different scenarios [33–35]:

- From a data-management perspective, special focus will be given to the data interchange and long-term preservation aspects, in order to allow data interchange with catalogue data from other libraries and make the managed data readable and usable also on a long-term basis [36];
- Intelligent and AI-based techniques will also have a prominent role: intelligent assistance features will be designed and implemented in order to bring new levels of assistance to the cataloguing process. Data entering will be supported by suggestions derived from user feedback and previously entered data, thus integrating and extending the author-/title-identification techniques we described in this paper; supervised machine-learning models will enable automatic publication-type recognition and provide a systematisation and classification of data according to the topographic design of the La Pira library;
- The design of incremental ML algorithms will ensure that the tool can "learn" and become more and more automated and effective with use. Both classic and deep-learning algorithms will be considered, deployed on parallel architectures for faster execution. Special attention will be given to interpretable machine-learning algorithms, following the recent interpretable machine-learning trend in different fields [37], with the aim of going beyond the black-box nature of ML suggestions and explaining them, also in the library cataloguing/cultural heritage context, where this has seldom been performed;
- Ee plan to extend the scope of the experimental tests by considering incrementally larger portions of the use case library.

AI techniques will be integrated in order to create a reproducible and reusable web tool enabling a simple cataloguing workflow overcoming language and field obstacles.

5. Conclusions

In this paper, we presented the first steps for designing a novel tool for the automatic extraction/sensing of knowledge from documents written in multiple non-Latin languages.

As shown in the preliminary tests, the performance of the currently developed techniques are encouraging. We will conclude the paper by briefly discussing the expected advantages of the final tool we are aiming to create, both from technical and non-technical perspectives.

From a technical point of view, many foreseen advantages are auspicable in the scientific domain:

- *Overcoming the limitations of current text-extraction tools*: non-uniform OCR with different characters, backwardness of OCR for Arabic, poor automation and extraction of additional metadata;
- *Faster pipeline*: To date, the cataloguing of documents has been performed manually, document by document, opening them one at a time, searching inside the information to be catalogued and then moving to the next one. By employing the tool being developed, we can drastically reduce the time spent cataloguing simpler documents (since the suggested output will need fewer modifications from the user), while for the most complex ones, the system can still provide useful suggestions. In both cases, this system will speed-up the entire procedure;
- *Greater consistency and fewer errors*: In fact, with the automatic suggestions supplied to the user, it will become easier to avoid mistakes while inserting the document's information, making the extracted data less error-prone. Moreover, the system will constantly perform checks on previously catalogued data in order to avoid inconsistencies, which can be very common in manual cataloguing (e.g., an author surname inserted with/without name, abbreviated name, etc.);
- *Consistently better system output through time*: by means of the machine-learning/intelligent features, the system will provide an output that will be better and better as the system is used, exploiting its "training" on previously catalogued, similar documents;
- *Flexibility of data output/exchange*: being a system based on a complete data-management system on which all the user and system output is stored, the system can at any time access the stored data and convert the bibliographic information toward most desired outputs, thus facilitating data exchange;
- *Efficiency and Explainability*: advanced data-management and machine-learning techniques will enable high efficiency levels for significant data-cataloguing needs, while the explainability of the models will make the intelligent assistance tools more usable.

Finally, from a broader standpoint, benefits from this research and the use of the tool are expected on several innovative fronts:

- Advancement of studies on cataloguing in multi-literate environments without leaning exclusively on confusing transliteration systems;
- Exchange of IT, humanist and library personnel, enhancement of professional skills, training activities extended to realities with similar needs;
- Strengthening of library services thanks to shared international standards, expanding library heritage, databases integration, maximum access to the heritage, possibility of using the language of the document without the mediation of other languages.

We are also aware of some limitations of our current research: in particular, achieving complete automation might be, in some ways, limited by the quality of the input images, which is not always satisfying; moreover, we are aware that text sensing on Arabic-script documents is an area where available research results and tools are not as developed as the ones for Latin scripts. This poses some obstacles but it is also a further motivation for what we are performing. Finally, machine-learning/intelligent features will only be possible with large amounts of data already loaded into the system, which will require a significant initial amount of manual work. In any case, we are confident that these limitations will be dealt with in our future research work.

In short, we are confident that this research will ultimately help in preserving and conserving culture, a crucial task, especially in the challenging scenario we consider.

Author Contributions: Conceptualization, R.M., F.R.; methodology, R.M., L.S., M.V.; software, L.S., M.V.; data curation and investigation, L.S., M.V., R.A.V.; writing—original draft preparation, all authors; writing—review and editing, R.M.; supervision, R.M.; project administration, S.D.N., F.R.; funding acquisition, S.B., S.D.N., R.M., F.R. All authors have read and agreed to the published version of the manuscript.

Funding: This research was funded by: MIM.fscire startup; Protocol 2020–2024 between The Italian Ministry for University and Research (MUR) and Fondazione per le Scienze Religiose (FSCIRE); Programme agreement 2021–2025 between The italian Ministry for University and research (MUR) and Fondazione per le Scienze Religiose (FSCIRE).

Institutional Review Board Statement: Not applicable.

Informed Consent Statement: Not applicable.

Data Availability Statement: At present the dataset of the La Pira Library is not publicly available, all items were donated from different contributors to the La Pira Library in PDF format and are property of the Fondazione per le Scienze Religiose (FSCIRE) and "Giorgio La Pira" Library. It is worth to remind that the main goal of the present research is to enable the creation of a digital library that will eventually enable to make those big amounts of data available, thus providing access to a wide array of users, from students and researchers to the general public.

Conflicts of Interest: The authors declare no conflict of interest.

Abbreviations

oq	manually assessed OCR output quality
iq	manually assessed OCR input quality
$qdiff$	resulting quality difference from oq-iq
$qscore$	overall quality of the OCR system result, given the input quality
v-$dist$	vertical distance
h-$dist$	horizontal distance
h-$diff$	text boxes height difference
$th_{v\text{-}dist}$	Threshold on vertical distance
$th_{h\text{-}diff}$	Threshold on height difference
$th_{h\text{-}dist}$	Threshold oh horizontal distance
h	boxes height
g	box(es) group

References

1. Nasir, I.M.; Khan, M.A.; Yasmin, M.; Shah, J.H.; Gabryel, M.; Scherer, R.; Damaševičius, R. Pearson Correlation-Based Feature Selection for Document Classification Using Balanced Training. *Sensors* **2020**, *20*, 6793. [CrossRef] [PubMed]
2. Kyamakya, K.; Haj Mosa, A.; Machot, F.A.; Chedjou, J.C. Document-Image Related Visual Sensors and Machine Learning Techniques. *Sensors* **2021**, *21*, 5849. [CrossRef] [PubMed]
3. Miller, M.T.; Romanov, M.G.; Savant, S.B. Digitizing the Textual Heritage of the Premodern Islamicate World: Principles and Plans. *Int. J. Middle East Stud.* **2018**, *50*, 103–109. [CrossRef]
4. Kitab Project. Available online: https://kitab-project.org/about/ (accessed on 10 March 2022).
5. Persian Digital Library, Roshan Institute for Persian Studies, University of Maryland. Available online: https://persdigumd.github.io/PDL/ (accessed on 10 March 2022).
6. Bergamaschi, S.; Martoglia, R.; Ruozzi, F.; Vigliermo, R.A.; De Nardis, S.; Sala, L.; Vanzini, M. Preserving and Conserving Culture: First Steps towards a Knowledge Extractor and Cataloguer for Multilingual and Multi-Alphabetic Heritages. In Proceedings of the Conference on Information Technology for Social Good, GoodIT '21, Rome, Italy, 9–11 September 2021; Association for Computing Machinery: New York, NY, USA, 2021; pp. 301–304. [CrossRef]
7. DAR Project. Available online: http://dar.bibalex.org/webpages/aboutdar.jsf (accessed on 10 March 2022).
8. Perseus Project. Available online: http://www.perseus.tufts.edu/hopper/research (accessed on 10 March 2022).
9. Shamela Library. Available online: http://shamela.ws/ (accessed on 10 March 2022).
10. Shiaonline Library. Available online: http://shiaonlinelibrary.com (accessed on 10 March 2022).
11. Ganjoor. Available online: https://ganjoor.net/ (accessed on 10 March 2022).
12. Hazm, Baray-e Pardazesh-e Zaban-e Farsi. Available online: https://www.sobhe.ir/hazm/ (accessed on 10 March 2022).
13. Steingass, F.J. *A Comprehensive Persian-English Dictionary, Including the Arabic Words and Phrases to be Met with in Persian Literature*; Routledge & K.Paul: London, UK, 1892.
14. ACO—Arabic Collections Online. Available online: https://dlib.nyu.edu/aco/ (accessed on 10 March 2022).

15. The British Library Projects: Arabic collection. Available online: https://www.bl.uk/collection-guides/arabic-manuscripts (accessed on 16 February 2022).
16. The British Library Projects: Persian collection. Available online: https://www.bl.uk/projects/digital-access-to-persian-manuscripts (accessed on 16 February 2022).
17. QDL—Qatar Digital Library. Available online: https://www.qdl.qa/en/about (accessed on 16 February 2022).
18. Iran Heritage. Available online: https://www.iranheritage.org/ (accessed on 16 February 2022).
19. Al-Muhtaseb, H.A. Arabic Text Recognition of Printed Manuscripts: Efficient Recognition of Off-Line Printed Arabic Text Using Hidden Markov Models, Bigram Statistical Language Model, and Post-processing. Ph.D. Thesis, University of Bradford, Bradford, UK, 2010.
20. Obaid, A.M. A New Pattern Matching Approach to the Recognition of Printed Arabic. In Proceedings of the Computational Approaches to Semitic Languages, SEMITIC@COLING 1998, Montreal, Canada, 16 August 1998; Université de Montréal: Montréal, QC, Canada, 1998.
21. Amin, A. Recognition of printed Arabic text using machine learning. In Proceedings of the Document Recognition V, San Jose, CA, USA, 24 January 1998; Lopresti, D.P., Zhou, J., Eds.; SPIE: Bellingham, WA, USA, 1998; Volume 3305, pp. 62–71.
22. Mohammad, K.; Qaroush, A.; Ayyesh, M.; Washha, M.; Alsadeh, A.; Agaian, S. Contour-based character segmentation for printed Arabic text with diacritics. *J. Electron. Imaging* **2019**, *28*, 043030. [CrossRef]
23. Mohamad, R.A.; Mokbel, C.; Likforman-Sulem, L. Combination of HMM-Based Classifiers for the Recognition of Arabic Handwritten Words. In Proceedings of the 9th International Conference on Document Analysis and Recognition (ICDAR 2007), Curitiba, Paraná, Brazil, 23–26 September; IEEE: Piscataway, NJ, USA, 2007; pp. 959–963.
24. Aghbari, Z.A.; Brook, S. HAH manuscripts: A holistic paradigm for classifying and retrieving historical Arabic handwritten documents. *Expert Syst. Appl.* **2009**, *36*, 10942–10951. [CrossRef]
25. Hamdani, M.; Doetsch, P.; Kozielski, M.; Mousa, A.E.; Ney, H. The RWTH Large Vocabulary Arabic Handwriting Recognition System. In Proceedings of the 11th IAPR International Workshop on Document Analysis Systems, Tours, France, 7–10 April 2014; IEEE: Piscataway, NJ, USA, 2014; pp. 111–115. [CrossRef]
26. Dehghan, M.; Faez, K. Farsi Handwritten Character Recognition with Moment Invariants. In Proceedings of the 13th International Conference on Digital Signal Processing, Santorini, Greece, 2–4 July 1997; Volume 2, pp. 507–510. [CrossRef]
27. Mowlaei, A.; Faez, K.; Haghighat, A.T. Feature extraction with wavelet transform for recognition of isolated handwritten Farsi/Arabic characters and numerals. In Proceedings of the 14th International Conference on Digital Signal Processing, DSP 2002, Santorini, Greece, 1–3 July 2002; IEEE: Piscataway, NJ, USA, 2002; pp. 923–926. [CrossRef]
28. Soleymani-Baghshah, M.; Shouraki, S.B.; Kasaei, S. A Novel Fuzzy Approach to Recognition of Online Persian Handwriting. In Proceedings of the Fifth Int. Conference on Intelligent Systems Design and Applications (ISDA 2005), Wroclaw, Poland, 8–10 September 2005; IEEE: Piscataway, NJ, USA, 2005; pp. 268–273. [CrossRef]
29. Ghadikolaie, M.F.Y.; Kabir, E.; Razzazi, F. Sub-Word Based Offline Handwritten Farsi Word Recognition Using Recurrent Neural Network. *ETRI J.* **2016**, *38*, 703–713. Available online: http://xxx.lanl.gov/abs/https://onlinelibrary.wiley.com/doi/pdf/10.4218/etrij.16.0115.0542 (accessed on 10 March 2022). [CrossRef]
30. Bonyani, M.; Jahangard, S. Persian Handwritten Digit, Character, and Words Recognition by Using Deep Learning Methods. *CoRR* **2020**. Available online: http://xxx.lanl.gov/abs/2010.12880 (accessed on 10 March 2022).
31. Zhang, K.; Li, H. Fusion-based recommender system. In Proceedings of the 2010 13th International Conference on Information Fusion, Edinburgh, UK, 26–29 July 2010; pp. 1–7. [CrossRef]
32. Martoglia, R. Facilitate IT-Providing SMEs in Software Development: A Semantic Helper for Filtering and Searching Knowledge. In Proceedings of the SEKE, Miami Beach, FL, USA, 7–9 July 2011; pp. 130–136.
33. Furini, M.; Mandreoli, F.; Martoglia, R.; Montangero, M. A Predictive Method to Improve the Effectiveness of Twitter Communication in a Cultural Heritage Scenario. *ACM J. Comput. Cult. Herit.* **2022**, *15*, 21. [CrossRef]
34. Martoglia, R. Invited speech: Data Analytics and (Interpretable) Machine Learning for Social Good. In Proceedings of the IEEE International Conference on Data, Information, Knowledge and Wisdom (DIKW), Haikou, China, 17–19 December 2021; IEEE: Piscataway, NJ, USA, 2021.
35. Martoglia, R.; Pontiroli, M. Let the Games Speak by Themselves: Towards Game Features Discovery Through Data-Driven Analysis and Explainable AI. In Proceedings of the IEEE International Conference on Data, Information, Knowledge and Wisdom (DIKW), Haikou, China, 17–19 December 2021; IEEE: Piscataway, NJ, USA, 2021.
36. Borghoff, U.M.; Rödig, P.; Scheffczyk, J.; Schmitz, L. *Long-Term Preservation of Digital Documents: Principles and Practices*; Springer: Berlin/Heidelberg, Germany, 2006.
37. Ahmad, M.A.; Eckert, C.; Teredesai, A.; McKelvey, G. Interpretable Machine Learning in Healthcare. *IEEE Intell. Inform. Bull.* **2018**, *19*, 1–7.

MDPI

Article

LIPSHOK: LIARA Portable Smart Home Kit

Kévin Chapron, Florentin Thullier *, Patrick Lapointe, Julien Maître, Kévin Bouchard and Sébastien Gaboury *

Laboratoire d'Intelligence Ambiante pour la Reconnaissance d'Activités, Département d'Informatique et de Mathématiques, Université du Québec à Chicoutimi, 555 Bd. de l'Université, Saguenay, QC G7H 2B1, Canada; kevin.chapron1@uqac.ca (K.C.); patrick.lapointe@hotmail.ca (P.L.); julien1_maitre@uqac.ca (J.M.); kevin_bouchard@uqac.ca (K.B.)
* Correspondence: florentin.thullier1@uqac.ca (F.T.); sebastien_gaboury@uqac.ca (S.G.)

Abstract: Several smart home architecture implementations have been proposed in the last decade. These architectures are mostly deployed in laboratories or inside real habitations built for research purposes to enable the use of ambient intelligence using a wide variety of sensors, actuators and machine learning algorithms. However, the major issues for most related smart home architectures are their price, proprietary hardware requirements and the need for highly specialized personnel to deploy such systems. To tackle these challenges, lighter forms of smart home architectures known as smart homes in a box (SHiB) have been proposed. While SHiB remain an encouraging first step towards lightweight yet affordable solutions, they still suffer from few drawbacks. Indeed, some of these kits lack hardware support for some technologies, and others do not include enough sensors and actuators to cover most smart homes' requirements. Thus, this paper introduces the LIARA Portable Smart Home Kit (LIPSHOK). It has been designed to provide an affordable SHiB solution that anyone is able to install in an existing home. Moreover, LIPSHOK is a generic kit that includes a total of four specialized sensor modules that were introduced independently, as our laboratory has been working on their development over the last few years. This paper first provides a summary of each of these modules and their respective benefits within a smart home context. Then, it mainly focus on the introduction of the LIPSHOK architecture that provides a framework to unify the use of the proposed sensors thanks to a common modular infrastructure capable of managing heterogeneous technologies. Finally, we compare our work to the existing SHiB kit solutions and outline that it offers a more affordable, extensible and scalable solution whose resources are distributed under an open-source license.

Keywords: ambient intelligence; smart home in a box; architecture; framework

Citation: Chapron, K.; Thullier, F.; Lapointe, P.; Maître J.; Bouchard K.; Gaboury S. LIPSHOK: LIARA Portable Smart Home Kit. *Sensors* **2022**, *22*, 2829. https://doi.org/10.3390/s22082829

Academic Editors: Pietro Manzoni, Claudio Palazzi and Ombretta Gaggi

Received: 23 February 2022
Accepted: 3 April 2022
Published: 7 April 2022

Publisher's Note: MDPI stays neutral with regard to jurisdictional claims in published maps and institutional affiliations.

1. Introduction

Over the years, various implementations of smart homes have been developed in laboratories or real habitations built for research purposes [1–8]. These works mainly focus on using sensors, effectors and learning algorithms to enable the use of ambient intelligence (Am.I.) as an empirical method in order to support older people's autonomy and health monitoring for medical purposes. For instance, enhanced homes with Am.I. allow one to improve the safety of residents who suffer from cognitive impairments. At the same time, they help healthcare professionals track a resident's condition so they are able to adapt their decisions. While each study in the field of smart homes has been developed to address these concerns and more specifically, the activity recognition problem [9], they all suggest different methods and infrastructures to achieve these objectives. Regardless, the major issues for most related smart home architectures are their expensive price, their lack of scalability due to proprietary hardware requirements that are rarely compatible with wearable devices [10] and the highly specialized personnel needed to deploy such systems [6]. The high costs involved in a number of traditional smart home implementations are primarily related to the prices of the required networking hardware and servers rather than the prices of the sensors and actuators themselves, since they remain relatively inexpensive.

In order to tackle these challenges, lighter implementations of smart home architectures known as smart home in a box (SHiB) have been proposed [5,11]. These kits have been designed mainly to offer cheap and easy to install solutions that do not compromise the capabilities of traditional smart home infrastructures. Moreover, these kits do not require any formal training to be operated. While existing SHiB kits remain an encouraging first step towards lightweight yet affordable smart home infrastructures, they still suffer from a few drawbacks. Indeed, these solutions do not always enable the use of various types of sensors or actuators, such as static and wearable devices with high or low sampling rates. Moreover, most of the kits that have been adopted so far are essentially built from a single technology (e.g., a ZigBee mesh network), thereby limiting the support for several other technologies that may also be involved in smart homes.

This paper introduces the LIARA Portable Smart Home Kit (LIPSHOK), a new SHiB kit that is intended to be a better alternative to existing smart home in a box implementations. The main advantage of LIPSHOK is that it is able to manage heterogeneous sensors and actuators, and several wireless communication technologies through a unified modular architecture. Moreover, all the modules that are part of the kit have been designed with the aim of providing a low-cost, self-contained and easy-to-install solution. In addition, since these components are open-source, it is possible for our SHiB kit to be extended with more features than the distributed ones to best suit specific needs, a task that may be cumbersome to achieve with proprietary hardware.

As of now, LIPSHOK comes with four specialized sensors that have been developed by our laboratory to fulfill modern smart home needs and related research. All the details as regards the four hardware components, including bathroom modules [12], a gait speed module [13], PIR-BLE-RSSI modules [14] and a smart wristband module [15], are provided in the next sections.

The rest of the paper is structured as follows: Section 2 presents the current state of smart home in a box kits and provides more insightful details about the motivation for the development of LIPSHOK. Section 3 briefly reports the sensors included in the kit. Next, Section 4 describes the design of the LIPSHOK architecture while Section 5 discusses the benefits of using LIPSHOK. Finally, Section 6 draws conclusions and Section 7 mentions future work we will accomplish.

2. Current State of Smart Home in a Box Kits

To the best of our knowledge, very few works have presented self-sufficient, affordable and easy to install smart home in a box kits.

The most popular one is CASAS: the smart home in a box [5]. This kit was developed at Washington State University with the main objectives of being easy to install and affordable. The physical layer offered by this kit is a mesh network composed of both sensors and actuators that communicate with each other through the ZigBee protocol. Each device can be powered by simple batteries while providing long-term functioning. Moreover, new devices may be added to the mesh network on demand, allowing the entire infrastructure to scale seamlessly and automatically. Indeed, the physical layer is linked to a ZigBee bridge that allows each device to communicate with a publish/subscribe manager that composes the messaging service middleware. Finally, an application layer is hosted on a small server where additional computing is performed for data storage and activity recognition.

The implementation proposed by the CASAS kit yields several advantages. First, the use of low-power communication through battery-powered sensors and actuators considerably reduces the cost and simplifies the cumbersome installation process required to make homes smart. Moreover, by its design the CASAS kit remains efficient, simple, scalable and particularly well suited for sensors with low data rates, including binary sensors and actuators (i.e., on/off values), such as a PIR motion sensor. However, the main limitations of this infrastructure is its inability to handle both non-ZigBee-enabled sensors—this technology is mandatory in CASAS—and sensors with high data throughput. Examples are inertial measurement units (IMUs) and positioning sensors based on received signal strength indication (RSSI), as they rely mainly on Wi-Fi or Bluetooth communication technologies.

In addition, ref. [16] have introduced the SPHERE smart home architecture. This smart home kit was designed to enable the use of environmental sensors, wearable devices and real-time video analysis. However, since it has been developed for long-term usage, its installation within existing homes remains challenging and requires qualified personnel. In addition, although special attention has been placed by the authors on the use of affordable consumer hardware, its specifications concerning the use of relatively powerful gateways make it a more expensive architecture than CASAS, although it also offers more possibilities. In that sense, as the SPHERE infrastructure may be seen as a more traditional smart home setup, it will not be discussed here in further detail.

A different team in the same research group was working, at the same time, on the development of an independent SHiB solution: SPHERE in a Box [11]. As a subset of the more comprehensive SPHERE smart home infrastructure, the SPHERE in a Box kit was designed, based on the same two main objectives of the CASAS kit: easy installation and affordability. The SPHERE in a Box kit is more oriented towards wearable devices than CASAS, as specific emphasis was placed on the integration of such devices inside smart homes throughout its presentation paper. Thus, it is capable of working with high-data-rate devices. In this case, a Bluetooth Low Energy (BLE) wristband is used with an embedded IMU. This device then communicates with several gateways strategically placed in the environment to locate the wearer through RSSI and to collect IMU data at a frequency of 25 Hz. Finally, each gateway transmits compressed and encrypted data from one or more wristband once a day to a backup database server through a router that provides a Wi-Fi access point and a 3G/4G cellular link to the Internet.

The main drawback of that solution is the lack of diversity in sensors, due to the limitation of including only wristbands with the kit. That leads us to state that it cannot be used as a fully autonomous smart home kit in its current state. Nevertheless, if the kit were to also contain other high-data-throughput sensors and a few low-data-rate sensors, the gateways would have to be improved first. According to our point of view, other wireless communication technologies than BLE and Wi-Fi would have to be added to the gateways, such as ZigBee or Z-Wave. This would allow the use of less power-consuming sensors and actuators, but it would also make the infrastructure more generic. However, given the authors' proposed hardware system, such improvements in the SPHERE in a Box kit appear to be perfectly achievable without considerably increasing its costs.

Through the detailed analysis of these two main previously proposed smart home in a box kits, the elements of which are provided by Table 1, the first requirement we identified is the need for such systems to be capable of handling both low and high-data-rate sensors and actuators, since both of them coexist in most current smart home designs [10]. Furthermore, since devices included in these kits may rely on heterogeneous technologies often mandated by the specific data acquisition needs that are used by machine learning applications, it remains important that the proposed solution does not limit itself to the support of a single technology. For example, the CASAS solution requires the use of ZigBee-enabled devices only. In this regard, both commercial and industrial sensors also represent one of the main concerns when designing a SHiB kit. Indeed, commercial sensors, although relatively cheap, rarely offer full control over the data they provide: sometimes the sampling frequency is not adjustable; otherwise, it is impossible to have access to raw data in favor of already pre-processed data. On the other hand, industrial sensors offer, most of the time, absolute control on the output data. However, their prices are not compatible with the constraint of proposing a low-cost smart home in a box kit.

Table 1. Elements of existing SHiB kits.

	CASAS [1]	**SPHERE in a Box** [11]
Main Technology	ZigBee	BLE
Included Modules	IR motion/light sensors (×24) Door sensor (×1) Temperature sensors (×2) Relays (×2)	Wristband (×1)
Ease of Installation	Yes no wiring required: battery-powered modules only.	Yes evaluated through a survey conducted by the authors.
Affordability	Moderate	High
Extensibility	No only compatible with ZigBee enabled modules.	Yes BLE and WiFi compatible out of the box. Possibility to add support for other technologies relatively easily.
Scalability	Yes	Yes
Sensor Data Rates	only low data rates	both low and high data rates out of the box

This paper presents LIPSHOK: a framework for a generic smart home in a box kit. The kit includes a total of four custom-made sensors that have been designed by our research laboratory to best address the modern challenges of today's smart homes requirements [17,18]. The main benefit of LIPSHOK is that it respects all the essential constraints for a SHiB kit which we stated previously. Indeed, it has been designed to be easily deployable in various existing environments without depending on a single technology, and the included devices are all heterogeneous. Some of them are static sensors and others are wearable devices, each of them having its specific hardware and software design, so the kit provides support for a wide variety of technologies. In addition, since the whole kit is distributed under an open-source license, the features are easily customizable and extendable. While each of these four sensors has been presented independently, this paper aims to describe their integration with the LIPSHOK framework in the same controlled environment within our laboratory. All details and materials required to reproduce such a deployment are provided.

3. Custom Sensors in LIPSHOK

As LIPSHOK is built on integrating several sensors that have been designed by our laboratory and introduced individually, we consider it important to first provide a brief overview of each one before presenting the framework in more detail.

3.1. Bathroom Modules

Corporal hygiene is a particularly useful indicator for detecting the impairment of the cognitive function of an individual living in a smart home. Indeed, it is known that people affected by a cognitive disorder will spend less and less time taking care of their own hygiene [19,20]. Moreover, in the context of personal hygiene, persons affected by a physical disability take longer to complete activities of daily living (ADLs), such as showering and going to the toilet [21,22], supporting that corporal hygiene remains a reliable indicator of physical condition for residents of smart homes. Hence, the first previously proposed module included in the LIPSHOK SHiB kit is a combination of two devices to be placed in the bathroom [12]. As shown in Figure 1, the two devices are placed both on the bathtub and over the toilet in the bathroom of the smart home.

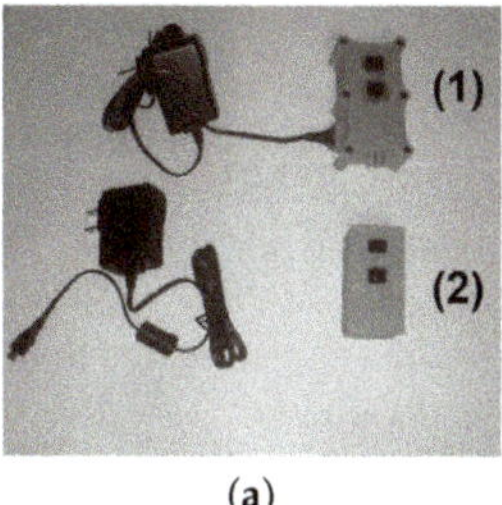 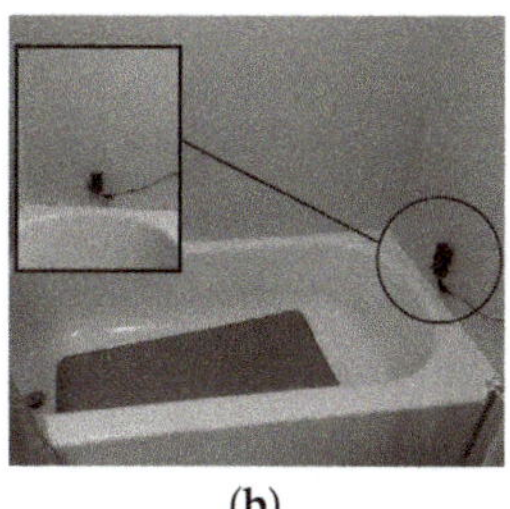 

(a)   (b)   (c)

Figure 1. The bathroom modules previously introduced by ref. [12]. (**a**) The bathroom modules where (**1**) refers to the bathtub device and (**2**) is the device for the toilets. (**b**) The first IRPS bathroom device placed in the bathtub. (**c**) The second IRPS bathroom device positioned over the toilets. Reprinted with permission from ref. [12] 2020 IEEE.

The bathtub device was roughly waterproofed to ensure minimal safety during the experiments, but it was not made to comply with both dust and water protection standards defined by ingress protection (IP), such as IP67 as defined by the EN-60529 standard (https://keystonecompliance.com/en-60529/ accessed on 22 February 2022). The two modules are based on the same sensor, an infrared proximity sensor (IRPS) (https://www.sparkfun.com/products/8958 accessed on 22 February 2022), allowing each module to compute the distance between the device and a potential human standing in front. This sensor provided accurate presence detection that could not have been achieved with any other sensors that had been reported in the literature. One reason is that passive infrared (PIR) sensors can never be sensitive to someone not moving while standing in front of them [5], and they are not capable of providing the distance. Moreover, the current state-of-the-art reports other types of sensors that are not effective enough at recognizing ADLs adequately, or whether they are being performed by the monitored person or for how long. Among these works, microphone-based recognition [23] and contact-based recognition [24] have only demonstrated the ability to detect the end of an event, providing no further related information.

As detailed in our previous research [12], the reliability of the bathroom modules was evaluated by their ability to accurately identify the activities "taking a shower" and "going to the toilet", through the use of threshold-based decision algorithms. The data used for this experiment were recorded over a 59-day period by eight participants, all of whom were healthy adults without any motor or cerebral disability. Moreover, a qualitative survey regarding the ease of installation and the acceptability of the bathroom modules by the participants was also conducted. Table 2 exposes an overview of the results obtained for the assessments.

Table 2. Overview of the results obtained with the bathroom modules [12].

Evaluation	Toilets	Shower/Bathtub
Activity recognition (F-measure)	95.26%	98.62%
Duration differences (% of difference)	3.90%	6.48%
Ease of installation (% of agreement)	91.43%	91.43%
Acceptability (% of positive response)	84.38%	90.63%

3.2. Gait Speed Module

Gait speed is known to be an excellent predictor of diseases such as mild cognitive impairment (MCI) in the older population [25]. More precisely, ref. [26] have demonstrated in a 20-year longitudinal study that although the age has an impact, patients diagnosed with MCI have a further decline in gait speed when compared to the healthy population. However, a possible remediation to help slow and control the progression of such cognitive decline is to consult with occupational therapists for regular monitoring of the gait speed (i.e., each year). Thus, gait speed monitoring appears to be a relevant use case of applied ambient intelligence inside smart homes. In that sense, considering the importance of this predictor which is currently rarely addressed in the literature, our team have been focused on the development of a device to be included in the LIPSHOK SHiB kit as an autonomous module that automatically monitors residents' gait speed in real time [14].

The gait speed module was built based on three IRPS sensors, the same as for the bathroom module. Each sensor is placed at a height of 90 cm on a hallway wall at a fixed distance of 60 cm from the others, resulting in a module with a total length of 120 cm, as shown in Figure 2. This gait speed module was designed to be affordable, easy to install and to provide a relatively straightforward way of operating. In a nutshell, whenever a monitored person walks by the module, the detection is monitored in order to allow the computation of the gait speed.

Figure 2. Gait speed module deployed in our smart home laboratory introduced by ref. [14]. Reprinted with permission from ref. [14] 2020 Springer Nature.

The gait speed module was evaluated in a three-phase experimental procedure which was as close as possible to real use case situations. The procedure was completed by nine participants, all being healthy adults without any known issues in physical condition. The first phase of the experiment focused on the evaluation of the precision of the speed (A) following a 5-meter walk test (5MWT) as introduced by ref. [27]. Then, the second phase of the experiment consisted of the evaluation of the user identification through the BLE RSSI only (B). Finally, the last phase focused on the combination of the identification of monitored persons and activity recognition (C). The data for the activity recognition process were collected with the wristband presented in Section 3.4. Table 3 exposes an overview of the results obtained for such assessments.

Table 3. Overview of the results obtained with the gait speed module [14].

Evaluation	Result
(A) speed precision (% of precision)	93.38%
(B) raw identification of monitored persons (% of accuracy)	48.00%
(C) identification of monitored persons and activity recognition aggregated (% of accuracy)	84.00%

3.3. PIR-BLE-RSSI Modules

The next module provided in the LIPSHOK SHiB kit is the PIR-BLE-RSSI device (several ones are used). One is shown in Figure 3. This module was designed by our team [13] in order to better address the key challenge when it comes to achieving an accurate activity recognition process in a multi-resident smart home context, namely, effectively associating the sensor observations and the right individuals [28]. Indeed, each PIR-BLE-RSSI module combines a passive-infrared motion sensor and a BLE unit, allowing it to detect movements through more or less restricted fields of view (i.e., 120°, 80°or 15°)

and room-level positioning when paired with any wearable device, such as a smartwatch or a smartphone.

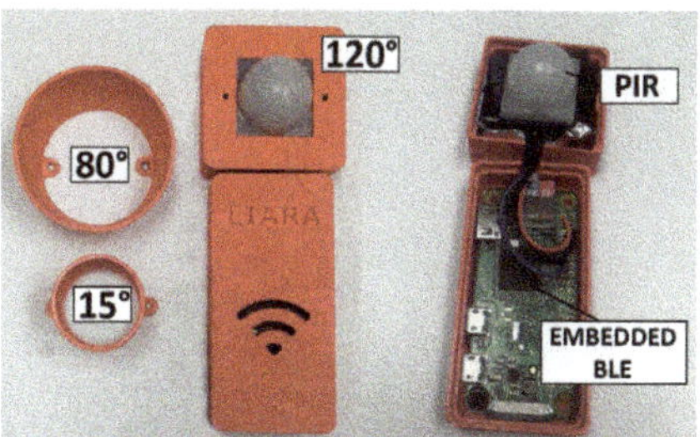

Figure 3. An example of the PIR-BLE-RSSI module introduced by ref. [13]. Reprinted with permission from ref. [13] 2020 Elsevier.

The evaluation was performed by installing multiple PIR-BLE-RSSI modules in our smart home laboratory of 43 m^2. Next, several realistic scenarios of ADLs were completed by eight participants, all being healthy adults, in a multi-resident setup. Firstly, two monitored persons (P_A and P_B) were equipped with one wearable device each and asked to perform ADLs simultaneously in different areas of the smart home (e.g., P_A: *washing hands* in the bathroom and P_B: *changing clothes* in the bedroom). Then, to collect some control data, a third monitored person was requested to stand in the same area without wearing a device for all scenarios in the experiments. Table 4 exposes an overview of the results obtained, outlining the robustness of our system when using multiple PIR-BLE-RSSI modules.

Table 4. Overview of the results obtained for the identification of monitored persons with the PIR-BLE-RSSI module [13].

Identification of Monitored Persons	Result
BLE only	90.22%
BLE and PIR aggregated	92.28%

3.4. Wristband Module

Over the past few years, wearable devices have been widely used in smart homes to address several research problems in various areas. Indeed, given their small size, their cheap price and their convenience of use and integration in intelligent environments, they have allowed researchers to suggest new systems for continuous health monitoring and the recognition of activities, gestures or falls [29–32]. In addition, wearable devices have also been shown to provide an excellent basis for addressing the multi-occupancy challenge, as they are capable of both tracking and identifying people, but also recognizing activities in an autonomous manner [33–35].

Our team focused on developing a wristband module to help with the rehabilitation of people affected by myotonic dystrophy type 1 (DM1) [15,36]—a hereditary neuromuscular disease that causes a variety of impairments, particularly muscle weakness—since it is a prevalent disease in the area of our University (i.e., Saguenay-Lac-Saint-Jean region of Quebec, Canada). As pictured in Figure 4, the proposed wristband hardware relies on a nRF52832 board (https://www.adafruit.com/product/3406 accessed on 22 February 2022) that embeds a native-Bluetooth chip to which a 9-degree-of-freedom (DoF) IMU has been added. However, three of the nine degrees of freedom corresponding to the magnetometer were deactivated, because in the context of activity recognition, it was determined that they did not provide enough relevant information, and therefore may be ignored, especially for devices with limited hardware [37].

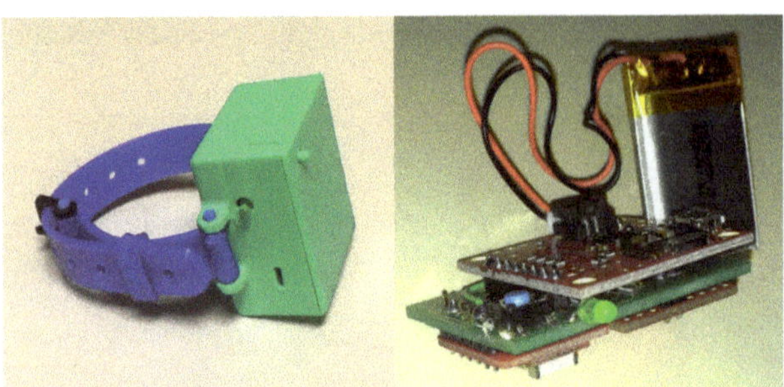

Figure 4. The wristband module introduced by ref. [15]. Reprinted with permission from ref. [15] 2021 IEEE.

In the context of LIPSHOK, the wristband is the most powerful module included in the SHiB kit. Indeed, the wristband module is the only one capable of accurately achieving the recognition of ADLs directly on its embedded hardware through a machine learning algorithm (i.e., a C4.5 decision tree) using the inertial data. However, while every other module presented in this paper is designed to work autonomously, as they are capable of performing simple activity detection on their own, combining their data with the data produced by the wristband can significantly improve the accuracy of the overall system for all the use cases that have been presented. For instance, the gait speed module, the PIR-BLE-RSSI, and the bathroom modules all have the ability to scan wristband identifiers and associate the readings with the nearest person based on an RSSI localization. As regards the recognition of ADLs through inertial data, an overall F-measure of 84.40% was obtained during the following experiment: Twenty participants were asked to perform, three times a week, a 5-activity training program (i.e., running, sit-to-stand, stand-to-sit, inactive and walking). The participants were divided into two equal groups of 10 people according to whether they had to complete the program with the wristband or without, in order to form a control group [15].

4. The Design of LIPSHOK

In this paper, we introduce LIPSHOK, LIARA Portable Smart Home Kit, a smart home in a box framework. We were motivated by the need to integrate the modules presented in the previous section into a standalone kit that is easy to install and inexpensive. As illustrated in Figure 5, the design of such a framework remains simple. Regarding our implementation, this architecture relies on a centralized unit that does not require powerful hardware to manage the flow of data generated by all the sensors a smart home may involve. Indeed, despite most of the implementations suggested in the literature [8], we have rather opted for the use of a Raspberry Pi 3B, one of the most well-known nanocomputers currently available on the market, since it offers an excellent price–performance ratio.

Moreover, as the proposed architecture is intended for a real-time usage of the data, it therefore avoids the need for expensive storage devices, and it also provides better privacy and security for the residents of smart homes. Nevertheless, the such a framework does not enforce the use real-time data. As it stands, it is possible to create a dedicated application within the LIPSHOK architecture in order to store incoming data in a database server located in another environment.

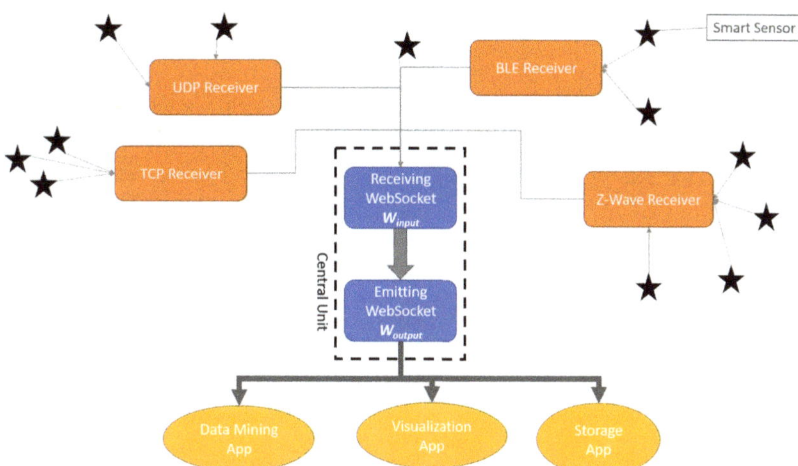

Figure 5. The detailed diagram of the suggested LIPSHOK framework.

4.1. Main Architecture

The central unit of the LIPSHOK framework hosts the core software component that mainly relies on the use of the websocket protocol (https://www.rfc-editor.org/rfc/inline-errata/rfc6455.html accessed on 22 February 2022). This technology is preferred, since it allows us to create a bidirectional communication channel that is kept open as long as required. In addition, websockets remain simple, well suited for real-time applications and allow one-to-many connections, which is particularly useful when it comes to supporting many smart home sensors.

In the LIPSHOK achitecture framework, there are two distinct websockets, the first one being required to acquire data transmitted by every receiver and the second one being required to emit aggregated data to external clients, such as dedicated applications (e.g., a visualization application) or other devices (e.g., a database server). In order to facilitate future explanations, it is first important to provide the definitions of some key concepts:

- **Smart sensor**: A smart sensor encompasses a physical sensor and a programmable chip, or a nanocomputer, that allows one to embed algorithms or any data processing software, and that provides either a wired or a wireless connectivity (e.g., BLE).
- **Receiver**: A receiver is an entity authorized to collect data of one or more sensors communicating with the same protocol (e.g., Z-Wave). Its only requirement is to be capable of connecting to both a websocket protocol and its related communication protocol. For example, the BLE receiver must be able to connect to any of the BLE smart sensors available and to send their data through the websocket protocol.
- W_{input} is the notation employed to refer the receiving websocket of the central unit. Its role is to combine and normalize data transmitted by the receivers. Moreover, it is also possible for this receiver to act as the default receiver for any smart sensor using a websocket as its only communication protocol.
- W_{output} is the notation employed to refer the emitting websocket of the central unit. Its role is to stream aggregated and normalized data obtained from smart sensors through W_{input} to every connected client. Therefore, if two interfaces are connected to W_{output}, they will both receive real-time data, regardless of the underlying communication protocol required by each sensor.

While our implementation suggests hosting the receivers within the same central unit, it is simple to see how the system may be improved by distributing them across multiple physical devices, as long as they have the proper requirements. The architecture should then benefit from better load distribution across the network and better robustness against single points of failure (SPoF).

4.2. Protocol Receivers

In order to provide a generic architecture capable of adapting to most technologies being used by sensors and actuators of smart homes, the suggested design of the LIPSHOK framework defines various software components as receivers. Every receiver operates on a specific network port. Each one is associated with a given technology, as detailed in Table 5. While these receivers cover most of the most popular technologies, including the ones required for every module included in the kit, more receivers may be added to the framework. Indeed, as an open-source project (https://github.com/kevinchapron/LIPSHOK-final accessed on 22 February 2022), the LIPSHOK framework was designed to be as extensible as possible. In addition, while the core software is written in GoLang (https://golang.org/ accessed on 22 February 2022), additional features may be developed in other languages without compromising its proper functioning, as long as they comply with the architectural guidelines related to port forwarding and encryption.

Table 5. Detailed receiver network configurations.

Label	Technology	Port
UDP Receiver	UDP	5010
TCP Receiver	TCP	5020
BLE Receiver	BLE	5030
Z-Wave Receiver	Z-Wave	5040
Main Receiver (W_{input})	Websocket	5001
Main Receiver (W_{output})	Websocket	5003

4.3. Security

The main requirement mandated by the LIPSHOK framework is the need for the data transmission process to implement a previously defined encryption layer. This is to prevent malicious users from acquiring sensible data that are easily usable through a man in the middle (MITM) attack [38]. Therefore, the suggested implementation is based on the advanced encryption standard (AES) algorithm. Thus, each message transmitted within the architecture must comply with the packet structure illustrated in Figure 6. Since the AES-128 master key is generated only one time, it is then stored securely in the central unit and in each module included in the kit. Moreover, to reduce the threats of potential packet interception, a unique AES initialization vector (IV) is generated for each new message.

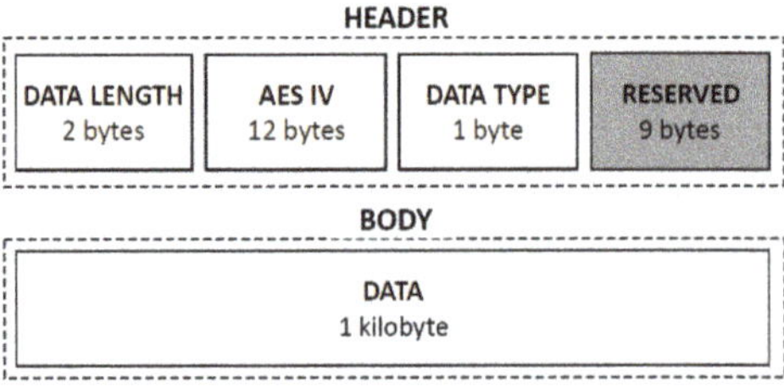

Figure 6. The structure of a message packet in the LIPSHOK architecture.

While AES is a powerful yet efficient encryption algorithm, some modules included in the kit, such as BLE-based devices (e.g., the wristband), limit the size reserved for the body of the message to 20 bytes, making them not suitable with AES encryption. To cope with such a problem, we therefore suggest using the PRESENT algorithm [39] since it is a lightweight block cipher encryption method advertised to be 2.5 times more cost-effective than AES (https://nieuws.kuleuven.be/en/content/2012/ultra-lightweight-encryption-method-becomes-international-standard accessed on 22 February 2022). The PRESENT algorithm uses an 8 byte block cipher. In that sense, as data transmitted by BLE-based devices within the architecture are 20 bytes long, the encryption algorithm is applied three times using the following bytes: $[0, 8]$, $[6, 14]$ and $[12, 20]$. A total of four bytes need to be overlapped to fit the algorithm. However, this overlap has no impact, since the decryption operation is applied in reverse order.

5. Why Use Lipshok?

From our point of view, the LIPSHOK SHiB kit detailed in this paper should benefit all stakeholders concerned with the use and development of intelligent environments. Inspired by the early proposal of ref. [5], the kit has also been designed to be easy to install in either new or already existing homes regardless of the available setup, for a very low price. For instance, Table 6 provides an evaluation of the cost for each module included in the LIPSHOK kit and the infrastructure costs. Furthermore, Table 7 offers a costs comparison of LIPSHOK with related SHiB solutions (i.e., CASAS [1] and SPHERE in a Box [11]) when deployed in a one-bedroom apartment. However, it must be noted that quoted prices for the LIPSHOK kit represent the costs for the production of the proofs of concept. Large-scale manufacturing of the different modules is expected to reduce significantly these costs.

Table 6. Summary of the cost for every module included in the LIPSHOK SHiB kit and for the hardware required to implement the architecture.

Module	Content	Qty.	Unit Price ($US)
LIPSHOK infrastructure	Raspberry Pi 3B+ board	1	35.00
	Minimal Raspberry equipment (Minimal equipment for the Raspberry Pi board includes a power cable and a class 10 micro SD memory card with a storage capacity of 32 GB.)	1	30.00
	ZigBee dongle	1	30.00
	Z-Wave dongle	1	60.00
	Total		155.00
Bathroom modules	Raspberry Pi Zero W board	2	10.00
	Minimal Raspberry equipment	2	20.00
	16-bit ADC	2	15.00
	IRPS sensor	2	15.00
	Total		120.00
Gait speed module	Raspberry Pi Zero W board	1	10.00
	Minimal Raspberry equipment	1	20.00
	16-bit ADC	1	15.00
	IRPS sensor	3	15.00
	Total		90.00
PIR-RSSI module	Raspberry Pi Zero W board	1	10.00
	Minimal Raspberry equipment	1	20.00
	PIR sensor	1	10.00
	Total		40.00
Wristband module	RedBear BLE Nano V2 board	1	15.00
	LSM9DS1 IMU sensor	1	17.50
	LiPo battery manager	1	21.50
	400 mAh LiPo battery	1	5.50
	Total		59.50

Table 7. Cost comparison of two related SHiB kits, CASAS [1] and SPHERE in a Box [11], with LIPSHOK.

SHiB kit	Content	Total Cost ($US)
CASAS [1]	Server (×1) IR motion/light sensors (×24) Door sensor (×1) Relays (×2) Temperature sensors (×2)	2765.00
SPHERE in a Box [11]	Cellular router (×1) WiFi gateways (×4) Wristband (×1)	500.00
LIPSHOK	Central unit (×1) Bathroom modules (×1) Gait speed module (×1) PIR-BLE-RSSI modules (×5) Wristband module (×1)	624.50

When compared to existing SHiB kits, such as CASAS and SPHERE in a Box, LIPSHOK is the most affordable when taking into account the number of sensors it provides out of the box. Furthermore, since the architecture has been made to allow integrating sensors and actuators based on various technologies and thus working at several data rates, the kit features better extensibility than SPHERE in a Box by default. In addition, having the entire LIPSHOK infrastructure (i.e., hardware blueprints and the firmware and algorithms) distributed under an open-source license also ensures enhanced extensibility, as it allows developers and researchers to easily upgrade the core features of the kit to best suit their needs.

Finally, LIPSHOK includes everything required from sensors to client applications enabling a fully operational smart home within a couple of hours of installation and configuration. Figure 7 shows the sensors-state-monitoring client application also provided as part of the kit (provisioned on the central unit by default). Additionally, it is important to note that the architecture may also be scaled-up to further meet high-availability requirements in order to improve fault tolerance in the same way as defined by refs. [8,10].

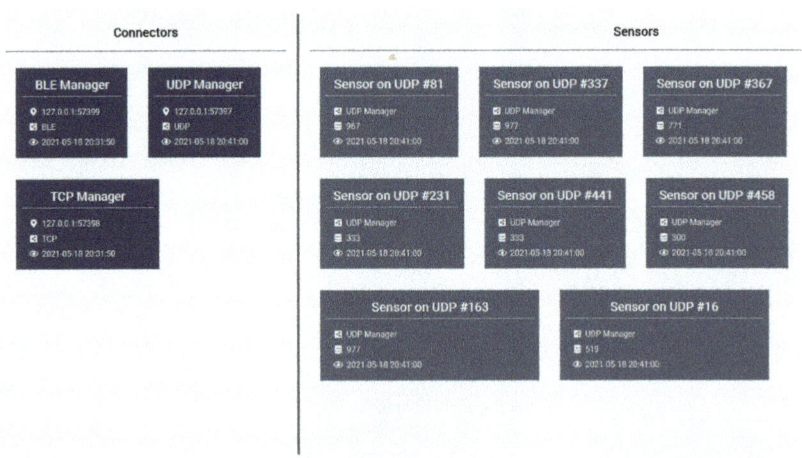

Figure 7. The sensors-state-monitoring application interface included with the LIPSHOK SHib kit.

6. Conclusions

In this paper, we have introduced the LIARA Portable Smart Home Kit (LIPSHOK), a smart home in a box (SHiB) kit capable of managing both heterogeneous sensors and actuators, along with several wireless communication technologies, in a unified modular architecture. In order to keep this architecture affordable, its design relies on a centralized unit that does not require powerful hardware to manage the flow of data generated by all the sensors in real time. Since it is a simple architecture that is based on inexpensive hardware, we have put our efforts toward offering a standalone and generic solution, in order to facilitate its installation. Moreover, four custom sensors are also featured in this kit. However, these devices do not exclude the possibility for researchers and developers to integrate other sensors or actuators, which may be either proprietary or custom-made in the LIPSHOK kit. While each of these four sensors has been presented independently, this is, to the best of our knowledge, the first time such a comprehensive SHiB kit has been presented.

7. Future Work

Future work will focus first on the refection and enhancement of the bathroom modules. Indeed, since these modules have not been designed to meet any of the IP standards, it seems important to offer modules that are IP67 considering the environment where they are installed. Moreover, the need for these modules to be powered through an electric cable, while it only outputs 5V, remains a drawback in their design that may raise concerns for smart home residents. Thus, we are working on making the bathroom modules capable of being powered by a button cell battery.

In addition, as a post-COVID-19 pandemic context is starting to emerge, our objective of deploying the complete LIPSHOK kit in real conditions, within a non-research laboratory environment, now appears achievable. Therefore, future work will focus on this essential step that is required to ensure the LIPSHOK SHiB kit works in residential settings.

Author Contributions: Conceptualization, K.C., P.L., K.B. and S.G.; methodology, K.C., F.T., P.L., S.G. and K.B.; software, K.C. and P.L.; experimentation and validation, K.C., P.L. and F.T.; writing—original draft preparation, K.C., F.T. and P.L.; writing—review and editing, S.G., J.M., K.B. and F.T. supervision, S.G., K.B. and J.M.; project administration, S.G. All authors have read and agreed to the published version of the manuscript.

Funding: This research received no external funding.

Institutional Review Board Statement: Not applicable.

Informed Consent Statement: Not applicable.

Data Availability Statement: Not applicable.

Acknowledgments: The authors would like to acknowledge all the people who participated in this study, as their involvement was essential to the success of this work.

Conflicts of Interest: The authors declare no conflict of interest.

Abbreviations

The following abbreviations are used in this manuscript:

5MWT	5-Meter Walk Test
ADL	Activities of Daily Living
AES	Advanced Encryption Standard
Am.I.	Ambient Intelligence
BLE	Bluetooth Low Energy
DM1	Myotonic Dystrophy type 1
DoF	Degrees-of-Freedom
IMU	Inertial Measurement Unit
IP	Ingress Protection
IRPS	Infrared Proximity Sensor
IV	Initialization Vector
LIPSHOK	LIARA Portable Smart Home Kit

MCI	Mild Cognitive Impairments
MITM	Man in the Middle
PIR	Passive Infrared
RSSI	Received Signal Strength Indication
SHiB	Smart Home in a Box
SPoF	Single Points of Failure
TCP	Transmission Control Protocol
UDP	User Datagram Protocol

References

1. Cook, D.; Youngblood, M.; Heierman, E.; Gopalratnam, K.; Rao, S.; Litvin, A.; Khawaja, F. MavHome: An Agent-Based Smart Home. In Proceedings of the First IEEE International Conference on Pervasive Computing and Communications, 2003. (PerCom 2003), Fort Worth, TX, USA, 26 March 2003; pp. 521–524. [CrossRef]
2. Helal, S.; Mann, W.; El-Zabadani, H.; King, J.; Kaddoura, Y.; Jansen, E. The Gator tech smart house: A programmable pervasive space. *Computer* **2005**, *38*, 50–60. [CrossRef]
3. Chen, L.; Nugent, C.; Mulvenna, M.; Finlay, D.; Hong, X. Semantic Smart Homes: Towards Knowledge Rich Assisted Living Environments. In *Intelligent Patient Management*; McClean, S., Millard, P., El-Darzi, E., Nugent, C., Eds.; Springer: Berlin/Heidelberg, Germany, 2009; pp. 279–296. [CrossRef]
4. Giroux, S.; Leblanc, T.; Bouzouane, A.; Bouchard, B.; Pigot, H.; Bauchet, J. The Praxis of Cognitive Assistance in Smart Homes. *BMI Book* **2009**, *3*, 183–211. [CrossRef]
5. Cook, D.J.; Crandall, A.S.; Thomas, B.L.; Krishnan, N.C. CASAS: A smart home in a box. *Computer* **2013**, *46*, 62–69. [CrossRef] [PubMed]
6. Bouchard, K.; Bouchard, B.; Bouzouane, A. Practical Guidelines to Build Smart Homes: Lessons Learned. In *Opportunistic Networking, Smart Home, Smart City, Smart Systems*, 1st ed.; CRC Press, Taylor & Francis: Boca Raton, FL, USA, 2014; pp. 1–34.
7. Lago, P.; Lang, F.; Roncancio, C.; Jiménez-Guarín, C.; Mateescu, R.; Bonnefond, N. The ContextAct@A4H Real-Life Dataset of Daily-Living Activities. In *Modeling and Using Context*; Brézillon, P., Turner, R., Penco, C., Eds.; Springer: Berlin/Heidelberg, Germany, 2017; Volume 10257, pp. 175–188. [CrossRef]
8. Plantevin, V.; Bouzouane, A.; Bouchard, B.; Gaboury, S. Towards a more reliable and scalable architecture for smart home environments. *J. Ambient. Intell. Humaniz. Comput.* **2019**, *10*, 2645–2656. [CrossRef]
9. Chen, L.; Hoey, J.; Nugent, C.D.; Cook, D.J.; Yu, Z. Sensor-Based Activity Recognition. *IEEE Trans. Syst. Man, Cybern. Part C (Appl. Rev.)* **2012**, *42*, 790–808. [CrossRef]
10. Thullier, F.; Hallé, S.; Gaboury, S. LE2ML: A Microservices-Based Machine Learning Workbench as Part of an Agnostic, Reliable and Scalable Architecture for Smart Homes. *J. Ambient. Intell. Humaniz. Comput.* **2021**, 1–22. [CrossRef]
11. Pope, J.; McConville, R.; Kozlowski, M.; Fafoutis, X.; Santos-Rodriguez, R.; Piechocki, R.J.; Craddock, I. SPHERE in a Box: Practical and Scalable EurValve Activity Monitoring Smart Home Kit. In Proceedings of the 2017 IEEE 42nd Conference on Local Computer Networks Workshops (LCN Workshops), Singapore, 9 October 2017; pp. 128–135. [CrossRef]
12. Chapron, K.; Lapointe, P.; Bouchard, K.; Gaboury, S. Highly accurate bathroom activity recognition using infrared proximity sensors. *IEEE J. Biomed. Health Inform.* **2020**, *24*, 2368–2377. [CrossRef]
13. Lapointe, P.; Chapron, K.; Bouchard, K.; Gaboury, S. A New Device to Track and Identify people in a Multi-Residents Context. *Procedia Comput. Sci.* **2020**, *170*, 403–410. [CrossRef]
14. Chapron, K.; Bouchard, K.; Gaboury, S. Real-time gait speed evaluation at home in a multi residents context. *Multimed. Tools Appl.* **2021**, *80*, 12931–12949. [CrossRef]
15. Chapron, K.; Lapointe, P.; Lessard, I.; Darsmstadt-Bélanger, H.; Bouchard, K.; Gagnon, C.; Lavoie, M.; Duchesne, É.; Gaboury, S. Acti-DM1: Monitoring the Activity Level of People With Myotonic Dystrophy Type 1 through Activity and Exercise Recognition. *IEEE Access* **2021**, *9*, 49960–49973. [CrossRef]
16. Woznowski, P.; Burrows, A.; Diethe, T.; Fafoutis, X.; Hall, J.; Hannuna, S.; Camplani, M.; Twomey, N.; Kozlowski, M.; Tan, B.; et al. SPHERE: A Sensor Platform for Healthcare in a Residential Environment. In *Designing, Developing, and Facilitating Smart Cities: Urban Design to IoT Solutions*; Angelakis, V., Tragos, E., Pöhls, H.C., Kapovits, A., Bassi, A., Eds.; Springer: Berlin/Heidelberg, Germany, 2017; pp. 315–333. [CrossRef]
17. Wilson, C.; Hargreaves, T.; Hauxwell-Baldwin, R. Smart Homes and Their Users: A Systematic Analysis and Key Challenges. *Pers. Ubiquitous Comput.* **2015**, *19*, 463–476. [CrossRef]
18. Bouchabou, D.; Nguyen, S.M.; Lohr, C.; LeDuc, B.; Kanellos, I. A Survey of Human Activity Recognition in Smart Homes Based on IoT Sensors Algorithms: Taxonomies, Challenges, and Opportunities with Deep Learning. *Sensors* **2021**, *21*, 6037. [CrossRef] [PubMed]
19. Folstein, M.; Anthony, J.C.; Parhad, I.; Duffy, B.; Gruenberg, E.M. The Meaning of Cognitive Impairment in the Elderly. *J. Am. Geriatr. Soc.* **1985**, *33*, 228–235. [CrossRef] [PubMed]
20. Fonseca, A.M.; Soares, E. The care taker's discourse about taking care of the elderly with alzheimer's disease. *Rev. Rene.* **2008**, *9*.
21. van Gelder, B.M.; Tijhuis, M.A.; Kalmijn, S.; Giampaoli, S.; Nissinen, A.; Kromhout, D. Physical activity in Relation to Cognitive Decline in Elderly Men. *Neurology* **2004**, *63*, 2316–2321. [CrossRef] [PubMed]

22. Milanović, Z.; Pantelić, S.; Trajković, N.; Sporiš, G.; Kostić, R.; James, N. Age-Related Decrease in Physical Activity and Functional Fitness Among Elderly Men and Women. *Clin. Interv. Aging* **2013**, *2013*, 549–556. [CrossRef]

23. Chen, J.; Zhang, J.; Kam, A.; Shue, L. An Automatic Acoustic Bathroom Monitoring System. In Proceedings of the 2005 IEEE International Symposium on Circuits and Systems (ISCAS), Kobe, Japan, 23–26 May 2005; Volume 2, pp. 1750–1753. [CrossRef]

24. Tapia, E.M.; Intille, S.S.; Larson, K. Activity Recognition in the Home Using Simple and Ubiquitous Sensors. In *Pervasive Computing*; Ferscha, A., Mattern, F., Eds.; Springer: Berlin/Heidelberg, Germany, 2004; Volume 3001, pp. 158–175. [CrossRef]

25. Marquis, S.; Moore, M.M.; Howieson, D.B.; Sexton, G.; Payami, H.; Kaye, J.A.; Camicioli, R. Independent Predictors of Cognitive Decline in Healthy Elderly Persons. *Arch. Neurol.* **2002**, *59*, 601–606. [CrossRef]

26. Buracchio, T.; Dodge, H.H.; Howieson, D.; Wasserman, D.; Kaye, J. The Trajectory of Gait Speed Preceding Mild Cognitive Impairment. *Arch. Neurol.* **2010**, *67*, 980–986. [CrossRef]

27. Wilson, C.M.; Kostsuca, S.R.; Boura, J.A. Utilization of a 5-Meter Walk Test in Evaluating Self-selected Gait Speed during Preoperative Screening of Patients Scheduled for Cardiac Surgery. *Cardiopulm. Phys. Ther. J.* **2013**, *24*, 36–43. [CrossRef]

28. Benmansour, A.; Bouchachia, A.; Feham, M. Multioccupant Activity Recognition in Pervasive Smart Home Environments. *ACM Comput. Surv.* **2015**, *48*, 1–36. [CrossRef]

29. Gao, L.; Bourke, A.; Nelson, J. Evaluation of Accelerometer Based Multi-Sensor Versus Single-Sensor Activity Recognition Systems. *Med Eng. Phys.* **2014**, *36*, 779–785. [CrossRef] [PubMed]

30. Adib, F.; Mao, H.; Kabelac, Z.; Katabi, D.; Miller, R.C. Smart Homes that Monitor Breathing and Heart Rate. In Proceedings of the 33rd Annual ACM Conference on Human Factors in Computing Systems—CHI'15, ACM, Seoul, Korea, 18–23 April 2015; pp. 837–846. [CrossRef]

31. Davis, K.; Owusu, E.; Bastani, V.; Marcenaro, L.; Hu, J.; Regazzoni, C.; Feijs, L. Activity Recognition Based on Inertial Sensors for Ambient Assisted Living. In Proceedings of the 19th International Conference on Information Fusion (FUSION), Heidelberg, Germany, 5–8 July 2016; pp. 371–378.

32. Khan, Y.; Ostfeld, A.E.; Lochner, C.M.; Pierre, A.; Arias, A.C. Monitoring of Vital Signs with Flexible and Wearable Medical Devices. *Adv. Mater.* **2016**, *28*, 4373–4395. [CrossRef] [PubMed]

33. Mokhtari, G.; Anvari-Moghaddam, A.; Zhang, Q.; Karunanithi, M. Multi-Residential Activity Labelling in Smart Homes with Wearable Tags Using BLE Technology. *Sensors* **2018**, *18*, 908. [CrossRef] [PubMed]

34. Arrotta, L.; Bettini, C.; Civitarese, G.; Presotto, R. Context-Aware Data Association for Multi-Inhabitant Sensor-Based Activity Recognition. In Proceedings of the 2020 21st IEEE International Conference on Mobile Data Management (MDM), Versailles, France, 30 June–3 July 2020; pp. 125–130. [CrossRef]

35. Alam, M.A.U.; Roy, N.; Misra, A. Tracking and Behavior Augmented Activity Recognition for Multiple Inhabitants. *IEEE Trans. Mob. Comput.* **2021**, *20*, 247–262. [CrossRef]

36. Chapron, K.; Plantevin, V.; Thullier, F.; Bouchard, K.; Duchesne, E.; Gaboury, S. A More Efficient Transportable and Scalable System for Real-Time Activities and Exercises Recognition. *Sensors* **2018**, *18*, 268. [CrossRef]

37. Thullier, F.; Plantevin, V.; Bouzouane, A.; Hallé, S.; Gaboury, S. A Comparison of Inertial Data Acquisition Methods for a Position-Independent Soil Types Recognition. In Proceedings of the 2018 IEEE SmartWorld, Ubiquitous Intelligence Computing, Advanced Trusted Computing, Scalable Computing Communications, Cloud Big Data Computing, Internet of People and Smart City Innovation (SmartWorld/SCALCOM/UIC/ATC/CBDCom/IOP/SCI), Guangzhou, China, 8–12 October 2018; pp. 1052–1056. [CrossRef]

38. Bhushan, B.; Sahoo, G.; Rai, A.K. Man-in-the-middle Attack in Wireless and Computer Networking—A Review. In Proceedings of the 2017 3rd International Conference on Advances in Computing, Communication & Automation (ICACCA) (Fall), Dehradun, India, 15–16 September 2017; pp. 1–6. [CrossRef]

39. Bogdanov, A.; Knudsen, L.R.; Leander, G.; Paar, C.; Poschmann, A.; Robshaw, M.J.B.; Seurin, Y.; Vikkelsoe, C. PRESENT: An Ultra-Lightweight Block Cipher. In *Cryptographic Hardware and Embedded Systems—CHES 2007*; Paillier, P., Verbauwhede, I., Eds.; Springer: Berlin/Heidelberg, Germany, 2007; pp. 450–466. [CrossRef]

sensors

Article

A Digital Twin Decision Support System for the Urban Facility Management Process [†]

Armir Bujari *, Alessandro Calvio, Luca Foschini, Andrea Sabbioni and Antonio Corradi

Department of Computer Science and Engineering (DISI), University of Bologna, 40136 Bologna, Italy; alessandro.calvio2@unibo.it (A.C.); luca.foschini@unibo.it (L.F.); andrea.sabbioni5@unibo.it (A.S.); antonio.corradi@unibo.it (A.C.)
* Correspondence: armir.bujari@unibo.it
† This Paper is an extended version of our paper published in: Bujari, A.; Calvio, A.; Foschini, L.; Sabbioni, A.; Corradi, A. IPPODAMO: A Digital Twin Support for Smart Cities Facility Management. In Proceedings of the Conference on Information Technology for Social Good, New York, NY, USA, 9–11 September 2021; pp. 49–54.

Abstract: The ever increasing pace of IoT deployment is opening the door to concrete implementations of smart city applications, enabling the large-scale sensing and modeling of (near-)real-time digital replicas of physical processes and environments. This digital replica could serve as the basis of a decision support system, providing insights into possible optimizations of resources in a smart city scenario. In this article, we discuss an extension of a prior work, presenting a detailed proof-of-concept implementation of a Digital Twin solution for the Urban Facility Management (UFM) process. The Interactive Planning Platform for City District Adaptive Maintenance Operations (IPPODAMO) is a distributed geographical system, fed with and ingesting heterogeneous data sources originating from different urban data providers. The data are subject to continuous refinements and algorithmic processes, used to quantify and build synthetic indexes measuring the activity level inside an area of interest. IPPODAMO takes into account potential interference from other stakeholders in the urban environment, enabling the informed scheduling of operations, aimed at minimizing interference and the costs of operations.

Keywords: Digital Twin; big data; geographic information system; smart city; Urban Facility Management; Apache Spark

Citation: Bujari, A.; Calvio, A.; Foschini, L.; Sabbioni, A.; Corradi, A. A Digital Twin Decision Support System for the Urban Facility Management Process. *Sensors* **2021**, *21*, 8460. https://doi.org/10.3390/s21248460

Academic Editor: Antonio Puliafito

Received: 13 November 2021
Accepted: 17 December 2021
Published: 18 December 2021

Publisher's Note: MDPI stays neutral with regard to jurisdictional claims in published maps and institutional affiliations.

1. Introduction

Digital transformation is an essential element for urban governance, enabling the efficient coordination and cooperation of multiple stakeholders involved in the urban environment. In the last decade, the concept of the smart city has gained tremendous importance and public/private institutions have started thinking about innovative ways to develop solutions tackling the complexity of the various phenomena [1]. At their core, these solutions rely on IoT and ICT to realize the perception, control and intelligent services aimed to optimize resource usage, bettering the services for the citizens [1,2]. To this aim, real-time access to accurate and open data is central to unlocking the economic value of the smart city potential, opening a rich ecosystem to suppliers for the development of new applications and services [3].

Digital Twin (DT) technology has revived the interest in the smart city concept, understood as a digital replica of an artefact, process or service, sufficient to be the basis for decision making [4]. This digital replica and the physical counterpart are often connected by streams of data, feeding and continuously updating the digital model, used as a descriptive or predictive tool for planning and operational purposes. While the concept of Digital Twin is by no means new, recent advances in 5G connectivity, AI, the democratization of sensing technology, etc., provide a solid technological basis and a new framework for investment. As an example, ABI Research predicts that the cost benefits deriving from the adoption of urban DTs alone could be worth USD 280 billion by 2030 [5].

Recognizing these benefits, different public and private institutions have already started investing in the technology, and this fact is attested by several urban DT initiatives, aimed at addressing different and complex problems in the smart city context. In the H2020 DUET project, the Flanders region aims at building a multi-purpose DT platform addressing the impact of mobility on the environment, to reduce the impact on human health [6]. In the same project, the city of Pilsen is building a proof-of-concept DT focusing on the interrelation between transport and noise pollution. The pilot aims to demonstrate the concept of the technology across transport and mobility, urban planning and the environment and well-being limits [7].

Recently, Bentley Systems and Microsoft announced a strategic partnership to advance DT for city planning and citizen engagement [8]. This partnership materialized in 2020 in a collaboration with the city of Dublin, pursuing the development of a large-scale urban DT used as a support tool for citizens to engage, from the safety of their own homes, in new development projects in their local communities.

It is evident that the success of an urban Digital Twin is directly impacted by the quality and quantity of the data sources that it relies on to model the physical counterpart. The modeling component, being either visual, e.g., based on a dashboard showing a consolidated view of the data, or data-driven, e.g., AI or statistical, is another core element of the framework. In this work, we discuss a concrete DT solution for the Urban Facility Management (UFM) process. The process under scrutiny comprises multiple stakeholders acting on a shared environment, e.g., different companies involved in the maintenance of various urban assets, each having a local view of the overall maintenance process. Their view on the activities is periodically consolidated in a joint conference called by the municipality, where mid-to-long-term operational details are discussed and a global, coarse-grained view of the process is established.

The scope and aim of the project is to showcase the use of DT technology as a decision support system, guiding UFM operators in their activity through the use of a rich set of correlation tools depicting the activities inside an area of interest. The system is equipped with a scheduling functionality, consulted to find feasible schedules for maintenance interventions, while minimizing some predetermined indexes, e.g., disturbance on mobility, interferences with other planned city events, etc. From a technological viewpoint, the Interactive Planning Platform for City District Adaptive Maintenance Operations (IPPODAMO [9]) is a proof-of-concept DT consisting of a (distributed) multi-layer geographical system, fed with heterogeneous data sources originating from different urban data providers. The data are initially staged at ingestion points, dedicated ingress machines, where they undergo syntactic and semantic transformations, and are successively forwarded to a big data processing framework for further refinements. The data are subject to different algorithmic processes, aimed at building a coherent view of the dynamics inside an area of interest, exploited by the UFM operated to make informed decisions on potential future maintenance operations. This work builds on a prior work [10], extending the study with a detailed discussion of some core system components, showcasing the advanced capabilities of the decision support system.

The article is organized as follows. Section 2 provides a concise background on the concepts and technological ecosystem adopted in this work. Section 3 briefly describes some data sources considered in the project, followed by a high-level functional overview of the multi-layer geographical system. Section 4 presents two distinct big data processing pipelines, providing some insights into their implementation and performance trends. While preserving the general aspect of our study and without loss of generality, in Section 5, we present the use cases and functionalities currently targeted by the platform. Finally, Section 6 draws the conclusions, delineating some future work.

2. Background

This section provides a brief overview of the Digital Twin concept, presenting its main building blocks and relating them to IPPODAMO. Next, we provide a concise and contextual survey of big data computing frameworks, part of the technological ecosystem adopted in this work.

2.1. Digital Twin Concept

Grieves is recognized to be the first that coined the term Digital Twin, using it to describe the digital product lifecycle management process [11]. Since then, thanks to the availability of modern computing platforms, engineering and management paradigms and practices, the term has gained in popularity, promoting DT technology as a key enabler of advanced digitization.

To date, many such applications have been successfully implemented in different domains, ranging from manufacturing [12] and logistics [13] to smart cities [14], etc. However, currently, there is no common understanding of the term, and in this respect, a taxonomy would help to demarcate the conceptual framework.

To this end, the authors in [4] provide a classification of the individual application areas of Digital Twins. As part of a comprehensive literature review, they develop a taxonomy for classifying the domains of application, stressing the importance and prior lack of consideration of this technology in information systems research. Building on this, the authors of [15] undertake a structured literature review and propose an empirical, multi-dimensional taxonomy of Digital Twin solutions, focusing on the technical and functional characteristics of the technological solution. As an add-on, the proposal allows for the classification of existing Industry 4.0 standards enabling a particular characteristic.

What unites the various DT approaches is the presence of some key technological building blocks that need to be considered and are crucial to its success.

1. Data link: the twin needs data collected from its real-world counterpart over its full lifecycle. The type and granularity of the data depend on the scope and context of deployment of the technology.

   IPPODAMO relies on a multitude of heterogeneous data sources, available at different granularities. Each data source has a dedicated (automatic) ingestion process used to transform the data, enriching the IPPODAMO data layer. For more information on the data sources and some computational aspects, we refer to the reader to Sections 3 and 4, respectively.

2. Deployment: the twin can be deployed and span the entire cloud-to-thing continuum, starting from the thing (IoT devices), the edge and/or the cloud. The specific deployment criteria depend on the scenario requirements, typically based on latency/bandwidth and/or security/privacy constraints.

   IPPODAMO is not directly involved in the raw data collection process. The system relies on third-party data providers, which collect, extract and fetch the data in dedicated ingress, cloud-backed machines. The raw data are anonymized and temporally retained in the system.

3. Modeling: the twin may contain different and heterogeneous computational and representational models pertaining to its real-world counterpart. This may range from first-principle models adhering to physical laws; data-driven; geometrical and material, such as Computer-Aided Design/Engineering (CAD, CAE); or visualization-oriented ones, such as mixed-reality.

   Our Digital Twin solution provides a consolidated view of heterogeneous urban data (descriptive Digital Twin), while at the same time, relying on historical and (near-)real-time data to perform near-to-mid-term predictions on the activity indexes (predictive Digital Twin). This allows operators to perform simulation scenarios aimed at minimizing predetermined indexes during routine and scheduled interventions. More information on the use cases and, in particular, the UFM scheduling is provided in Section 5.

4. APIs: the twin needs to interact with other components, e.g., twins in a composite system or an operator consulting the system. To facilitate these interactions, various APIs must be available to allow for information collection and control between the twin and its real-world counterpart.

    IPPODAMO presents the UFM operator an intuitive, high level user interface abstracting low-level system interfaces. At the same time, IPPODAMO has a programmatic interface, exposing well-defined ReST APIs through which other systems and operators could integrate and interact. The rationale behind this choice is to enable future potential integrations of IPPODAMO into a larger, federated ecosystem of smart city platforms. This aspect of the study is beyond the scope of this article.

5. Security/privacy: considering the role and scope of the DT, physical-to-digital interactions require security/privacy mechanisms aimed at securing the contents of the twin and the interaction flows between the twin and its physical counterpart.

    The solution is deployed on a state-of-the-art virtualized environment equipped with all the necessary security features. To this end, different administrative levels are provisioned, for accessing both the virtualized system and the system functionalities.

In the following, we provide a concise survey on some big data computing frameworks, motivating the choice of the identified technological ecosystem.

2.2. Big Data Computing

Smart cities generate and require, more than often, the collection of massive amounts of geo-referenced data from heterogeneous sources. Depending on the process and purpose of the system(s), these data must be quickly processed and analyzed to benefit from the information. To this end, big data computing frameworks have gained tremendous importance, enabling complex and online processing of information, allowing us to gain insights about phenomena in (near-)real time.

Cluster computing frameworks such as Apache Hadoop, based on the MapReduce compute model, are optimized for offline data analysis and are not ideal candidates for fast and online data processing due to the overhead of storing/fetching data at intermediate computation steps [16]. Current efforts in the relevant state-of-the-art have shifted toward promoting a new computing paradigm referred to as Stream Processing [17]. This paradigm adheres to the dataflow programming model, where computation is split and modeled as a directed acyclic graph and data flow through the graph, subjected to various operations [18]. Micro-batching represents a middle ground in this continuum: the idea is to discretize the continuous flow of data as a set of continuous sequences of small chunks of data, delivered to the system for processing. In this context, frameworks such as Apache Spark adopt this philosophy [19]. While Apache Spark is not a purely streaming approach, it introduces a powerful abstraction—the resilient distributed dataset (RDD [20])—allowing for distributed and in-memory computation. This programming abstraction supports efficient batch, iterative and online micro-batching processing of data, and it is a perfect match for the dynamics of our UFM scenario. Moreover, the cluster computing platform offers a wide range of libraries and computation models, as well as the geographical extensions needed to handle spatial data [21].

3. The IPPODAMO Platform

In this section, we start by presenting the data sources used by the Digital Twin, discussing their spatial and temporal characteristics. Next, we provide a high-level overview of the functional components comprising the technical solution.

3.1. Data Sources

IPPODAMO relies on a multitude of heterogeneous data sources provided, in part, by the project partners, including a telco operator and a company operating in the UFM sector.

Referring to Figure 1, the twin relies on (anonymized) vehicular and human presence data, combined to extract a measure of the activity inside an area of interest. The data sources are geo-referenced, embodying different levels of granularity, subject to different data processing steps and aggregation procedures aimed at building a composable activity level index (refer to Section 5). At the time of writing, the system has processed and stores two years of historical data and actively processes (near) real-time updates from each data source. These data have important value and enable IPPODAMO to gain (near-)real-time insights into the activities, but also to simulate near-to-mid-term evolutions of the activity level by exploiting the historical data.

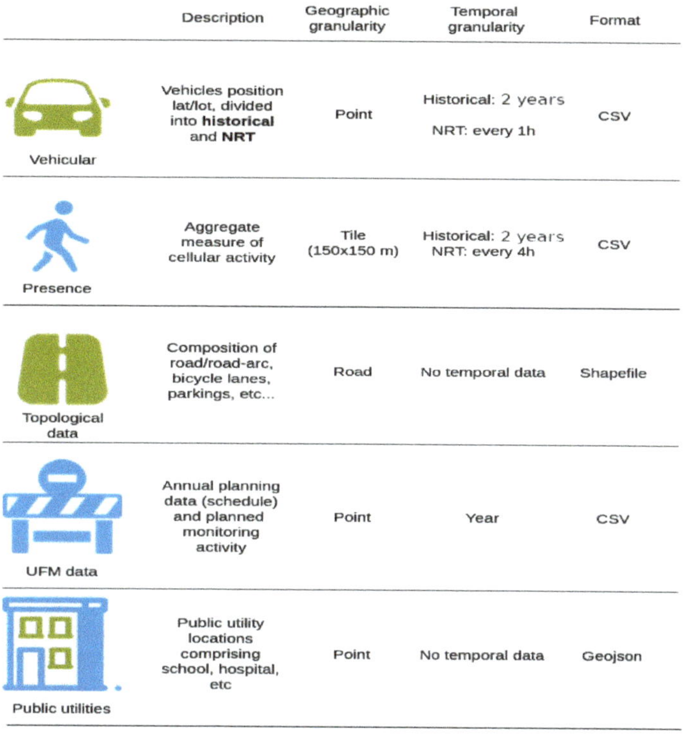

	Description	Geographic granularity	Temporal granularity	Format
Vehicular	Vehicles position lat/lot, divided into **historical** and **NRT**	Point	Historical: 2 years NRT: every 1h	CSV
Presence	Aggregate measure of cellular activity	Tile (150x150 m)	Historical: 2 years NRT: every 4h	CSV
Topological data	Composition of road/road-arc, bicycle lanes, parkings, etc…	Road	No temporal data	Shapefile
UFM data	Annual planning data (schedule) and planned monitoring activity	Point	Year	CSV
Public utilities	Public utility locations comprising school, hospital, etc	Point	No temporal data	Geojson

Figure 1. IPPODAMO data sources.

Other important data relate to the UFM process itself, which comprises data generated from: (i) the urban monitoring activity, e.g., the status of an urban asset assessed periodically through field inspections, (ii) annual planning and scheduled operations, e.g., repair interventions, (iii) geographical data concerning public utilities such as hospitals, schools, cycling lanes, etc. The data are provided and updated by the company operating in the UFM sector, having a vested interest in accurately depicting and monitoring the status of the urban assets.

Additional data sources are available and extracted from the open data portal, curated and maintained by the municipality of Bologna, Italy, and these include: (i) city events and (ii) other public utility maintenance operations.

All the above-mentioned data sources undergo dedicated processing steps and are stored in a logically centralized system, providing a consolidated and multi-source data layer. A rich set of visualizations can be built, guiding the UFM operator in their work. At the same time, more advanced functionalities, relying on the historical data to predict future evolutions of the phenomena inside an area of interest, are possible, and this topic is discussed in Section 5.

3.2. Technological Ecosystem

The platform (Figure 2) is structured in four main conceptual layers: (i) the ingestion layer, which interacts with the data providers, continuously acquiring new data, performing syntactic transformations, pushing them upwards for further refinements; (ii) the big data processing layer, which performs semantic transformation and enrichment of raw data fed from the ingestion points; (iii) the storage layer, providing advanced memorization and query capability over (near-)real-time and historical data, and (iv) the analytics layer, presenting to the customer an advanced (near-)real-time layer with query capabilities, aggregate metrics and advanced representations of data.

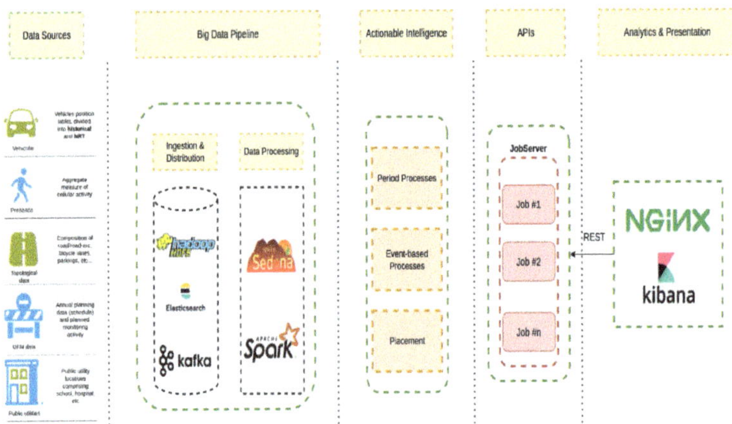

Figure 2. Technological components of the IPPODAMO platform.

For each data source used by the Digital Twin and, in particular, for the vehicular and presence data, there is a custom ingestion process tasked with reading the raw data, performing some syntactic transformations and pushing them towards the big data cluster. The various data sources are retrieved and pushed in parallel to specific Kafka topics, which identify also the semantic processing pipeline. Indeed, to enable the reliable and fast delivery of data from the ingestion points to our analytics and storage platform, we rely on Apache Kafka, an open-source, distributed, message-oriented middleware [22]. This choice is driven by the capabilities of this platform, including, but not limited to, its capability to gracefully scale the processing in the presence of high-throughput and low-latency ingress data. This matches the requirements of our domain, where data are constantly and periodically collected from many heterogeneous sources, e.g., vehicle black boxes, cellular, public transport, etc.

To provide advanced and fast processing capabilities for spatial data, the technological stack integrates and relies on Apache Sedona [23]. Apache Sedona is a distributed processing library, which builds on the in-memory computation abstraction of the Apache Spark framework and is able to provide advanced and fast queries over spatial data. Thanks to this spatial processing framework, we are able to blend and elaborate different data sources, creating rich representations of various phenomena in an area of interest.

Once the data have been processed, they are stored in Elasticsearch, a fast and scalable no-SQL database with advanced capabilities of indexing and querying key-value data [24]. Thanks to the advanced integration between Spark and Elasticsearch, we can use the solution as both a data provider and as a distributed storage system. The data are subject to further refinements and algorithmic processes aimed at creating different layers of aggregations, calculating synthetic indexes, etc., which are then used by the planner functionality to identify suitable time intervals during which to schedule urban operations. The raw data and information extracted from the data are presented to the end-users in different forms through advanced visualization dashboards available through Kibana, part of the Elasticsearch ecosystem. Last is the JobServer component, an optimized RESTful interface for submitting and managing Apache Spark tasks [25]. A job request, e.g., for a suggestion on a maintenance operation schedule, is a JSON-formatted request, triggering the execution of a specific algorithm resulting in a JSON-formatted output visualized through a web-based user interface. This component composes part of the IPPODAMO programmatic API, which could be used to integrate the solution into a larger, federated ecosystem of platforms in a smart city context.

4. Big Data Processing

In this section, we provide some technical details on the big data processing pipelines dedicated to the transformation and enrichment of incoming raw data. Next, we present and analyze the performance trend of two distinct data processing pipelines, providing in-depth insights into some system mechanisms.

4.1. Processing Pipeline(s)

The data sources introduced in Section 3 are subject to different processing pipelines due to the inherent syntactic and semantic differences that they embody. Concerning the vehicular and presence data, of relevance to the UFM process is the measure of the activity level inside a particular area in time. To this end, both pipelines are finalized to implement a counting technique measuring the activity volume. The geographical granularity of the presence data is accounted on a tile basis, a square-shaped geographical covering an area of 150 m $\times$ 150 m, while the vehicular data are point data—latitude and longitude—and can be accounted for at any meaningful granularity. These data resolutions are imposed by the data provider. It is important to note that in scenarios where both vehicular and presence data need to be accounted for, the granularity of this aggregation operation can be a tile or a multiple of tile entities.

Referring to Figure 3, both vehicular and presence data sources are initially subjected to some syntactic transformations before being forwarded from the Kafka producer, simplifying the ingestion process on the Spark cluster. From here on, the data are subject to semantic transformation processes, enriching the source data with relevant geographical information. In particular, the last operation in the pipeline accounts for the activity index information, retaining it in a distributed memory support. This operation should be done in a fast and efficient way; otherwise, it risks becoming a bottleneck for the overall system, delaying the ingestion performance.

To this end, we rely on a spatial partitioning mechanism, dividing the interest area among the cluster nodes. The area of interest to the project is initially saved in the underlying distributed file system and programmatically loaded and partitioned using the GeoHash spatial partitioning scheme [26]. This allows us to distribute topological data and index thereof among the cluster nodes. Indeed, in scenarios where ingress data are uniformly distributed inside the area of interest, this allows us, on average, to equally distribute and scale the computation among available cluster nodes. In particular, the data are accounted for by performing distributed point-in-polygon (PiP) and k-Nearest-Neighbor (kNN) operations, and these operations are enclosed by the *GeoRDDPresence* and *GeoRDDVehicular* functional components, respectively, for the presence and vehicular data. The data, once retained, are then subjected to additional processes, aggregating and slicing the data in the time and space domains (refer to Section 5). Other sources of information are those containing topological information concerning urban assets such as cycling and bus lanes, hospitals, schools, etc., and open data offered by the municipality, e.g., city events. Topological data have a dedicated batch processing pipeline; at its crux is a geo-join operation, aimed at enriching the IPPODAMO baseline topological map with additional information on urban assets. A dedicated update procedure is available whereby old information is discarded and only the fresh information is considered.

Last are the data sources containing operational information on the facility management process. UFM data have spatial and time information and are currently handled by a similar processing pipeline to the topological data.

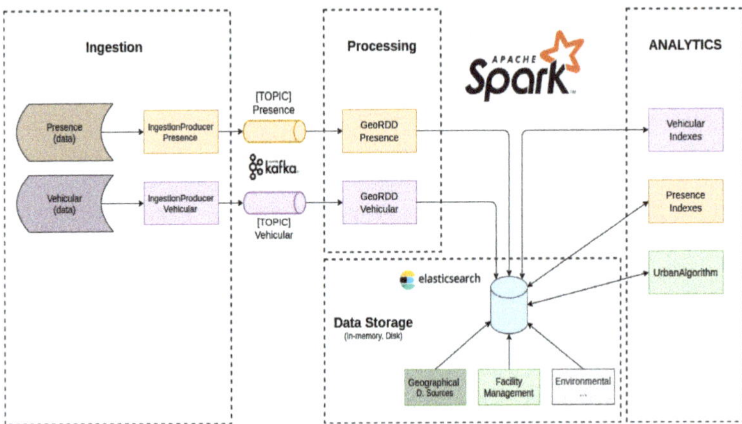

Figure 3. IPPODAMO (big) data processing pipelines and information flow.

4.2. Performance Analysis

The periodicity of the individual data sources imposes some operational constraints on the data processing pipelines, and this consideration applies to the vehicular and presence data. As an example, when a new batch of vehicular data enters the system, the corresponding end-to-end processing pipeline comprising the various syntactic and semantic transformations should take no more than 60 min (refer to Figure 1). Failure to do so would create a backlog of data that increases over time, slowing down the ingestion performance.

In particular, among the operations shown in Figure 3, the one embodying the highest computational burden is the *GeoRDDVehicular* operation, which requires the execution of a kNN operation for each data point present in the hourly dataset. We report that the end-to-end sequential processing of an hourly vehicular dataset containing, on average, 15,000 records (trips) often results in a violation of its operational constraint.

To address the issue, we leverage some specific constructs of the Apache Sedona framework, aimed at distributing the computational effort among the various cluster nodes. As anticipated, the proposed solution makes use of the (i) GeoHash spatial partitioning scheme, allowing the partitioning of a geographical area of interest among cluster nodes, and the (ii) broadcast primitive implementing a distributed, read-only shared memory support. In particular, the ingress vehicular dataset is enclosed as a broadcast variable, shared among worker nodes, where each node is responsible for the computation of a kNN operation over a subset of the points contained in the original dataset. This approach allows the parallel computation of individual kNN operations, whose outcome is later merged and retained in ElasticSearch. Figure 4 shows the performance trend of the optimized vehicular data processing pipeline. In particular, Figure 4a plots the processing time with a varying number of input records. The resulting trend is monotonic, allowing for the timely processing of the input dataset, adhering to the operational constraints imposed by the data source periodicity. Figure 4b puts the processing time into a greater context, accounting for additional operations occurring before and after the *GeoRDDVehicular* processing step, comprising data (de)serialization and output communication to the driver node, with the kNN processing step accounting for nearly 91% of the processing time.

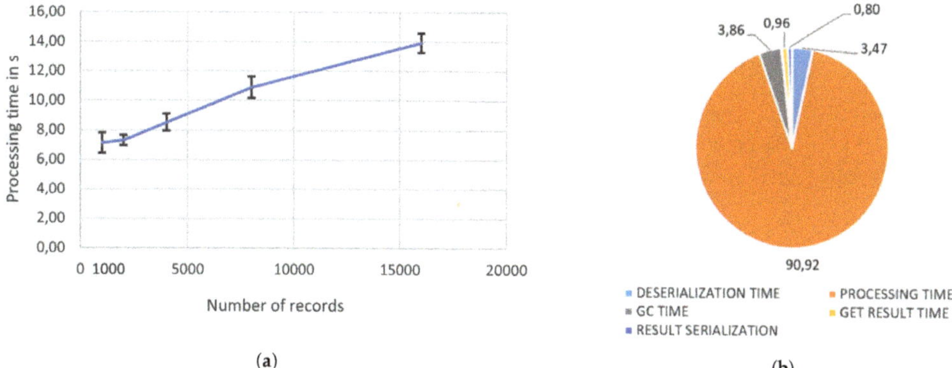

(a)   (b)

Figure 4. Vehicular processing pipeline performance. The experiments were carried out in a testbed comprising 4 VMs—1 driver and 3 workers—each equipped with 8 vCores, 32 GB vRAM and 150 GB data SSD support. (**a**) Overall processing time under varying number of ingress records. (**b**) Processing time decomposition for the 15,000 record configuration.

5. A Decision Support System

At its core, the IPPODAMO platform serves as a decision support system, aiding UFM operators in their daily activity. In this section, we start by discussing a set of identified use cases best targeting the needs of the UFM operator. Next, we discuss the concept behind the activity level index, a synthetic index used by some underlying system functionalities. Then, we provide a brief description of the algorithmic details of the scheduler functionality, along with a validation study showcasing its capabilities.

5.1. Use Cases

While preserving the general aspect of our study and without loss of generality, we identified two broad use cases, which showcase the capabilities of the solution in the following directions:

- UC#1: Rich (comparative) analysis by providing a set of configurable visualizations, used to quantify and visualize the activities inside an area of interest.
- UC#2: UFM scheduling functionality guiding the placement of maintenance operations in time.
- UC#3: Quantitative evaluation of the annual planning interventions.

Concerning the first use case, the system provides a rich set of configurable visualizations, allowing the UFM operator to consult and confront the historical and current trend of data inside an area of interest. The second use case aims to provide the UFM planner with a proactive decision tool, guinding the scheduling decisions. The last use case allows the platform administrator to perform an *a posteriori* evaluation of the annual planning schedule, dictated by the data that IPPODAMO has ingested. Through this functionality, we would like also to be able to perform a qualitative evaluation of the algorithmic decisions made by IPPODAMO, confronting them with the knowledge of the UFM specialist. At the core of these use cases is the activity level index, which is discussed in the following section.

5.2. Activity Level Index

Once the data are processed, they are stored in ElasticSearch and are subjected to further periodic and event-based algorithmic processes, aggregating and slicing the data in the time and space domains. Intuitively, the activity index is a measure of the activity level inside an area of interest.

At first, vehicular data are stored at their finest granularity, contributing to the traffic volume in a specific point of the underlying road topology. These point-wise data are aggregated and accounted on a road-arch basis, a constituent of the road topology. The data are also sliced in the time domain, accounting for the daily traffic volume and the traffic volume on some configurable rush hours. Once aggregated, the data are normalized and stored at different scales, e.g., for better visualization. This computed quantity constitutes the activity level index derived from the vehicular data.

The human presence data are subjected to a similar workflow. The difference is in the granularity (tile entity) that the data are accounted for. Recall that this granularity is imposed on us by the data provider. Once the individual indexes are computed, they can be composed and weighted according to some criteria. The current implementation allows an operator to simulate different scenarios by specifying the weights accordingly.

These indexes are periodically updated and maintained, and they constitute the input for the, e.g., UFM scheduler. Currently, three implementations for the activity level index are available, and the planners' behavior is parametric on the index type:

1. Raw index: scaled activity level computed from the raw data. This index has only look-back capabilities and is used by the UFM specialist to compare the goodness of the performed annual activities. This index is connected to the functionality of UC#3, via which we would like to obtain a qualitative evaluation of the platforms' operations.
2. Smoothed index: a window-based, weighted average index adopted to filter noise and potential erratic behavior of the raw index. Similarly, this index has only look-back capabilities, serving as a starting point for more sophisticated types of indexes.
3. Predictive index: this index has look-ahead capabilities, leveraging the past to predict near-to-mid-term, e.g., monthly-based evolution, activity levels in a specified geographical area. This index is used to create future hypotheses for scheduling operations.

Concerning the predictive index, we rely on state-of-the art algorithms, capable of inferring seasonality and local phenomena from the data. Currently, we are evaluating some practical design considerations that can occur in a dynamic and time-varying scenario such as ours.

5.3. UFM Scheduling Algorithm

This functionality is used in the first use case, and aids the UFM operator in searching for a suitable timeframe during which schedule a maintenance intervention. A maintenance operation may consist of a minor/major repair operation of an urban asset; it has fixed coordinates in space, a predicted duration, and an optional timeframe in which it needs to be scheduled. The scheduling criteria vary, and, depending on the objective, one would like to avoid or minimize disturbance to nearby activities. To this aim, the system has the capability to express and consider all these and other constraints when performing the search for a suitable timeframe.

Once a request is issued, the system receives all the constraints expressed by the operator, including the list of attributes, e.g., coordinate, expected duration, etc., for each intervention. The algorithm then exploits the geographical coordinates to compute the activity level index inside an area of interest and gathers all the potential interferences with nearby ongoing activities. To this aim, the functional component relies on the Spark SQL library and geographical primitives available in Apache Sedona. Recall that the algorithms' behavior is parametric on the index type, and different types of indexes are available. The final outcome of this computation is the construction of the index history and its hypothetical evolution in time.

Once the activity level index is computed, and all interferences have been retrieved, a final index is processed for the time horizon under consideration served as input to the scheduling algorithm. The exact algorithmic details and final index composition are beyond the scope of this article.

5.4. Discussion

Figure 5 shows a comparative view through which the UFM operator can assess the vehicular activity in configurable timeframes and areas. These views fall inside the functionalities provisioned in the second use case, and are all configurable via dedicated high-level graphical interactions. Once the configuration has been set up, a Kibana view is generated and mirrored in the frontend.

Figure 6 shows some sample outputs generated by the UFM scheduler functionality provisioned in the second use case. For this assessment, the algorithm relies on the predictive index and co-locality information identifying nearby interferences, e.g., city events, other public utility maintenance operations announced by the municipality, etc. A maintenance operation has a specific location in space, and a duration in time that could span several days or hours (Figure 6). Indeed, as already discussed, the data are sliced in the time domain: in Figure 6a, the algorithm exploits the daily predictive index to position a maintenance operation in time, while, in Figure 6d, the maintenance operation is, generally, positioned in a series of consecutive (pre-configured) time intervals. The IPPODAMO interface allows an operator to specify this additional search criterion.

In all the charts, the lines denoted in green identify the minimum cost schedules along with potential identified interferences, gathered and reported via the user interface. It is noteworthy to point out that, in the current implementation, interferences do not contribute to the index, but rather serve as additional information guiding the UFM operator to make an informed decision (Figure 6b,c). In addition, the interface allows the UFM operator to customize the weights of the individual parameters, e.g., as shown in Figure 6c, where the only quantity contributing to the index is that derived from the vehicular data.

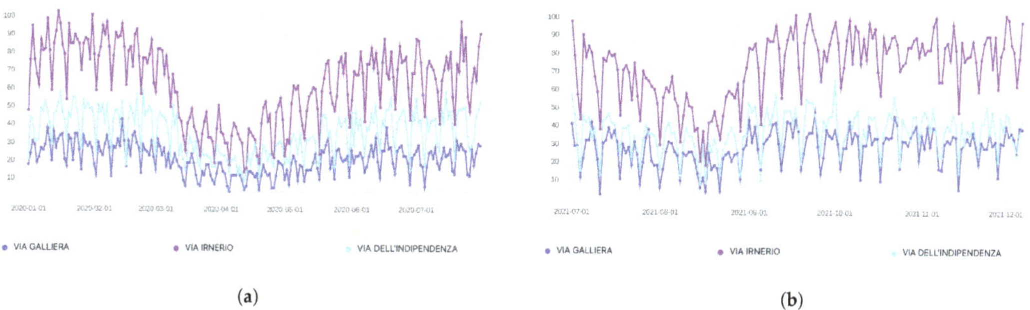

(a)   (b)

Figure 5. Trend of the vehicular index—scale [0, 100] for better visualization—in the pre-COVID-19, during and post-COVID-19 period. (**a**) Timeline comprising the COVID-19 period. (**b**) Post-COVID-19 period.

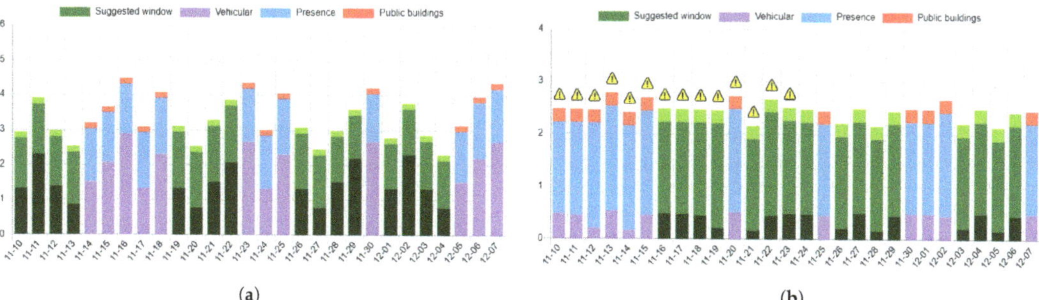

(a)   (b)

Figure 6. *Cont.*

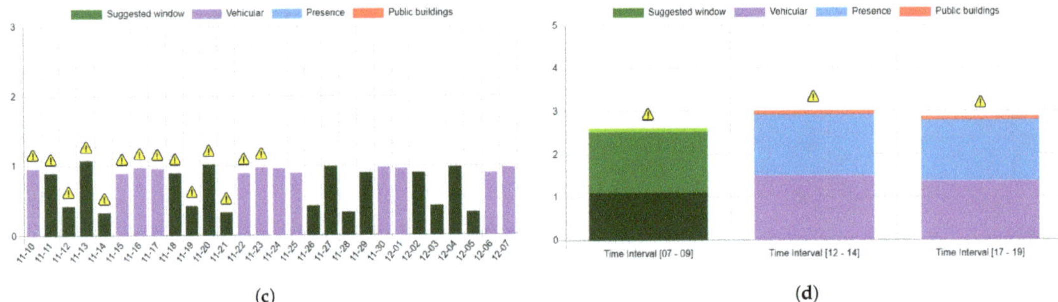

(c) (d)

Figure 6. The UFM scheduler result, providing the operator with potential timeframes (green color) during which to schedule a specific maintenance operation. The final activity index is decomposed in all its constituent values, contributing to the final index. The analysis considers a period of one month, starting from 10 November 2021. (**a**) Equal cost timeframes proposed by the scheduler for a maintenance operation in Via Indipendenza X, Bologna, Italy. (**b**) Identified non-binding interference for a maintenance operation in Via Zamboni, Bologna, Italy, in the interval [10, 23]. (**c**) Equal cost timeframes, vehicular data only, proposed by the scheduler for a maintenance operation in Via Zamboni. (**d**) Scheduling of an urgent intervention in Via Zamboni, relying on the next-day prediction of the activity index.

6. Conclusions

In this work, we presented a Digital Twin solution for the Urban Facility Management process in a smart city context. IPPODAMO is a multi-layer, distributed system making use of a multitude of heterogeneous data sources to accurately depict and predict the dynamics inside a geographical area of interest. The decision support system consists of a wide variety of visualizations, including a scheduler functionality, aiding UFM operators in their maintenance placement activity.

Currently, the solution is being tested in a real operational scenario, and we are studying emergent software behavior, identifying near-to-mid-term directions to extend the software. Of paramount importance is the capability to quantify the benefits of the solution through measurable KPIs. To this end, we are collaborating with the private sector and structuring a qualitative data gathering process that could serve as a basis for the value proposition of the proposal.

Author Contributions: Conceptualization, A.B., A.C. (Alessandro Calvio) and L.F.; Data curation, A.B., A.C. (Alessandro Calvio) and A.S.; Investigation, A.B., A.C. (Alessandro Calvio), A.S., L.F. and A.C. (Antonio Corradi); Resources, A.B., L.F. and A.C. (Antonio Corradi); Software, A.B., A.C. (Alessandro Calvio) and A.S.; Validation, A.B., A.C. (Alessandro Calvio) and A.S.; Writing—original draft, A.B., A.C. (Alessandro Calvio) and A.S.; Writing—review and editing, L.F. and A.C. (Antonio Corradi). All authors have read and agreed to the published version of the manuscript.

Funding: This work has been partially funded by the Interactive Planning Platform fOr city District Adaptive Maintenance Operations (IPPODAMO) project (CUP: C31J20000000008), a BI-REX Industry 4.0 Competence Center project.

Institutional Review Board Statement: Not applicable.

Informed Consent Statement: Not applicable.

Data Availability Statement: Not applicable.

Conflicts of Interest: The authors declare no conflict of interest.

References

1. Andrisano, O.; Bartolini, I.; Bellavista, P.; Boeri, A.; Bononi, L.; Borghetti, A.; Vigo, D. The Need of Multidisciplinary Approaches and Engineering Tools for the Development and Implementation of the Smart City Paradigm. *Proc. IEEE* **2018**, *106*, 738–760. [CrossRef]
2. Bujari, A.; Bergamini, C.; Corradi, A.; Foschini, L.; Palazzi, C. E.; Sabbioni, A. A Geo-Distributed Architectural Approach Favouring Smart Tourism Development in the 5G Era. In Proceedings of the EAI International Conference on Smart Objects and Technologies for Social Good, Antwerp, Belgium, 14–16 September 2020.
3. Ferretti, S.; Furini, M.; Palazzi, C.E.; Roccetti, M.; Salomoni, P. WWW Recycling for a Better World. *ACM Com.* **2010**, *53*, 139–143. [CrossRef]
4. Enders, M.R.; Hoßbach, N. Dimensions of Digital Twin Applications-A Literature Review. In Proceedings of the Americas Conference on Information Systems (AMCIS), Cancun, Mexico, 15–17 August 2019.
5. ABI Research. The Use of Digital Twins for Urban Planning to Yield US$280 Billion in Cost Savings by 2030. 2021. Available online: https://www.abiresearch.com/press/use-digital-twins-urban-planning-yield-us280-billion-cost-savings-2030/ (accessed on 1 December 2021).
6. EU H2020 DUET. Digital Urban European Twins-Flanders Twin. 2020. Available online: https://www.digitalurbantwins.com/flanderstwin (accessed on 1 December 2021).
7. EU H2020 DUET. Digital Urban European Twins-Pilsen Twin. 2020. Available online: https://www.digitalurbantwins.com/pilsen-twin (accessed on 1 December 2021).
8. Sarah Wray, Cities Today. Bentley Systems and Microsoft Team up on City Digital Twins. 2020. Available online: https://cities-today.com/bentley-systems-and-microsoft-team-up-on-digital-twins/ (accessed on 1 December 2021).
9. IPPODAMO. Interactive Planning Platform for City District Adaptive Maintenance Operations. 2020. Available online: https://www.ippodamoproject.it/ (accessed on 1 December 2021).
10. Bujari, A.; Calvio, A.; Foschini, L.; Sabbioni, A.; Corradi, A. IPPODAMO: A Digital Twin Support for Smart Cities Facility Management. In Proceedings of the ACM Conference on Information Technology for Social Good (GoodIT), Rome, Italy, 9–11 September 2021; pp. 49–54. [CrossRef]
11. Grieves, M.; Vickers, J., Digital Twin: Mitigating Unpredictable, Undesirable Emergent Behavior in Complex Systems. In *Transdisciplinary Perspectives on Complex Systems: New Findings and Approaches*; Springer International Publishing: Cham, Switzerland, 2017; pp. 85–113.
12. Tao, F.; Zhang, H.; Liu, A.; Nee, A.Y.C. Digital Twin in Industry: State-of-the-Art. *IEEE Trans. Ind. Inform.* **2019**, *15*, 2405–2415. [CrossRef]
13. Korth, B.; Schwede, C.; Zajac, M. Simulation-ready Digital Twin for Realtime Management of Logistics Systems. In Proceedings of the IEEE Big Data, Seattle, WA, USA, 10–13 December 2018.
14. White, G.; Zink, A.; Codecá, L.; Siobhán, C. A Digital Twin Smart City for Citizen Feedback. *Cities* **2021**, *110*, 103064. [CrossRef]
15. van der Valk, H.; Haße, H.; Möller, F.; Arbter, M.; Henning, J.L.; Otto, B. A Taxonomy of Digital Twins. In Proceedings of the Americas Conference on Information Systems (AMCIS), Salt Lake City, UT, USA, 10–14 August 2020.
16. Ding, M.; Zheng, L.; Lu, Y.; Li, L.; Guo, S.; Guo, M. More Convenient More Overhead: The Performance Evaluation of Hadoop Streaming. In Proceedings of the ACM Symposium on Research in Applied Computation, Miami, FL, USA, 19–22 October 2011; pp. 307–313.
17. Hirzel, M.; Soulé, R.; Schneider, S.; Gedik, B.; Grimm, R. A Catalog of Stream Processing Optimizations. *ACM Comput. Surv.* **2014**, *46*, 1–34. [CrossRef]
18. Fu, X.; Ghaffar, T.; Davis, J.C.; Lee, D. EdgeWise: A Better Stream Processing Engine for the Edge. In Proceedings of the USENIX Annual Technical Conference, Renton, WA, USA, 10–12 July 2019; pp. 929–946.
19. Zaharia, M.; Xin, R.S.; Wendell, P.; Das, T.; Armbrust, M.; Dave, A.; Meng, X.; Rosen, J.; Venkataraman, S.; Franklin, M.J.; et al. Apache Spark: A Unified Engine for Big Data Processing. *Commun. ACM* **2016**, *59*, 56–65. [CrossRef]
20. Zaharia, M.; Chowdhury, M.; Das, T.; Dave, A.; Ma, J.; McCauly, M.; Franklin, M.J.; Shenker, S.; Stoica, I. Resilient Distributed Datasets: A Fault-Tolerant Abstraction for In-Memory Cluster Computing. In Proceedings of the USENIX NSDI, San Jose, CA, USA, 25–27 April 2012.
21. Yu, J.; Zhang, Z.; Sarwat, M. Spatial Data Management in Apache Spark: The GeoSpark Perspective and Beyond. *Geoinformatica* **2019**, *23*, 37–78. [CrossRef]
22. Kafka, A. A Distributed Event Streaming Platform. Available online: https://kafka.apache.org/ (accessed on 1 November 2021).
23. Sedona, A. Cluster Computing System for Processing Large-Scale Spatial Data. Available online: https://sedona.apache.org/ (accessed on 1 November 2021).
24. Elastic. Elasticsearch, Logstash, and Kibana. Available online: https://www.elastic.co/what-is/elk-stack (accessed on 1 November 2021).
25. JobServer. REST Job Server for Apache Spark. Available online: https://github.com/spark-jobserver/ (accessed on 1 November 2021).
26. Lee, K.; Ganti, R.; Srivatsa, M.; Liu, L. Efficient Spatial Query Processing for Big Data. In Proceedings of the ACM SIGSPATIAL International Conference on Advances in Geographic Information Systems, Dallas, TX, USA, 4–7 November 2014; pp. 469–472.

Article

A 5G-Enabled Smart Waste Management System for University Campus †

Edoardo Longo [1], Fatih Alperen Sahin [1], Alessandro E. C. Redondi [1,*], Patrizia Bolzan [2], Massimo Bianchini [2] and Stefano Maffei [2]

[1] Dipartimento di Elettronica, Informazione e Bioingegneria, Politecnico di Milano, 20133 Milano, Italy; edoardo.longo@polimi.it (E.L.); fatihalperensahin@gmail.com (F.A.S.)
[2] Polifactory-Dipartimento di Design, Politecnico di Milano, 20133 Milano, Italy; patrizia.bolzan@polimi.it (P.B.); massimo.bianchini@polimi.it (M.B.); stefano.maffei@polimi.it (S.M.)
* Correspondence: alessandroenrico.redondi@polimi.it
† This paper is an extended version of our paper published in ACM International Conference on Information Technology for Social Good (GoodIT 2021), 9–11 September 2021, Rome, Italy.

Abstract: Future university campuses will be characterized by a series of novel services enabled by the vision of Internet of Things, such as smart parking and smart libraries. In this paper, we propose a complete solution for a smart waste management system with the purpose of increasing the recycling rate in the campus and provide better management of the entire waste cycle. The system is based on a prototype of a smart waste bin, able to accurately classify pieces of trash typically produced in the campus premises with a hybrid sensor/image classification algorithm, as well as automatically segregate the different waste materials. We discuss the entire design of the system prototype, from the analysis of requirements to the implementation details and we evaluate its performance in different scenarios. Finally, we discuss advanced application functionalities built around the smart waste bin, such as optimized maintenance scheduling.

Keywords: Smart Campus; smart waste management; waste classification; multi access edge computing

Citation: Longo, E.; Sahin, F.A.; Redondi, A.E.C.; Bolzan, P.; Bianchini, M.; Maffei, S. A 5G-Enabled Smart Waste Management System for University Campus. *Sensors* **2021**, *21*, 8278. https://doi.org/10.3390/s21248278

Academic Editors: Claudio Palazzi, Ombretta Gaggi and Pietro Manzoni

Received: 8 November 2021
Accepted: 9 December 2021
Published: 10 December 2021

Publisher's Note: MDPI stays neutral with regard to jurisdictional claims in published maps and institutional affiliations.

1. Introduction

The Internet of Things vision is becoming a reality, transforming the way we live and interact with the environment. Many conventional places are acquiring smart characteristics thanks to a multitude of small, low cost and connected computing devices able to sense, process and communicate data from the environment to cloud/Internet services. Smart Cities and Smart Buildings, to name a few, are all different realizations of such a vision and are nowadays of great interest to academic researchers as well as having great potential for the industrial world. Smart Campuses are of particular interest in this scenario, and they can be seen as the perfect place for initial steps towards the realization of large-scale projects targeting Smart Cities [1]. Indeed, university campuses mimic cities in many aspects: they generally extend on a vast urban area, they are composed of many buildings of different types (administrative buildings, research laboratories, classrooms, residences, bar/restaurants) and populated by different types of people (students, teachers, administrative and technical staff, etc.). At the same time, the management is somehow more flexible than what is found in proper cities and municipalities since universities are by nature more open at accepting innovations and new technologies, even if still not completely mature. Several solutions have been recently proposed in association with the concept of Smart Campus [2]: smart parking systems [3,4], microgrids [5], smart libraries [6,7], systems for classroom monitoring and occupancy estimation [8,9] as well as sustainable solutions [10,11], are all examples of smart applications implemented in university campuses.

One area that received particular attention in the last decade is the efficient management of university solid waste (USW) [12]. Recycling such waste is crucial from several points of view: from an economic perspective, turning solid waste into a resource is fundamental to the realization of a circular economy, where one industry's waste becomes another raw material. At the same time, efficient and sustainable management of waste helps reduce health and environmental problems: recycling materials helps cut emissions from landfills and from new extraction/processing sites, and mitigates environmental issues such as water/air pollution and littering.

Several works in the literature have addressed the analysis of how much waste is produced in a university campus, with estimates ranging from 50 to 150 g/user/day (i.e., 20–50 Kg per user each year) [13–15]. Considering that, according to recent statistics [16], each person in Europe produce half a tonne of municipal waste per year, USW alone account for about 1 tenth of the total waste produced in cities.

Moreover, some studies [17,18] have analysed the composition of waste produced in university campuses, concluding that the majority of USW is composed of organic waste suitable for composting, followed by recyclable materials such as plastic, glass and paper. As different types of waste require different recycling processes, segregating and separating waste at its source is key to the effective management of the recycling chain. While industrial waste is generally treated with large-scale segregators, the task of waste separation is much more challenging at the municipal or campus level, as it is solely based on the goodwill of people and the level of readiness of the recycling infrastructure available.

To facilitate the separate collection of waste starting from the beginning of the sorting chain, that is, from public waste bins, information technologies may come to help. In particular, embedding different types of sensors and actuators into such waste bins, connecting them to the Internet and driving them through intelligent algorithms (i.e., following the vision of the Internet of Things (IoT)) may give an incredible boost to the recycling performance.

Motivated by these reasons, this paper extends our previous paper [19] and describes the realization of a complete solution for the efficient management of USW. The key building block of our proposal is a novel prototype of a Smart Waste Bin (SWB), a smart object able to automatically sort different types of trash directly at the place of generation using a multi-sensor approach, thus easing the management of the entire trash cycle. Peculiar features characterize the system: the SWB adopts a hybrid scalar/visual sensor waste classification algorithm that allows for accurate waste recognition as well as an innovative dual-motor design for automatic waste segregation. Moreover, the SWB and the management system designed around it are integrated with the recently introduced 5G networking architecture, particularly for what concerns the advantages of using a Multi-access Edge Computing (MEC) server. Indeed, the intelligence driving the SWB resides at the edge of a 5G cellular network, rather than in a cloud server or locally on the object itself. This approach brings several benefits, such as reduced delay in waste recognition and reduced energy consumption, making the SWB more appealing to everyday use.

In detail, the contributions of this work are the following: first, we illustrate the design and implementation of the smart waste bin, detailing the steps made for its creation from the analysis of the requirements to the physical realization of its external and internal parts. Second, we give details on the algorithms governing the SWB, including the main functioning logic as well as the multi-sensorial artificial intelligence used for recognizing and sorting different types of trash. For the latter, we propose different ways of fusing information coming from the different sensors, evaluating the performance obtained. Third, we evaluate a fully working prototype of the SWB in different scenarios, showing through experiments on a real 5G network that moving the artificial intelligence on the MEC is beneficial under both latency and energy consumption perspectives. Finally, we showcase the potential of a management system built around a multitude of (simulated) smart waste bins, allowing for, e.g., easy and optimized maintenance.

The remainder of this paper is structured as follows: Section 2 briefly reviews the main related works in the field of smart waste management. Section 3 details the physical realization of the proposed smart waste bin prototype, focusing on the main working logic and the offered functionalities. Section 4 provides a detailed description of the hybrid scalar/visual machine learning algorithm used to perform waste classification and evaluates the performance obtained when such intelligence is run locally on the SWB, in the Cloud or on a 5G MEC. Section 5 focuses on the management backend server and the advanced features offered by the provided user application. Finally, Section 6 concludes the paper.

2. Related Work

Facilitated by the wide commercial availability of low-cost sensors, microcontrollers and communication modules, several research works focusing on prototyping smart waste management systems have appeared in the last few years.

A class of these works focus mainly on monitoring the amount of trash and the fill level of waste bins in order to send alerts and optimize the emptying procedures [20–22]. Generally, ultrasonic sensors are used to estimate the fill level by measuring the distance from the lid on top of the bin to the trash in the compartment. Sometimes a load cell sensor is incorporated at the bottom of the bin to measure the weight of the waste [23,24]. As an example, in [25] authors propose a system with ultrasonic sensors connected to a Microcontroller Unit (MCU) that sends an SMS message to the municipality if the waste level is above a certain threshold. Knowing the waste levels and the locations of the corresponding bins, the routing and scheduling of the garbage picking procedures can be optimized; as a result, authors claim that the service cost can be cut by 50%.

The second class of works focus on techniques for recognizing and sorting different types of trash, with several approaches. Some works use scalar sensors, such as electromagnetic sensors (capacitive or inductive sensors), which can be utilized for detection of nonferrous metal fractions based upon electrical conductivity of the sample [26,27]. Alternatively, photoelectric sensors (obtained coupling a Light Emitter Diode (LED) source and a photodiode as a receiver) can be used to recognize the type of material (especially in presence of transparent wrappings) [28]. Other works focus on Radio-frequency identification (RFID) technology to sort the different categories of waste, assuming that each piece of trash is equipped with a smart RFID tag containing the information on the particular type of material [29,30].

With the success and popularity of machine learning, and in particular of Convolutional Neural Networks (CNN) in the field of computer vision, a considerable amount of works tackle the problem of image-based waste recognition [31–34]. A common approach is to use already existing CNN models (pre-trained over very large image databases, such as ImageNet [35]), which are known to provide excellent results in terms of image classification (e.g., AlexNet [36] or VGG16 [37]), and fine tune their last layers of the neural network with datasets containing images of pieces of trash [38]. All these works report excellent performance in the task of trash classification, reporting accuracies generally above 90% when four target classes of glass, paper, metal and plastic are concerned.

Some works also propose prototypes not only to recognize different pieces of trash, but also to move them in proper compartments after recognition. The operation is typically performed through the use of Direct Current (DC) or stepper motors [39,40]. As an example, in [41] waste is placed on a conveyor belt and classified in different categories via image-based recognition and a trained CNN. After classification, an automatic hand hammer is used to push the waste into a specifically labelled bucket.

For what concerns communication technologies, most of the aforementioned works contemplate the use of radio technology to communicate application data such as the bin fill levels or other information to a remote management server. Often, a GSM module is used [23–25], although recent works explored the possibility of using other types of communication such as LoRa/LoRaWAN [42,43].

This paper proposes a complete solution for waste management that comprises most of the features encountered in the recent literature. The proposed Smart Waste Bin offers accurate waste classification through a hybrid scalar/visual sensor system, as well as automatic waste segregation with an innovative dual-motor setup and waste level tracking. In addition, the entire system makes efficient use of 5G connectivity and the availability of MEC technology to increase recycling rates while providing reduced operation costs, response time and energy consumption.

3. Building the Smart Bin

3.1. Requirement Analysis

Before designing the system, we conducted an analysis to understand (i) how people interact with the traditional waste bins currently available in the campus premises and (ii) what is the composition of the waste produced, two pieces of information that are key for building an effective yet user-friendly prototype. The analysis was conducted in the Bovisa Campus of Politecnico di Milano university, which hosts departments and classrooms for both the Engineering and Design schools and hosts roughly 10,000 people considering students, faculty and administrative staff. For one week, we filmed the behaviour of people during the lunchtime break (12:30–13:30), collecting statistics on the type of trash produced as well as studying the behaviour of each person when handling the trash in front of the existing waste bins. The area analysed is an area generally used by students for consuming lunch. Two trash collecting points are present in the area, both equipped with four coloured bins collecting different types of trash according to the regulation of the municipality of Milan (paper, plastic/aluminium, glass, unsorted trash) (Figure 1). We observed that the most recurring behavior of a person after lunch is to collect all pieces of trash, move to one of the waste collecting points and then manually sorting all pieces of trash in the correct bins, one at a time. Another observed behavior consists of throwing all the different pieces of trash in the unsorted bin. Although such a latter behavior happens less frequently, it is detrimental for recycling purposes. In total, we analysed about 400 interactions between humans and trash bins: the average amount of time spent by the first group of users, the ones sorting the trash in the correct bins, is 5.3 s. The composition of the waste produced is observed as it follows: 24% plastic/aluminium, 22% paper, 2% paper and 52% residual waste (unsorted). Such percentages are in line with other studies conducted in university campuses [18]. Based on such observations, we designed a Smart Waste Bin able to (i) accurately classify and segregate trash while requiring minimal effort to the users and (ii) keep the required interaction time below the average observed during the requirement analysis. The realized bin is the central element of a more general Smart Waste Management System, illustrated in Figure 2, and detailed in the following Sections.

(**a**) Correct waste disposal   (**b**) Incorrect waste disposal

Figure 1. Two frames of the recordings used for analysing the student's behaviour. The average interaction time is estimated from the video.

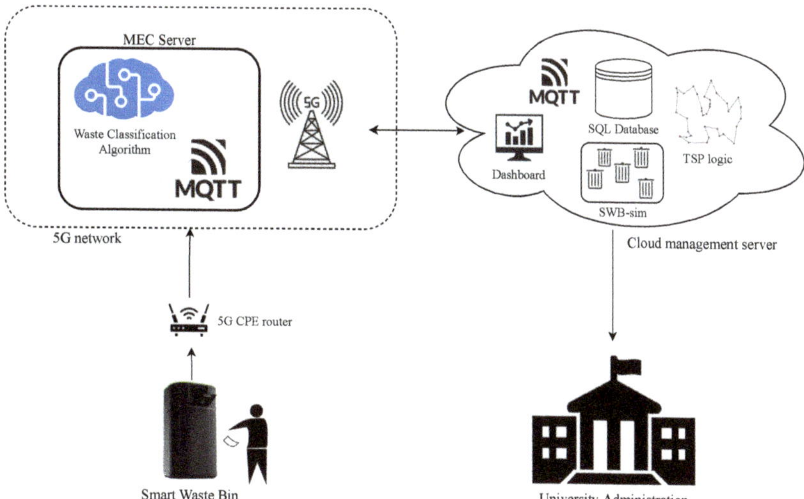

Figure 2. Overview of the Smart Waste Management System: the Smart Waste Bin leverages 5G MEC architecture to accurately classify trash in order to automatically segregate it. Usage data from the smart waste bin is also transmitted to a central backend server, which allows the university administration to provide optimized waste management.

3.2. Prototype Design

The proposed Smart Waste Bin (SWB) is composed of a unique solid body measuring 90 cm in height, with a diameter of 62 cm. The external body is digitally fabricated with a large size FDM (Fused Deposition Modeling) 3D printer, using a thermoplastic material. A washable protective varnish is applied on the whole exterior prior to the final colouring process. Figure 3 shows a digital render of the prototype, while Figure 4 shows the final realized version. The 3D-printed body hides an aluminium structure, which gives solidity to the entire prototype and is used for supporting all the hardware and the electronics needed, as well as the Garbage Unit (GU). The GU contains four circular aluminium structures, used for holding four standard 110 L bags for collecting glass, paper, plastic/metal and residual waste. We opted to maintain the same type of garbage bags already used for traditional bins in the campus for all type of waste, although the requirement analysis clearly showed different usages among the four different type of waste, in order not to modify the supplying operations of the waste management service of the campus and facilitate a transition between already existing bins and smart bins. To ease the tasks of garbage bag replacement, cleaning and other maintenance activities, the entire GU can be easily opened through sliding guides placed at the bottom of the SWB (as shown in Figure 2, right).

A convenient flap door is placed on the front side of the Smart Bin, easily accessible through a metal handle mounted on its top. The door embeds a LED matrix, covered with a laser-cut semi-transparent plastic material, which is used to signal if the SWB is correctly functioning (with a green arrow) or not (with a red cross). In the latter case, an automatic lock avoids opening the door. In normal conditions, opening the door reveals the Waste Disposal Unit (WDU), where objects to be thrown away are deposited and eventually recognized, one at a time. The user deposits a piece of waste in the WDU, which contains a rotating circular shelf with an aperture surrounded by a semicircular structure connected to a couple of servo motors (Figure 5). This area is used for taking measurements from the piece of trash using a hybrid scalar/visual sensor system, which are subsequently fed to a waste classification algorithm (detailed in Section 4). After the waste is recognized in one of the four trash classes, it is automatically moved in the proper bag, thanks to the servo motors.

The data acquired by such sensors is then passed to the waste classification algorithm, which is detailed in Section 4.

3.3.2. Automatic Waste Segregation

Trash segregation is obtained through a pair of servo motors, which allow for precision control of the movement and rotors position. The two motors are located in the central spindle of the bin, one on top of the other and allow to move a piece of trash in the proper bag. One motor controls the plastic shelf rotation in the WDU, while the second is attached to the semicircular structure. Disposal of a piece of waste happens in two steps: first, the shelf and the semicircular structure rotate in the same direction so that the piece of trash is located on top of the right bin. Then, the shelf rotates in the opposite direction so that its aperture let the piece of trash fall into the bin (Figure 5). Both motors are wired to the Raspberry Pi and controlled through the GPIO pins. In order to ensure the correct positioning of the two motors, they are automatic calibrated during every boot of the SWB thanks to specific magnets located on the motors' hardware.

3.3.3. Fill Levels Engine

Each bag in the garbage unit is equipped with a Time-of-Flight (ToF) VL53L0X distance sensor, similar to the one used in the Waste Sensing module. Thanks to a laser, the sensors can accurately measure the distance between the top of the GU (Figure 6 and the garbage inside the correspondent bin bag, providing the estimated fill level of each trash bag. Then, the fill level is used as user feedback displaying the percentage level on the upper surface of the smart waste bin through LED strips, and transmitted to a remote server for advanced functionalities and management purposes.

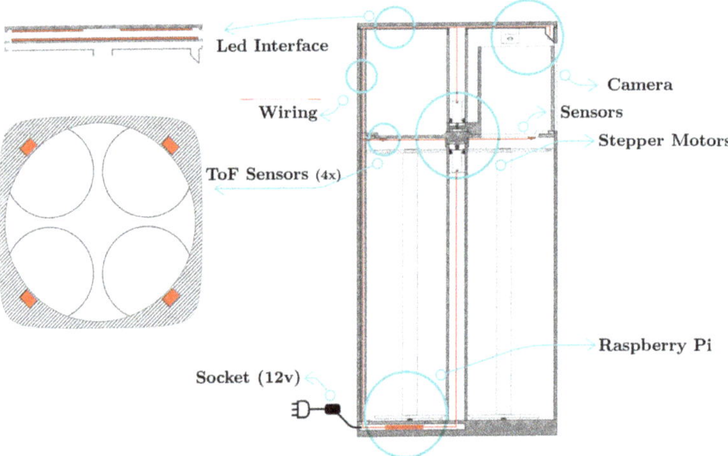

Figure 6. SWB vertical section and internal details.

3.4. Standard Operating Procedure

Figure 7 illustrates the functional flow diagram of the smart waste bin. Upon activation, the smart waste bin performs the following operations:

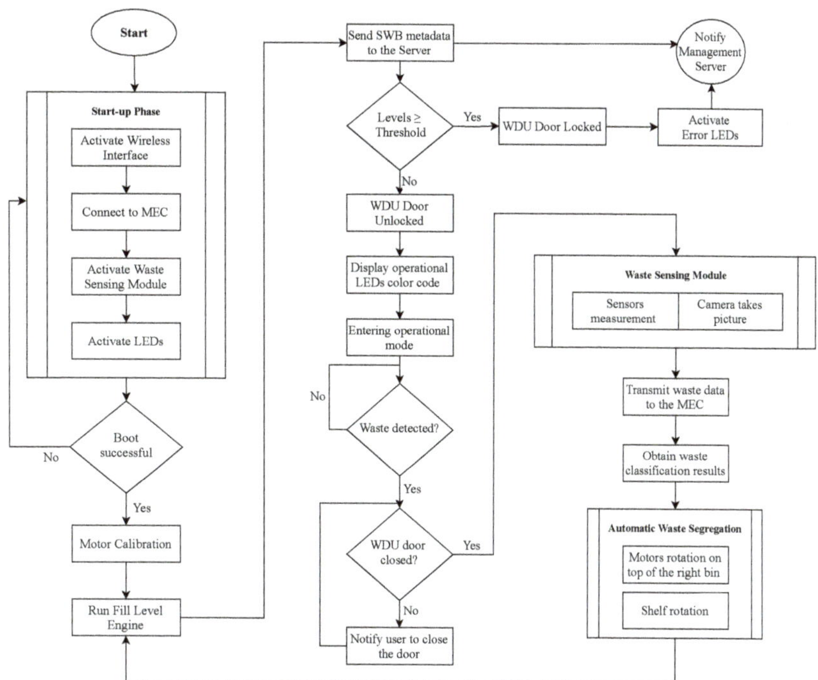

Figure 7. Functional flow diagram of the smart waste bin.

1. *Start-up routine*: during this phase, the firmware executes a series of checks for all the sensors and actuators, as well as for the wireless connectivity with an external server. If all the checks are passed, the SWB can be started; otherwise, the flap door is locked and all LEDs are turned to red color.

2. *Motors synchronization and calibration*: after the start-up, the motors that control the automatic waste segregation need to align with the waste disposal unit to ensure correct disposal of trash. Such regulation allows the motors to set their starting position and subsequently compute the positions (in terms of degrees of rotation) of the four trash bags. For this purpose, the motors perform one complete 360 degrees start-up spin: we use a magnet and a Hall effect sensor to mark the starting position of both the shelf and the semicircular structure, thus calibrating the system.

3. *Fill Levels Engine*: after the controlling operations, the SWB verify that it has enough room to store new pieces of trash, using the Fill Levels Engine. The current level of each bag is estimated and transmitted via MQTT to an external server. The topic used for such signalling is `smartbin/swb_id/fle/material` where `swb_id` and `material` are the strings controlling the SWB identifier and the waste material corresponding to the sensed bag. If the levels exceed a specified threshold (75% in our case), the waste management administrator is promptly notified, in order to empty the bin before it can saturate the bag size. Moreover, when one or more bags fill levels reach 100% of the capacity, the bin activation is interrupted, the door is locked, and all LEDs are turned red, waiting for an operator to take action.

When all the operations above are completed, the smart waste bin enters the idle state and the LEDs on the top (as in Figure 4a) become green, indicating it is ready to accept recycling items. The operations are as follows:

4. *Waste insertion*: when the SWB is active, a user willing to throw a piece of trash can open the lid of the waste disposal unit, insert an object on the shelf as in Figure 5(1) and, finally, close the lid to activate the classification process.

5. *WSM activation*: upon the closure of the WDU, the Smart Waste Bin, thanks to the Time-of-Flight sensors, detects that an object is ready to be analysed. At this point, in order to avoid any interference from outside, the SWB securely closes the lid and activates the waste sensing module, gathering measurements from the sensor as well as taking an image with the installed camera.

6. *Waste Classification and Segregation*: the sensed data is passed to the waste classification algorithm (Section 4), which returns, as a result, the estimated type of the piece of trash. Finally, the motors are activated and the object is moved in its correspondent bag.

7. *Release*: after the object is disposed correctly, the motors come back to the starting position, the fill levels engine is again activated, and the SWB is ready for another operation.

4. Waste Classification Algorithm

The smart waste bin implements a hybrid waste classification algorithm that leverages data from both the scalar sensors and the camera installed in the WDU to distinguish the specific type of waste inserted. We train the algorithm to distinguish among four different classes according to the rules of the municipality of Milan: glass, paper, plastic/metal, and unsorted.

4.1. Dataset

We created a dataset for training the waste classifier, collecting the most frequent waste items found in our university campus' bins and surveying the students about the most common garbage objects thrown into the trash. We collected about 65 different waste items, which were inserted into the smart waste bin for data collection. Since waste objects are not always in their pristine forms when being thrown away but are often dirty, distorted, torn, or crumpled, each item was inserted multiple times into the SWB. Each time, we changed the position of the object inside the WDU as well as applied physical deformations to modify its shape. From the initial 65 items, we collected 3125 data observations, each one composed of one image acquired by the camera sensor as well as a vector of measurements collected by the other scalar sensors. Finally, we grouped objects of the same type together: as an example, all different observations of beverages in aluminium cans (e.g., Coke, Fanta Orange, Red Bull) are grouped in the class `metal can`. After this operation, the final dataset is composed of 40 classes, each one with roughly 80 observations. As a last step, each item in the dataset is labelled with one of the five classes of trash: glass, paper, plastic, metal, and unsorted. We obtained 7 objects in the glass class, 9 in the paper class, 13 for the plastic class, 4 for the metal class and 7 for the unsorted class (see Table 1 for a complete list). Figure 8 shows a sample of the pictures used for the training dataset taken by the bin's camera, while Table 2 reports the summary statistics for the data gathered by the scalar sensors, divided by class. As one can see from Table 2, the scalar sensors allow capturing some characteristics of specific materials such as the conductivity of metals or the different transparency between paper and plastic. The complete dataset is made publicly available at https://tinyurl.com/SWB-dataset (accessed on 22 May 2021).

Table 1. List of objects contained in each waste class of the dataset.

Glass	Coke bottle, Beck's Beer, Aperol Bottle Heineken Beer, Jar, Red Beer, Water Bottle
Paper	Business Card, Candy Box, Cup, Flyers Paper Bag, Juice Box, Magazine, Paper Napkins, Newspaper
Plastic	Blue Bottle, White Bottle, Green Bottle Coffee Capsule Packet, Transparent Glass White Dish, Green Dish, Red Dish Cutlery, Tea bottle, Fiesta Snack, Yogurt Cup, Plastic Bag
Metal	Aluminium can, Metal Box, Aluminium Foil, Jar Lid
Unsorted	Backing Paper, Bic Pen, CD, Cigarettes, Lighter, Marker, Receipt

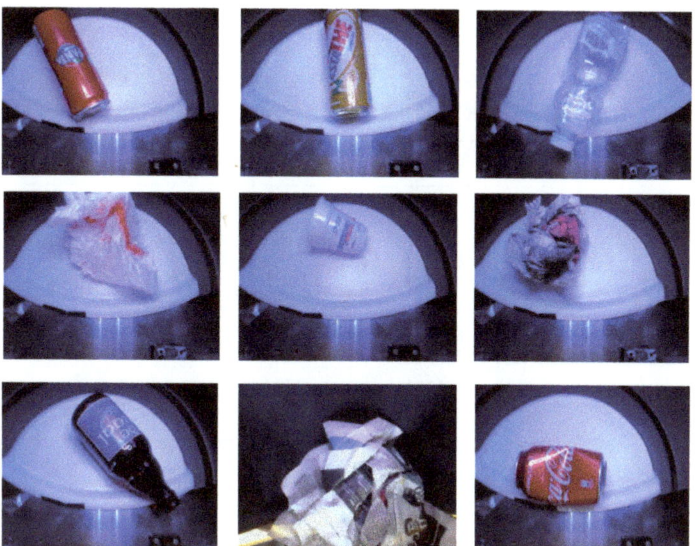

Figure 8. A sample of the pictures used as training dataset. The objects were acquired by the waste sensing module directly on the white shelf of the waste disposal unit.

Table 2. Summary table reporting the per class average and standard deviation obtained by the Inductive Sensor (IS), Capacitive Sensor (CS) and Photoelectric sensors (PS).

	IS	CS	PS
Glass	0 ± 0.0	0.96 ± 0.23	14.28 ± 8.9
Paper	0 ± 0.0	0.12 ± 0.11	0.68 ± 0.6
Plastic	0 ± 0.0	0.12 ± 0.12	17.00 ± 8.7
Metal	0.93 ± 0.25	0.98 ± 0.12	4.35 ± 2.3
Unsorted	0 ± 0.0	0.18 ± 0.13	7.12 ± 3.2

4.2. Waste Classification

We observe that the type of data returned by the two different types of sensors is very different: the inductive and capacitive sensors return a binary value, the photoelectric sensors return a real value and the camera produces an image. In the following, we will first derive two different classification models, leveraging either the scalar sensor data or the images from the camera. Then, we will explore two strategies to effectively fuse all this information in a single classification algorithm, which differ in terms of *where* data integration happens: at learning time or at prediction time.

4.2.1. Classification from Scalar Data

As a first step, we trained a classifier to leverage the data retrieved by the scalar sensors only. As a preprocessing step, each sensor data was normalized in order to have zero mean and unit variance. We split the available data into train and test subsets, according to stratified k-fold cross-validation with $k = 5$. The training data was given as input to a logistic regression classifier, using as labels the object materials. The performance of the resulting 5-class model is evaluated on the test folds, and we report in Table 3 the results obtained in the form of a confusion matrix, considering all test folds. As one can see from the Table, waste classification starting from the scalar sensors only allows to already reach a good starting point, with an average accuracy of about 89.6%. Some classes have very high recognition accuracy: indeed, the glass, metal and paper classes are recognized with accuracy higher than 95% given the unique property of the materials and the way they interact with the available sensors (i.e., inductive and capacitive sensors).

Table 3. Test confusion matrix obtained for sensor-based classification.

		Predicted Class				
		Glass	Paper	Plastic	Metal	Unsorted
	Glass	530 (97%)	6 (1%)	2 (<1%)		8 (1%)
True	Paper		697 (99%)			5 (1%)
Class	Plastic	1 (1%)	44 (4%)	846 (83%)		123 (12%)
	Metal	2 (1%)			310 (99%)	
	Unsorted	1 (<1%)	35 (6%)	96 (18%)		414 (76%)

4.2.2. Classification from Images

As a second step, we build an image-based waste classifier. We base our approach on the use of a Convolutional Neural Network (CNN) classifier, thanks to its proven effectiveness in image classification tasks. Training a CNN classifier from scratch, avoiding the issue of overfitting, generally requires a massive amount of training images. Due to the relatively small size of our dataset, we rely on the concept of *transfer learning*: we start from a CNN image classifier pre-trained on the ImageNet dataset [44], and re-train only its last layers on our dataset in order to specialize it to the task of classifying trash. Since each CNN layer learns filters of increasing complexity, the earlier layers learn to detect basic features such as edges, corners, textures whilst later layers detect patterns, object parts, tags, and the final layers detect objects. Therefore, fine-tuning the last layers on our dataset while keeping the previous layers enables us to reach an accurate model without needing a huge image dataset as input. Several pre-trained CNN models, differing in structure (number of layers, number of neurons per layer, etc.) are already available: in order to select the one that best fits our purposes, we performed fine-tuning and studied the resulting model accuracy as well as Single Forward Pass (SFP) time (that is, the time it takes for the CNN to process an image and return the classification result). The following CNN models were considered for comparison: NASNet-A-Mobile, MobileNet-v2, MobileNet-v3-large, MobileNet-v3-small, ResNet-18, ResNet-34, ResNet-50, ResNet-101, GoogleNet, ShuffeNetv2-1.0, SqueezeNet-v1.1 and Inception-v3. Due to the large variability of the appearance of objects inside each waste material class, tests were performed in the following way: we first split the dataset in train and test folds according to the same 5-fold cross-validation procedure used for the scalar sensor-based classifier. This time, however, we trained the CNNs using the object labels rather than the material labels. At inference time, we mapped back each object to its material class. All tests were performed on an Intel Core i7-6700HQ CPU, equipped with a NVIDIA GeForce GTX 950M GPU, and 16 GB RAM. The accuracies obtained are illustrated in Figure 9: we select the ResNet-18 model as the best compromise between accuracy and SFP time. Table 4 shows the confusion matrix obtained with the fine-tuned ResNet-18 model. As one can see, the average accuracy hits about 93%, with no material class having accuracy higher than 95%.

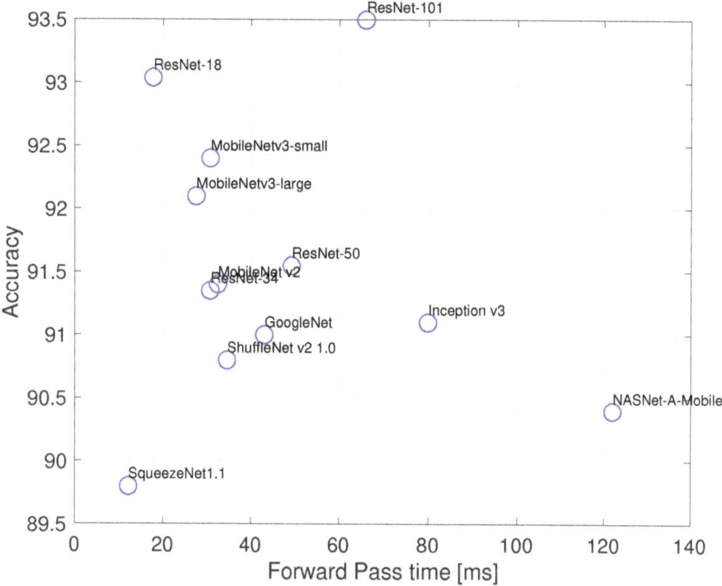

Figure 9. Material class accuracy vs. mean forward pass time.

Table 4. Test confusion matrix obtained for image-based classification.

		Predicted Class				
		Glass	**Paper**	**Plastic**	**Metal**	**Unsorted**
	Glass	486 (89%)		39 (7%)		21 (4%)
True	**Paper**	21 (3%)	653 (93%)	22 (3%)		6 (1%)
Class	**Plastic**	40 (4%)		963 (95%)		11 (1%)
	Metal	3 (1%)	4 (1%)		283 (91%)	22 (7%)
	Unsorted	11 (2%)	4 (1%)	6 (1%)	7 (1%)	518 (95%)

4.2.3. Hybrid Classification

Looking at the results obtained classifying waste with scalar sensor or image data, it is clear that each method has its pros and cons. Scalar sensor data outperforms image-based classification for some materials (e.g., metal), while image-based classification obtains similar results for each class. In the following, we propose two different strategies to exploit the best features of the two different approaches.

1. *Integration at prediction time:* a first approach consists of running the two classifiers in parallel and then taking a decision considering the lowest (training) classification error (Figure 10). Let y be the output of the two classifiers, taking qualitative value C = [glass, paper, plastic, metal, unsorted],, i.e., the output class. For each classifier, we compute the a posteriori misclassification error probability $P(x \neq C|y = C)$, being x the true class. To do this, we use the Bayes' theorem:

$$P(x \neq C|y = C) = \frac{P(y = C|x \neq C)P(y = C)}{P(x \neq C)},\qquad(1)$$

where $P(y = C|x \neq C)$, $P(y = C)$ and $P(x \neq C)$ are the likelihood of misclassification for class C, the prior output and the prior class probabilities, respectively. We estimated such quantities from the (training) confusion matrix of each classifier. In case the two classifiers agree on the output class, the method obviously returns the same class C; in case the two classifiers disagree, the class C having the lowest misclassification error is selected.

As an example, let y_s = plastic and y_i = glass be the output of the sensor-based and image-based classifiers, respectively. Assuming the values contained in Table 3 and 4 as the learnt probabilities during training we have:

$$P(x \neq \text{plastic}|y_s = \text{plastic}) = \frac{0.1038 \times 0.303}{0.675} = 4.65\%,\qquad(2)$$

while

$$P(x \neq \text{glass}|y_i = \text{glass}) = \frac{0.1337 \times 0.179}{0.825} = 2.9\%;\qquad(3)$$

The system will therefore select y_i as final class, since its associated error is lower.

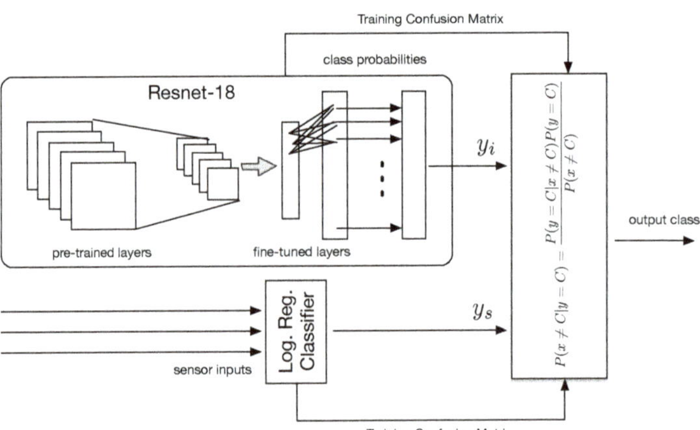

Figure 10. Integration at prediction time.

2. *Integration at learning time:* A second approach is to train a new classifier, where input features come from all available sensors. To do this, we note that the last layer of the fine-tuned CNN consists of 40 nodes, where each node outputs a value between 0 and 1 that represents the probability that the input image belongs to one of the 40 object classes. We treat such values as new features, which are fed to a regularized logistic regression classifier together with the scalar sensor measurements (Figure 11. The classifier is again trained according to k-fold cross-validation using as ground truth labels the waste materials.

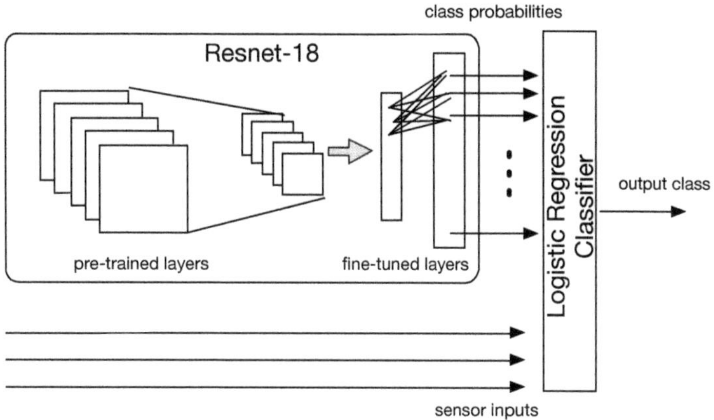

Figure 11. Integration at learning time.

The results obtained on the test set for the two strategies are contained in Tables 5 and 6, for the integration at prediction and learning cases, respectively. As one can see, both approaches allow to increase performance compared to solely using the scalar sensor-based or image-based approaches. In particular, integration at prediction time allows obtaining an accuracy of 96.12%, while the best result is obtained with the integration at learning time approach (97.37%). This is particularly promising, especially to cope with specific waste objects such as that of shattered glass. In this case (fortunately rare, according to our survey) relying solely on an image-based recognition would be very difficult given the high variance associated with images of glass fragments. Indeed, using also scalar sensors in the system may greatly improve the recognition accuracy.

Table 5. Test confusion matrix obtained for hybrid classification with integration at prediction.

		Predicted Class				
		Glass	Paper	Plastic	Metal	Unsorted
	Glass	535 (98%)	3 (<1%)	2 (<1%)	0	6 (<1%)
True	**Paper**	2 (<1%)	700 (99%)			
Class	**Plastic**	7 (1%)	20 (2%)	951 (94%)		36 (3%)
	Metal	2 (<1%)			310 (99%)	
	Unsorted	11 (2%)	9 (2%)	11 (2%)	12 (2%)	503 (92%)

Table 6. Test confusion matrix obtained for hybrid classification with integration at learning.

		Predicted Class				
		Glass	Paper	Plastic	Metal	Unsorted
	Glass	542 (99%)		2 (<1%)		2 (<1%)
True	**Paper**		701 (99%)			1 (<1%)
Class	**Plastic**	1 (<1%)	14 (1%)	963 (95%)		36 (3%)
	Metal				312 (100%)	
	Unsorted	6 (1%)	7 (1%)	13 (3%)		520 (95%)

4.3. Waste Classifier Location

The hybrid model with integration at learning time has been exported for being tested in three different scenarios, differing in where the classification takes place.

1. *Local recognition:* first, we run the classifier on the Raspberry PI controlling the SWB. In this case, the SWB does not require any connection to an external server as all decisions are taken locally.
2. *Cloud-based recognition:* as a second test, we move the classifier on a cloud-based server hosted on Amazon Web Services EC2, located in Ireland. Data gathered from the WSM is transmitted to a listening process on the server: upon reception, the classifier is run and the response is transmitted back to the SWB. We used SCP to transfer data from the SWB to the server, while the MQTT protocol was used to reply from the server to the SWB.
3. *MEC-based recognition:* finally, we move the classifier on a multi-access edge computing server, provided by Vodafone Italia S.p.A, located in the core Vodafone network in Milan and running an Ubuntu Server machine with the same characteristics of the AWS EC2 instance. Access to the MEC is enabled by using the 5G connection through the Huawei 5G CPE router, which allows for a low-latency and high-bandwidth connection.

For every scenario, we tested the total recognition time of a waste item and the overall energy consumption of the SWB.

4.3.1. Recognition Time

The total recognition time is composed of CNN execution time and the picture transfer time from the Raspberry Pi to the server. (Image acquisition time is assumed constant and thus discarded.) For the local scenario, since the picture is processed internally on the Raspberry Pi, the total time equals the execution time of the CNN, which is around 3 s. For the cloud and MEC server scenarios, the total time also includes the transfer time of the picture from the Bin to the server. In these cases, the Image Acquisition Module of the Raspberry Pi takes a picture of the trash, sends it to the cloud or MEC server using Secure Copy Protocol (SCP); then, the server feeds the picture to the CNN, and the resulting label along with the confidence level is sent back to the SWB as an MQTT publish message. The time measurement summary is given in Table 7.

Table 7. Total waste recognition time.

	Local	Cloud Server	MEC Server
Avg. Data Transfer Time (ms)	-	343.3	191.3
Avg. Classification Time (ms)	3159.2	123.9	123.9
Avg. Total Recognition Time (ms)	3159.2	467.1	315.2

As one can see, the total time on the Raspberry Pi is 5–6 times longer than the others taking over 3 s due to the low computational power available. Since the cloud and the MEC server have equivalent hardware specifications, the CNN recognition time is identical on the two machines. However, as the MEC server is located closer to the Smart Waste Bin compared to the cloud server, the data transfer time is greatly reduced. For this reason, we can see a clear improvement for the MEC approach in the Average Total Time. In any case, note that the total time is well below the average time of 5.3 s spent with the traditional bins and estimated from the requirement analysis. This means that the use of the SWB speeds up an average interaction with a human, also reducing waste misplacement.

4.3.2. Energy Consumption

To measure the energy consumption of the SWB, we used an Adafruit INA219 High Side DC Current Sensor wired to an Arduino Uno and connected in series to the Raspberry

Pi of the SWB. The method calculates the integral of power over the execution time, i.e., the sum of instant power samples taken by the sensor unit, illustrated in Figure 12. Even though the unit continuously takes measurements, since the samples are discrete, the exact energy consumption is not measured but estimated. As one can see in Table 8, the energy consumption reflects the total recognition time. In particular, using the MEC, we can save up to 15% of energy compared to the cloud version.

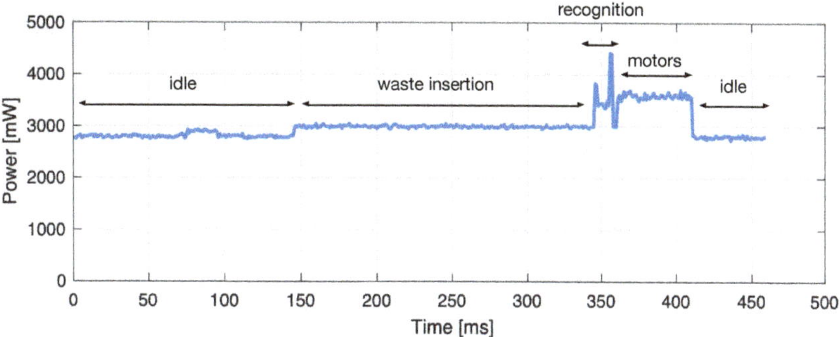

Figure 12. Energy measurement during one normal operation cycle.

Table 8. Energy consumption of the Raspberry Pi when the waste classifier is run locally, on the MEC or on the cloud server.

	Local	Cloud Server	MEC Server
Bin Energy Consumption (J/object)	11.69	1.28	0.93

5. Management Application

The smart waste bin collects not only data relative to the waste classification but also a multitude of heterogeneous information such as time and frequency usage, bag filling levels, emptying time. Such additional metadata may be of enormous value for optimizing the waste collection task in a university campus, as well as larger scenarios such as a city. For these reasons, all the information collected by the bin (working status, filling level for each waste class, etc.) are periodically transmitted to a management server, hosted remotely, which stores the data for advanced uses. In the following, we provide a brief description of such a management server: to fully test the functionalities offered, we also provide a Smart Waste Bin simulator (SWB-sim), which allows simulating a multitude of SWB instances, therefore, providing enough data.

5.1. Smart Waste Bin Simulator (SWB-Sim)

To overcome the practical issues of physically realizing multiple prototypes, we propose a simulator that virtually creates thousands of bins with different usage profiles, such as frequency of interaction with people and distribution of waste produced. The simulation software is written in Python and replicates an arbitrary number of Smart Waste Bin devices in a simulated environment with adjustable parameters. At the program start-up, a user-specified number of Smart Waste Bins devices are simulated, placed in an area of interest either randomly or in specific positions. Then, the simulation system runs the engine for the process of waste generation. We leveraged Python capabilities to generate a discrete-time simulation scenario that can either run in real-time or in a speed-up fashion. Each bin's waste level at a certain time is modelled as a normal distribution, according to [45]. Indeed, the amount of waste deposited by each person in a bin can be represented as a stochastic variable. Therefore, according to the central limit theorem, the sum of many stochastic variables of arbitrary probability distributions approaches a normal distribution. The simulator allows to use five template distributions for each bin, according to different usage profiles from very low to very high usages, which in turn control the mean and variance of the associated normal distribution. Moreover, in order to adhere to real-world constraints, the simulator takes into consideration specific environment characteristics such as university closing time, holidays, and most expected waste type. Periodically, the data generated by each bin in the simulator is transmitted to a remote server via MQTT, using the same message format as the prototype. The smart waste bin simulation system is available at: https://tinyurl.com/SWB-sim (accessed on 22 May 2021).

5.2. Management Server

All data produced and transmitted by the SWB, real or simulated, are received by a server application running on a public server. The main tasks of the management server are the following:

1. *Data storage*: the server runs an MQTT broker that accepts messages from the smart waste bins. Each module publishes messages on specific MQTT topics: for example, in a scenario with two SWBs named `swb1` and `swb2`, the topic `smartbin/swb1/fle/glass` is used for publishing messages of the glass bag's filling levels; while the topic `smartbin/swb2/daily` is used for communicating the daily usage summary of the recycling bin as a `Json` file. Upon reception of a message, the server reads its content and saves the received information in a local SQL database.
2. *Data visualization*: the server also provides a web-based dashboard for data visualization and monitoring purposes. The dashboard is implemented with Node-RED, a framework built on top of Node.js that has recently become very popular in IoT application development. As shown in Figure 13, the dashboard shows aggregated information for each smart waste bin connected to the system: (i) on the top part, the fill levels for the four materials with their daily correspondent trend represented in a chart; and (ii) in the bottom part, a map summarizing the status of all the SWBs present in an area, with different colours according to the overall fill level of each bin, allowing to easily keep track of the status of the bin from the landfill operators.
3. *TSP for waste collection*: The management server also allows to calculate an optimized route for the operator in charge of the waste collection. The task is faced as a *Travelling Salesman Problem (TSP)*. In particular, the goal is to minimise the travelling time starting and finishing at a specific node (e.g., the landfill site) after visiting each other node exactly once. In particular, the nodes are represented by the bins and the weight on the links is the travel time of a specific road. Moreover, to avoid useless stops at an empty recycling bin, the SWB with a filling level lower than 75% of the total are automatically excluded by the TSP problem.

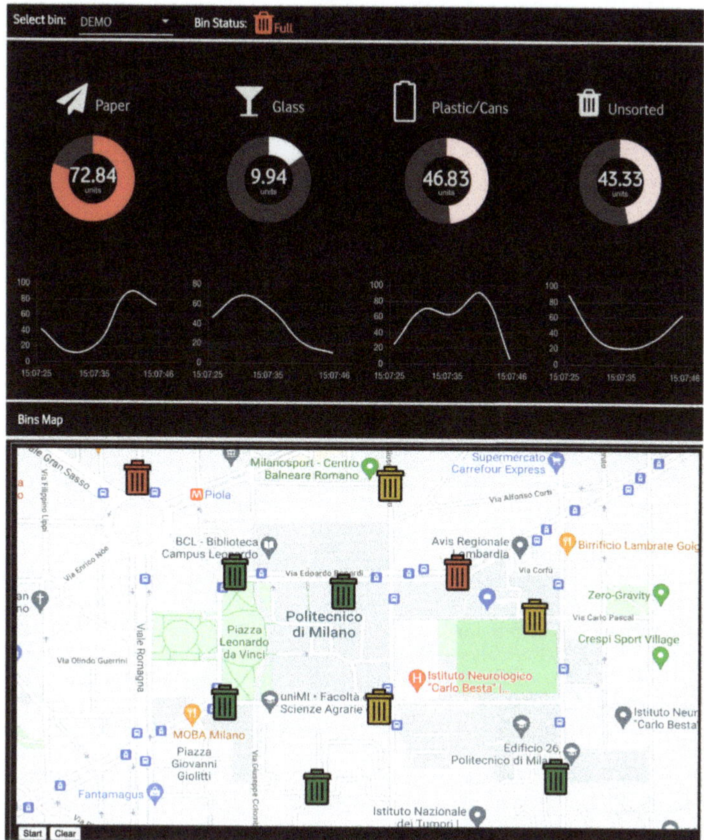

Figure 13. Smart waste bin backend dashboard implemented with Node-RED. The map shows the bins' position as well as a graphical summary of the fill levels. Respectively, red when at least one class is greater than 70%; yellow when at least one class is greater than 40% and green when all the classes are below or equal 40%.

6. Conclusions

We have presented the design and implementation of a waste management system for Smart Campuses. The system is based on a smart waste bin prototype, an innovative device that can be used for automatically recycling objects using a hybrid sensor/image-based classifier. Results showed that the proposed approach reached an accuracy of over 97% for waste classification. In addition, we evaluated the device in different network scenarios, including moving the artificial intelligence on the MEC of the 5G network, reducing the recognition latency and the energy consumption. Moreover, we presented an application server which is able to easily monitor the status of waste bins present in the campus, as well as optimizing the management procedures (e.g., waste collection). We believe such a system will be extremely useful in the near future, considering the increased environmental impact of waste generated by people, which requires correct recycling. For this reason, we plan to create many other smart waste bin devices and deploy them on the university premises. This will also enable the possibility to study the interaction between students and smart waste bins, paving the way for possible future system optimizations.

Author Contributions: Conceptualization, A.E.C.R., M.B. and P.B.; methodology, A.E.C.R., M.B. and P.B., software, E.L., F.A.S.; validation, E.L. and F.A.S.; writing—original draft preparation, E.L. and A.E.C.R.; writing—review and editing, P.B.; supervision, A.E.C.R. and M.B.; project administration, S.M. All authors have read and agreed to the published version of the manuscript.

Funding: This research work has been supported by project BASE5G (project id: 1155850) funded by Regione Lombardia within the framework POR FESR 2014-2020 and by Vodafone Italia S.p.A under the Italian 5G trial use case *Smart City and Smart Campus*.

Data Availability Statement: The waste dataset and the Smart Waste Bin simulator used in this work are available at https://tinyurl.com/SWB-dataset and https://tinyurl.com/SWB-sim, (accessed on 22 May 2021).

Acknowledgments: The authors would like to thank Angelo De Iesi, for their invaluable help in the product design of the Smart Waste Bin, and Sabrina Baggioni, Gianpiero Carocci, Giacoma Caruso, Andrea Ferrara and Stefano Bauro from Vodafone Italia S.p.A. for their fruitful ideas.

Conflicts of Interest: The authors declare no conflict of interest.

References

1. Min-Allah, N.; Alrashed, S. Smart campus—A sketch. *Sustain. Cities Soc.* **2020**, *59*, 102231. [CrossRef] [PubMed]
2. Abuarqoub, A.; Abusaimeh, H.; Hammoudeh, M.; Uliyan, D.; Abu-Hashem, M.A.; Murad, S.; Al-Jarrah, M.; Al-Fayez, F. A survey on internet of things enabled smart campus applications. In Proceedings of the International Conference on Future Networks and Distributed Systems, Cambridge, UK, 19–20 July 2017; pp. 1–7.
3. Mohandes, M.; Deriche, M.; Abuelma'atti, M.T.; Tasadduq, N. Preference-based smart parking system in a university campus. *IET Intell. Transp. Syst.* **2019**, *13*, 417–423. [CrossRef]
4. Nagowah, S.D.; Sta, H.B.; Gobin-Rahimbux, B.A. An ontology for an IoT-enabled smart parking in a university campus. In Proceedings of the 2019 IEEE International Smart Cities Conference (ISC2), Casablanca, Morocco, 14–17 October 2019; pp. 474–479.
5. Alrashed, S. Key performance indicators for Smart Campus and Microgrid. *Sustain. Cities Soc.* **2020**, *60*, 102264. [CrossRef]
6. Antevski, K.; Redondi, A.E.; Pitic, R. A hybrid BLE and Wi-Fi localization system for the creation of study groups in smart libraries. In Proceedings of the 2016 9th IFIP Wireless and Mobile Networking Conference (WMNC), Colmar, France, 11–13 July 2016; pp. 41–48.
7. Chan, H.C.; Chan, L. Smart library and smart campus. *J. Serv. Sci. Manag.* **2018**, *11*, 543–564. [CrossRef]
8. Longo, E.; Redondi, A.E.; Bianchini, M.; Bolzan, P.; Maffei, S. Smart gate: A modular system for occupancy and environmental monitoring of spaces. In Proceedings of the 2020 5th International Conference on Smart and Sustainable Technologies (SpliTech), Split, Croatia, 1–4 July 2020; pp. 1–6.
9. Tse, R.; Mirri, S.; Tang, S.K.; Pau, G.; Salomoni, P. Modelling and Visualizing People Flow in Smart Buildings: A Case Study in a University Campus. In Proceedings of the Conference on Information Technology for Social Good, Roma, Italy, 9–11 September 2021; pp. 309–312.
10. Villegas-Ch, W.; Palacios-Pacheco, X.; Luján-Mora, S. Application of a smart city model to a traditional university campus with a big data architecture: A sustainable smart campus. *Sustainability* **2019**, *11*, 2857. [CrossRef]
11. Ceccarini, C.; Mirri, S.; Salomoni, P.; Prandi, C. On exploiting Data Visualization and IoT for Increasing Sustainability and Safety in a Smart Campus. *Mob. Netw. Appl.* **2021**, *26*, 2066–2075. [CrossRef]
12. Smyth, D.P.; Fredeen, A.L.; Booth, A.L. Reducing solid waste in higher education: The first step towards 'greening'a university campus. *Resour. Conserv. Recycl.* **2010**, *54*, 1007–1016. [CrossRef]
13. de Vega, C.A.; Ojeda-Benitez, S.; Ramirez-Barreto, M.E. Mexican educational institutions and waste management programmes: A University case study. *Resour. Conserv. Recycl.* **2003**, *39*, 283–296.
14. Taghizadeh, S.; Ghassemzadeh, H.R.; Vahed, M.M.; Fellegari, R. Solid waste characterization and management within university campuses case study: University of Tabriz. *Elixir Pollut.* **2012**, *43*, 6650–6654.
15. Okeniyi, J.O.; Anwan, E.U. Solid wastes generation in Covenant University, Ota, Nigeria: Characterisation and implication for sustainable waste management. *J. Mater. Environ. Sci* **2012**, *3*, 419–424.
16. Eurostat Statistics Explained. Municipal Waste Statistics. Available online: https://ec.europa.eu/eurostat/statistics-explained/index.php?title=Municipal_waste_statistics. (accessed on 22 May 2021)
17. Ruiz Morales, M. Caracterizacion de residuos solidos en la Universidad Iberoamericana, Ciudad de Mexico. *Rev. Int. Contam. Ambient.* **2012**, *28*, 93–97.
18. Gallardo, A.; Edo-Alcón, N.; Carlos, M.; Renau, M. The determination of waste generation and composition as an essential tool to improve the waste management plan of a university. *Waste Manag.* **2016**, *53*, 3–11. [CrossRef]
19. Longo, E.; Sahin, F.A.; Redondi, A.E.; Bolzan, P.; Bianchini, M.; Maffei, S. Take the trash out... to the edge. Creating a Smart Waste Bin based on 5G Multi-access Edge Computing. In Proceedings of the Conference on Information Technology for Social Good, Roma, Italy, 9–11 September 2021; pp. 55–60.

20. Yusof, N.M.; Jidin, A.Z.; Rahim, M.I. Smart garbage monitoring system for waste management. In Proceedings of the MATEC Web of Conferences, Ho Chi Minh City, Vietnam, 5–6 August 2016; Volume 97, p. 01098.
21. Dubey, S.; Singh, M.K.; Singh, P.; Aggarwal, S. Waste Management of Residential Society using Machine Learning and IoT Approach. In Proceedings of the 2020 International Conference on Emerging Smart Computing and Informatics (ESCI), Pune, India, 12–14 March 2020; pp. 293–297.
22. Bano, A.; Ud Din, I.; Al-Huqail, A.A. AIoT-Based Smart Bin for Real-Time Monitoring and Management of Solid Waste. *Sci. Program.* **2020**, *2020*, 6613263. [CrossRef]
23. Aleyadeh, S.; Taha, A.E.M. An IoT-Based architecture for waste management. In Proceedings of the 2018 IEEE International Conference on Communications Workshops (ICC Workshops), Kansas City, MO, USA, 20–24 May 2018; pp. 1–4.
24. Pardini, K.; Rodrigues, J.J.; Diallo, O.; Das, A.K.; de Albuquerque, V.H.C.; Kozlov, S.A. A smart waste management solution geared towards citizens. *Sensors* **2020**, *20*, 2380. [CrossRef]
25. Kaushik, D.; Yadav, S. Multipurpose Street Smart Garbage bin based on Iot. *Int. J. Adv. Res. Comput. Sci.* **2017**, *8*, 1145–1149.
26. Mesina, M.; De Jong, T.; Dalmijn, W. Improvements in separation of non-ferrous scrap metals using an electromagnetic sensor. *Phys. Sep. Sci. Eng.* **1970**, *12*, 87–101. [CrossRef]
27. Rafeeq, M.; Alam, S. Automation of plastic, metal and glass waste materials segregation using arduino in scrap industry. In Proceedings of the 2016 International Conference on Communication and Electronics Systems (ICCES), Piscataway, NJ, USA, 21–22 July 2016; pp. 1–5.
28. Tiwari, C.; Nagarathna, K. Waste management using solar smart bin. In Proceedings of the 2017 International Conference on Energy, Communication, Data Analytics and Soft Computing (ICECDS), Chennai, India, 1–2 August 2017; pp. 1123–1126.
29. Chowdhury, B.; Chowdhury, M.U. RFID-based real-time smart waste management system. In Proceedings of the 2007 Australasian Telecommunication Networks and Applications Conference, Christchurch, New Zealand, 2–5 December 2007; pp. 175–180.
30. Sinha, A.; Couderc, P. Using Owl Ontologies for Selective Waste Sorting and Recycling. In Proceedings of the OWL: Experiences and Directions Workshop 2012, Heraklion, Greece, 27–28 May 2012; Volume 849.
31. Costa, B.S.; Bernardes, A.C.; Pereira, J.V.; Zampa, V.H.; Pereira, V.A.; Matos, G.F.; Soares, E.A.; Soares, C.L.; Silva, A.F. Artificial intelligence in automated sorting in trash recycling. In Proceedings of the Anais do XV Encontro Nacional de Inteligência Artificial e Computacional, São Paulo, Brazil, 22–25 October 2018; pp. 198–205.
32. Aral, R.A.; Keskin, Ş.R.; Kaya, M.; Hacıömeroğlu, M. Classification of trashnet dataset based on deep learning models. In Proceedings of the 2018 IEEE International Conference on Big Data (Big Data), Boston, MA, USA, 11–14 December 2018; pp. 2058–2062.
33. Srinilta, C.; Kanharattanachai, S. Municipal solid waste segregation with CNN. In Proceedings of the 2019 5th International Conference on Engineering, Applied Sciences and Technology (ICEAST), Piscataway, NJ, USA, 2–5 July 2019; pp. 1–4.
34. Shi, C.; Xia, R.; Wang, L. A Novel Multi-Branch Channel Expansion Network for Garbage Image Classification. *IEEE Access* **2020**, *8*, 154436–154452. [CrossRef]
35. Deng, J.; Dong, W.; Socher, R.; Li, L.J.; Li, K.; Fei-Fei, L. Imagenet: A large-scale hierarchical image database. In Proceedings of the 2009 IEEE Conference on Computer Vision and Pattern Recognition, Miami, FL, USA, 20–25 June 2009; pp. 248–255.
36. Krizhevsky, A.; Sutskever, I.; Hinton, G.E. Imagenet classification with deep convolutional neural networks. *Adv. Neural Inf. Process. Syst.* **2012**, *25*, 1097–1105. [CrossRef]
37. Simonyan, K.; Zisserman, A. Very Deep Convolutional Networks for Large-Scale Image Recognition. In Proceedings of the 3rd International Conference on Learning Representations, ICLR 2015, San Diego, CA, USA, 7–9 May 2015.
38. Thung, G. Trashnet GitHub Repository. Available online: https://github.com/garythung/trashnet (accessed on 22 May 2021).
39. Norhafiza, S.; Masiri, K.; Faezah, A.N.; Nadiah, A.N.; Aslila, A. The effectiveness of segregation recyclable materials by automated motorized bin. *J. Adv. Manuf. Technol. (JAMT)* **2018**, *12*, 409–420.
40. Kamarudin, N.H.; Abdullah, N.E.; Halim, I.S.A.; Hassan, S.L.M. Development of automatic waste segregator with monitoring system. In Proceedings of the 2019 4th International Conference on Information Technology, Information Systems and Electrical Engineering (ICITISEE), Piscataway, NJ, USA, 20–21 November 2019; pp. 190–195.
41. Gondal, A.U.; Sadiq, M.I.; Ali, T.; Irfan, M.; Shaf, A.; Aamir, M.; Shoaib, M.; Glowacz, A.; Tadeusiewicz, R.; Kantoch, E. Real Time Multipurpose Smart Waste Classification Model for Efficient Recycling in Smart Cities Using Multilayer Convolutional Neural Network and Perceptron. *Sensors* **2021**, *21*, 4916. [CrossRef]
42. Lozano, Á.; Caridad, J.; De Paz, J.F.; Villarrubia Gonzalez, G.; Bajo, J. Smart waste collection system with low consumption LoRaWAN nodes and route optimization. *Sensors* **2018**, *18*, 1465. [CrossRef]
43. Baldo, D.; Mecocci, A.; Parrino, S.; Peruzzi, G.; Pozzebon, A. A Multi-Layer LoRaWAN Infrastructure for Smart Waste Management. *Sensors* **2021**, *21*, 2600. [CrossRef]
44. Stanford Vision Lab; Stanford University; Princeton University. ImageNet Database Website. Available online: https://www.image-net.org/index.php (accessed on 22 May 2021).
45. Johansson, O.M. The effect of dynamic scheduling and routing in a solid waste management system. *Waste Manag.* **2006**, *26*, 875–885. [CrossRef]

Article

Algorithms for Smooth, Safe and Quick Routing on Sensor-Equipped Grid Networks

Giovanni Andreatta, Carla De Francesco * and Luigi De Giovanni *

Dipartimento di Matematica "Tullio Levi-Civita", Università degli Studi di Padova, 35122 Padova, Italy; giovanni.andreatta@unipd.it
* Correspondence: carla@math.unipd.it (C.D.F.); luigi@math.unipd.it (L.D.G.)

Abstract: Automation plays an important role in modern transportation and handling systems, e.g., to control the routes of aircraft and ground service equipment in airport aprons, automated guided vehicles in port terminals or in public transportation, handling robots in automated factories, drones in warehouse picking operations, etc. Information technology provides hardware and software (e.g., collision detection sensors, routing and collision avoidance logic) that contribute to safe and efficient operations, with relevant social benefits in terms of improved system performance and reduced accident rates. In this context, we address the design of efficient collision-free routes in a minimum-size routing network. We consider a grid and a set of vehicles, each moving from the bottom of the origin column to the top of the destination column. Smooth nonstop paths are required, without collisions nor deviations from shortest paths, and we investigate the minimum number of horizontal lanes allowing for such routing. The problem is known as fleet quickest routing problem on grids. We propose a mathematical formulation solved, for small instances, through standard solvers. For larger instances, we devise heuristics that, based on known combinatorial properties, define priorities, and design collision-free routes. Experiments on random instances show that our algorithms are able to quickly provide good quality solutions.

Keywords: automated transportation network; collision-free routing; grid network; optimization algorithm; integer linear programming; heuristics

Citation: Andreatta, G.; De Francesco, C.; De Giovanni, L. Algorithms for Smooth, Safe and Quick Routing on Sensor-Equipped Grid Networks. *Sensors* **2021**, *21*, 8188. https://doi.org/10.3390/s21248188

Academic Editor: Pietro Manzoni

Received: 15 November 2021
Accepted: 6 December 2021
Published: 8 December 2021

Publisher's Note: MDPI stays neutral with regard to jurisdictional claims in published maps and institutional affiliations.

1. Introduction

Modern transportation and handling systems greatly benefit from information technology (IT) and automation, as demonstrated by the consolidated use of sensor-equipped transport networks, automated guided vehicles (AGVs), self-moving robots, as well as the growing adoption of drones, in many industrial, logistic and public transportation environments. Typical examples can be found in railway transportation systems, where optic or acoustic sensors on trains and tracks, integrated by collision detection and avoidance logic, support safe and efficient operations. IT also supports taxiways operations in airports airside [1], where aircraft, passenger buses as well as many ground service vehicles (like baggage dollies, passenger steps, tow-tractors, follow-me cars, etc.) run intersecting routes between the boarding gates and the runways, and the risk of collisions or deadlocks has to be constantly monitored. Another application in logistic networks involves the use of AGVs in port terminals [2] to transport containers from the berths on the quay along the shoreline to dockside stacks and land access points. In a similar way, automated warehouses or factories adopt vehicles (like AGVs or drones) to transfer goods or materials from the depot shelves to the delivery docks or between production lines [3], or to perform other inventory, inspection or surveillance operations [4,5]. In all of these cases, the traffic load may be relevant and appropriate vehicle routes must be designed and operated, in order to mitigate the risk of collision while preserving the system efficiency in terms of transportation time and cost. To this end, IT provides hardware and software devices to support safe and efficient transport network operations. A sensing network, including sensors installed on the network infrastructure and/or the vehicles, collects information (position, direction, velocity, etc.), that is processed by software logic that schedules vehicle movements and detects possible conflicts or deadlocks. In this context, the availability of optimization algorithms can be determinant in reducing transportation time, cost and accident rates, with relevant economic as well as social benefits.

We focus on automated transportation systems where, for the sake of safety, potential collisions should be avoided in advance as much as possible. In particular, we envision a system where, given the initial and the goal positions of each vehicle, a set of collision-free *nominal* routes are determined, and the sensing network and related logic manage, at real-time, possible conflicts or deadlocks that may arise due to unpredicted events causing any vehicle or network disruptions that prevent following the predefined schedule. Moreover, in many cases, like, container port terminals (see, e.g., [6]), logistic and industrial warehouses, etc., automated transportation systems rely on a grid network topology, where the vehicle moves on intersecting horizontal and vertical lanes. This motivates us to consider a simplified, although realistic, setting, where vehicles are initially positioned on one side of the grid network (e.g., the berths along the shoreline in a port terminal, or the gates of an airport apron) and have to reach a destination on the opposite side (e.g., the land or the runways access points), and the time needed to move between any two consecutive lanes is the same for all vehicles. Under these settings, the most efficient way for a single vehicle to reach its destination is to follow a smooth *nonstop* path that starts from the origin and only contains moves on horizontal and vertical lanes in the same direction, the one towards the destination. Such kinds of shortest paths on the grid do not contain horizontal (or vertical) moves in opposite directions and are called *Manhattan paths*. Clearly, there exists more than one Manhattan path for each vehicle. If a fleet of two or more vehicles has to be routed, choosing Manhattan paths may cause collisions, since two vehicles may require to cross the same intersection or the same road segment between two lines at the same time. Collisions can be avoided by choosing different Manhattan paths rather than stopping vehicles along the path, in order to preserve efficiency. To this end, let us consider, without loss of generality, the case where vehicles have to move from the bottom side of the grid to the top side and observe that it is always possible to route vehicles, without stops, on a set of collision-free Manhattan paths where each vehicle performs all the required horizontal moves on a different dedicated horizontal lane. The drawback of such a solution is the possibly large number of required horizontal lanes, which corresponds to long displacement times and large infrastructural and operational costs, including, e.g., land consumption, sensing network installation and operation and transportation costs. On the other hand, a small number of available horizontal lanes may not be sufficient to guarantee the possibility of finding collision-free routes without stopping or deviating from the Manhattan paths.

The question of determining the smallest number of required horizontal lanes is the object of the fleet quickest routing problem on grids (FQRP-G), which can be stated as follows. We are given a grid network made of intersecting horizontal and vertical lanes and a set of vehicles. The time to move between consecutive lanes is constant and the same for every vehicle. Each vehicle is initially positioned at its origin at one side of the grid and has to reach its destination at the opposite side: without loss of generality, let origins be located at the bottom and destinations at the top of the grid, that is, the route of each vehicle starts at the bottom of a vertical lane and ends at the top of a (possibly different) vertical lane. We want to determine the minimum number of horizontal lanes that allow routing the vehicles on a set of collision-free nonstop Manhattan paths.

The scope of the paper is presenting an exact solution approach to FQRP-G, based on mathematical programming, and alternative fast heuristics that exploit relevant theoretical results presented in the literature, with the aim of assessing their computational performance and their impact on the design of time-cost efficient and safe routing systems. After reviewing the literature related to FQRP-G in Section 2, the general methodology adopted in this paper, based on modelling the problem on an undirected grid graph, is presented in

Section 3, together with previous theoretical results that are relevant for our work, and an integer linear programming formulation of FQRP-G. Fast heuristic algorithms are reported in Section 4, one corresponding to a more efficient implementation than that proposed in [7], and further greedy procedures that prioritize vehicles based on measures computed on a conflict graph, defined in Section 3.2. Section 5 reports on computational experiments on a benchmark of more than 200 random and on-purpose designed instances of different sizes up to 300 columns and vehicles. Results show that the proposed exact approach is able to solve instances up to about 150 vehicles in a few seconds, whereas running times become longer than one minute, and exponentially increase for larger instances. In any case, we show that the optimal solution, on average, would enable large per cent savings in terms of required horizontal levels. The tested heuristics always run in a blink. Moreover, even if the gap from the optimal solution may be, in theory, very large, the worst-case performance is just observed on on-purpose designed instances, whereas the performance on random instances, in particular for the first heuristic, shows just a few additional required horizontal lanes with respect to the optimal solution. This means that, as discussed in Section 6, the proposed heuristic can be used in realistic settings with relevant savings in terms of transportation and sensing infrastructure while preserving vehicle route safety. The concluding Section 7 summarizes the findings of the paper and draws some lines for further research.

2. Literature Review

Several works in literature are related to FQRP-G and, more generally, to designing a routing network and finding collision-free schedules for multiple vehicles.

In the collision-free route planning problem (also known as *multi-agent path finding* in artificial intelligence literature), a set of vehicles with a given origin and destination has to move in a given routing network, modelled as a directed graph. A first group of papers presents *static* approaches, where, for each vehicle, nominal routes are computed on the underlying routing network, taking load-balance factors into account, to prevent collisions as far as possible [8,9], or, if the application context allows, by dividing the routing area into non-intersecting zones, each occupied by one vehicle at a time, as in regional control models [9–12]. In general, such approaches cannot guarantee collision-free nominal routes, and additional methods are required during their execution to detect and resolve collisions and, in case, deadlocks, based on, e.g., Petri Net approaches [13,14], graph-theoretic models [15], queries on geospatial reference grid systems [16] and searching the space of possible deviations from nominal routes [17]. A specialized static approach for grid networks is presented in [18], where initial routes that minimize collisions are chosen from equivalent Manhattan paths, and selected collision avoidance rules, based on preliminary collisions classification, are applied during execution.

An improved static method is proposed in [19], where statically computed load-balanced paths are post-processed by resource reservation and deadlock prevention techniques inspired by [20], leading to collision-free routes.

Notice that, in general, collision and deadlock avoidance introduce deviations from nominal shortest paths as well as delays in the vehicle schedule, since stops may be required during routes operation. By considering deviations and delays already at the planning stage, *dynamic* approaches are able to directly determine optimized collision-free paths and schedules, by taking into account that the impact of vehicle routes on network resources changes over time. For a general network topology, the authors of [21] developed a heuristic based on a mathematical programming formulation and column generation, whereas exact algorithms are devised in [22] for the special case where the routing network consists of two horizontal lanes and vertical bridges between them, and in [23] for the special case of two vehicles on a grid network. The dynamic approach proposed in [24] for the general case, iteratively computes shortest paths on a time-expanded network, and it is suitable for online settings, where transportation requests may appear during operations. In [25], a time-expanded network allows dynamically modelling the problem as a multi-commodity network flow [26]: the corresponding integer linear programming formulation, solved by state-of-the-art off-the-shelf solvers, provides either cost- or time-optimal schedules for up to 50 vehicles to be routed on a grid network.

The literature also integrates collision avoidance methods and dynamic algorithms with general heuristic searching techniques (A*-based search, evolutionary algorithms, particle swarm optimization, neighbourhood search, etc. [27]) that efficiently explore different vehicle priorities and conflict-resolution policies, as well as alternative routes towards the destinations. For example, in [28], an improved A* algorithm searches the paths between vehicles origin and destination in a grid network related to a warehouse environment, also taking congestion measures into account, and grid specific priority rules are used to solve residual conflicts. In [29], the D* Lite search algorithm is run on a reachability graph obtained from a suitable coloured Petri net that models feasible multi-AGV trajectories. A Time Enhanced A* search is proposed in [30] to find collision-free route plans in a time-expanded network, and integrated with tabu search techniques to further improve the efficiency by changing the assignment of transport tasks between robots. The Conflict Based Search proposed in [31], and further enhanced in [32], explores a constraint tree whose nodes are evaluated through nominal shortest routes and, in case of collisions, branches are generated corresponding to alternative vehicle priorities. For the solution of a real ship traffic optimization problem, the authors of [33] integrate the dynamic collision-free routing algorithm proposed in [24] into a local search scheme that explores the space of possible alternative scheduling decisions related to precedence conflicts between ships that compete for traversing a waterway with limited capacity and equipped with sidings to allow ships stopping and passing each other, according to the chosen precedence strategy.

We remark that the collision-free routing methodologies described above allow for vehicle stops and deviations from the nominal shortest paths (Manhattan paths, in the case of grids), whereas our research focuses on smooth nonstop routes. With this respect, FQRP-G has relations to the design of at-grade traffic networks without conflicts, aiming at configuring and operating a routing network where all roads run at the same level (at-grade) and all vehicles can seamlessly move from their origin to their destination without stopping. A grid-shaped network is proposed in [34], where conventional four-leg lane intersections are replaced by a combination of suitable intersections with restrictions on the permitted lane exchanges, giving rise to paths where vehicles can safely move without stops between any two points of the grid, at the cost of an additional detour with respect to the Manhattan path, which may represent a good trade-off, especially for automated routing networks [35]. An alternative design is obtained in [36], by tiling together hexagon blocks with one-way or bidirectional links, able to avoid intersection conflicts between any nonstop paths.

In this context, a conflict-free routing system based on grid networks with alternating one-way lanes and no detour from nominal shortest paths is presented in [37]: platoons, each representing a virtual sequence of non-conflicting vehicles running on the same lane, are scheduled on a regular basis (rhythm), in such a way that, in each moment in time, just one virtual platoon crosses an intersection; each (real) vehicle is scheduled to join a synchronized nonstop sequence of virtual platoons to cover a Manhattan path from its origin to its destination. Notice that, due to the limited length of virtual platoons, a vehicle may need to wait at the border of the grid before joining the first platoon and proceeding to its destination, so that the problem is to optimize the vehicle entry times, which is modelled and solved in [37] with an integer linear programming formulation.

Even if the literature presented above shares common features with FQRP-G, it presents significant limitations in the scope of our research. In fact, as already observed, the reviewed routing algorithms may entail delays during the execution of the routes, as well as deviations from Manhattan paths. Even conflict-free routing systems involve either detours from optimal paths (like, e.g., [34]) or delays at the beginning of the schedules (like, e.g., [37]). As a consequence, the routes provided by previous methods are, in general, worse than the ones expected from the solution of FQRP-G, since we are looking for algorithms able to avoid collisions and deadlocks while preserving shortest paths with no initial nor intermediate delays. Moreover, under the hypothesis on the distribution of vehicles origins and destinations given in Section 1, the goal of FQRP-G is to minimize the grid size, which, instead, is given and fixed in previous literature, where different metrics are optimized.

Concerning works involving grid networks and nonstop collision-free routing on Manhattan paths and, hence, more strictly related to FQRP-G, a first heuristic approach is proposed in [38], where collision avoidance is guaranteed by one-way horizontal lanes and by prioritization of horizontal moves, based on the distance of a vehicle from its destination: the number of required horizontal lanes is equal to the number of vehicles, in the worst case, and smaller on average. Improved upper bounds and heuristic algorithms for FQRP-G are discussed in [7,39,40], based on the analysis of the potential conflicts arising between vehicles, and the related properties. In particular, thanks to theoretical results derived under the one-way lanes hypothesis and, as far as [7] is concerned, by restricting Manhattan paths to those containing only one horizontal leg, the number of required horizontal lanes is limited by roughly the number of vehicles divided by four, using the heuristic proposed in [7], whereas the bound claimed by the authors of [40] has to be amended, as observed in [39]. The results presented in [7,38] that are relevant for the analysis proposed in our work, will be reviewed in Section 3.

3. Methodology: A Mathematical Formulation

In this section, we describe a mathematical model for FQRP-G, based on a graph representation. It will be used to introduce notation and to review some relevant properties presented in the literature. By exploiting such properties, we then propose a mathematical programming formulation of FQRP-G, which will be the base for the exact approach proposed in this work.

3.1. Graph Model and Notation

The analysis and the development of solution methods for FQRP-G starts from modelling the grid network as an undirected grid graph $G = (N, E)$: the set N contains the vertices, each corresponding to the intersection of one horizontal and one vertical lane, and the set E contains the lane segments, each connecting two consecutive nodes on the same horizontal or vertical lane (see Figure 1a).

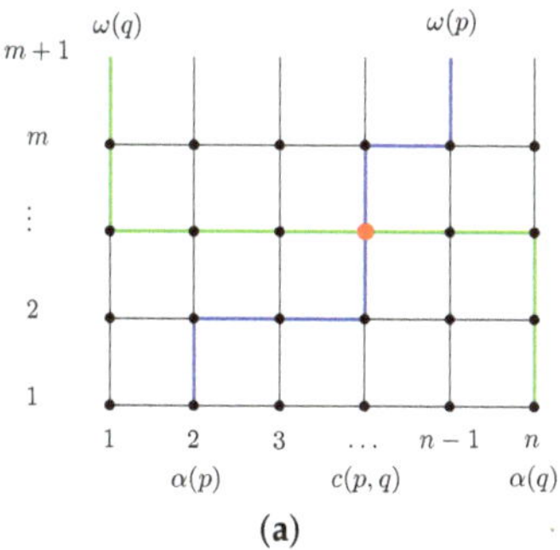
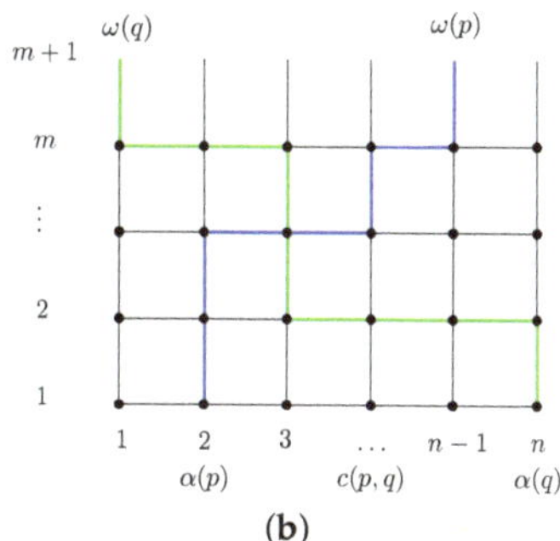

Figure 1. A sample grid graph with two conflicting vehicles routed on: (**a**) paths colliding in the red node; (**b**) collision-free paths.

We denote with n the number of vertical lanes (or *columns*) and with m the number of horizontal lanes (or *rows* or *levels*) in the grid. Columns are numbered from left to right from 1 to n, rows from bottom to top from 1 to m, so that each node representing the intersection between column i and row j is identified by the pair (i, j). Without loss of generality, we consider vehicle origins located at the bottom of the grid, in row 1, and, thus, vehicle destinations at the top of it, in row m. Moreover, the dummy level $m + 1$ in

Figure 1 at the top of the grid, simply represents the vehicle exit points from the grid. Let $\alpha(k)$ denote the starting column of vehicle k and $\omega(k)$ its destination column.

All vehicles start at the same time and, according to the FQRP-G definition, they never stop until they reach their destination. We recall that the time to cross an edge is constant and the same, for every edge and vehicle, and we can take it as the unit time. We can thus assume that the time is discrete and that, at each moment in time, each vehicle has to make a move, either vertically or horizontally.

We recall that, for the sake of efficiency, each vehicle has to reach its final destination using a *Manhattan path* on the grid, that is a (shortest) path that does not contain moves in opposite directions. Moreover, in order to avoid collisions, two different vehicles cannot use the same edge, or be in the same node, at the same time. In particular, no two vehicles can start from the same position, nor can share the same final position.

We also say that two vehicles i and j are in (or have) an *edge* (resp. a *node*) *conflict* between each other, if there exists a Manhattan path π_i of i and a Manhattan path π_j of j that use one same edge (resp. node) at the same time. If π_i and π_j are chosen, then a *collision* between i and j occurs. For example, vehicles p and q in Figure 1 have a node conflict, as shown, e.g., by the two Manhattan paths depicted in Figure 1a, that use the same node at time 4. Clearly, since we search for a set of pairwise collision-free paths, a solution to FQRP-G is feasible if and only if it does not contain any such a pair of paths. With reference to the example of Figure 1, notice that the node conflict between p and q can be avoided by, e.g., choosing the two Manhattan paths of Figure 1b.

We can divide the vehicles into three sets, S, R and L, in the following way:

$S = \{k : \omega(k) = \alpha(k)\}$,
$R = \{k : \omega(k) > \alpha(k)\}$,
$L = \{k : \omega(k) < \alpha(k)\}$.

Vehicles belonging to S have to proceed *straight* to their final destination, and they have no conflict with other vehicles.

Vehicle k_1 belonging to R will have to make $\omega(k_1) - \alpha(k_1)$ horizontal moves to the *right* and may have conflicts with vehicles belonging to L.

Vehicle k_2 belonging to L will have to make $\alpha(k_2) - \omega(k_2)$ horizontal moves to the *left* and may have conflicts with vehicles belonging to R.

Since, for each vehicle belonging to S, there is only one Manhattan path, we have to choose a path only for vehicles in R or L.

In the proposed modelling framework, the FQRP-G objective can be stated as follows: we want to find the minimum number of levels necessary for all the vehicles to complete all horizontal moves before reaching their final destination column without collisions. A conflict between two vehicles can exist only if one of them belongs to R and the other to L. Moreover, since we consider nonstop Manhattan paths, further necessary conditions can be established. To this end, given vehicles k_1 and k_2, let $c(k_1, k_2) = \lfloor \alpha(k_1) + \alpha(k_2) \rfloor / 2$.

An edge conflict between vehicles k_1 and k_2 exists if and only if the pair (k_1, k_2) belongs to the set

$$C_{odd} = \{(k_1, k_2) \in R \times L : \ \alpha(k_1) < \alpha(k_2), \ \alpha(k_1) + \alpha(k_2) \text{ is odd},$$
$$\omega(k_1) \geq c(k_1, k_2) + 1, \ \omega(k_2) \leq c(k_1, k_2)\}. \quad (1)$$

The conflict only occurs on a horizontal edge joining a node of column $c(k_1, k_2)$ with a node of column $c(k_1, k_2) + 1$. To avoid collisions related to edge conflicts between vehicles k_1 and k_2 with $(k_1, k_2) \in C_{odd}$, vehicles k_1 and k_2 have to cross the space between columns $c(k_1, k_2)$ and $c(k_1, k_2) + 1$ at different levels.

A node conflict between vehicles k_1 and k_2 exists if and only if the pair (k_1, k_2) belongs to the set

$$C_{even} = \{(k_1, k_2) \in R \times L : \ \alpha(k_1) < \alpha(k_2), \ \alpha(k_1) + \alpha(k_2) \text{ is even},$$
$$\omega(k_1) \geq c(k_1, k_2), \ \omega(k_2) \leq c(k_1, k_2)\}. \quad (2)$$

The conflict only occurs on a node of column $c(k_1, k_2)$.

Node conflicts can be further classified as (see [7]):

- *B-conflict*, if $\omega(k_1) > c(k_1, k_2)$ and $\omega(k_2) < c(k_1, k_2)$,
- *C-conflict*, if either $\omega(k_1) = c(k_1, k_2)$ or $\omega(k_2) = c(k_1, k_2)$.

To avoid collisions related to B-conflicts between vehicles k_1 and k_2 such that $(k_1, k_2) \in C_{even}$, the set of nodes in column $c(k_1, k_2)$ visited by vehicle k_1 has to be disjoint from the set of nodes in the same column visited by vehicle k_2.

Consider now vehicles k_1 and k_2 such that $(k_1, k_2) \in C_{even}$, subject to a C-conflict, and assume that $\omega(k_2) = c(k_1, k_2)$. In this case, any Manhattan path of vehicle k_2 reaches column $c(k_1, k_2)$ and then proceeds with vertical steps only, remaining in such column. Therefore, vehicle k_1 needs to visit and leave column $c(k_1, k_2)$ before k_2 reaches this column. Hence, as observed by the authors of [7], to avoid collisions related to such C-conflict, it is necessary and sufficient that vehicle k_1 leaves column $c(k_1, k_2)$ on a lower level than that on which vehicle k_2 reaches it.

Given two vehicles k_1 and k_2 subject to a C-conflict, we say that k_2 *has a C-conflict with* k_1 if $\omega(k_2) = c(k_1, k_2)$ and, vice versa, k_1 *has a C-conflict with* k_2 if $\omega(k_1) = c(k_1, k_2)$. Such relation is not symmetric: if k_2 has a C-conflict with k_1, then k_1 does not have a C-conflict with k_2. Furthermore, if k_2 has a C-conflict with k_1, then it cannot have any other C-conflict with any vehicle distinct from k_1.

3.2. Review of Relevant Previous Results

Andreatta et al. in [38] consider FQRP-G and propose a heuristic dispatching algorithm (DA) to solve it. The algorithm incrementally builds vehicle routes and its underlying idea is to give priority to the horizontal movement of the vehicles with higher numbers of remaining horizontal steps. DA provides collision-free Manhattan paths and its computational complexity is $\mathcal{O}(n^2)$. As observed in [38], the route generated by DA for any vehicle is, by construction, a *simple* Manhattan path, i.e., a Manhattan path such that all its horizontal moves are performed on one level only. Moreover, no level contains horizontal moves in opposite directions, that is, grid rows corresponds to one-way horizontal lanes. Concerning the objective function value, the number of necessary levels, i.e., the number of levels at which at least one vehicle moves horizontally, is bounded by the number of vehicles, hence by n, in the worst case, even if it can be significantly smaller for specific FQRP-G instances.

The minimum number of levels that ensures the existence of collision-free routes in any instance of FQRP-G for a given n, has been deeply investigated by Cenci et al. in [7]. They tackle FQRP-G defining *C-conflict paths*, i.e., sequences of vehicles such that each vehicle in the sequence has a C-conflict with the following one (we recall that the definition of C-conflict is not symmetric). They prove that the length of the longest C-conflict path that can be observed in any instance of FQRP-G on a grid with $n \geq 3$ columns is equal to

$$1 + \left\lfloor \frac{n-1}{4} \right\rfloor. \tag{3}$$

Then they assume that only simple Manhattan paths are feasible and that each grid level allows movements in one direction only (one-way horizontal lanes). These conditions exclude collisions related to edge and B-conflicts and restrict the attention to C-conflicts. Under such hypotheses, Ref. [7] proves that, for $n \geq 3$,

$$m^* = 3 + \left\lfloor \frac{n-1}{4} \right\rfloor \tag{4}$$

is the number of levels of the grid that guarantees the existence of a feasible solution to every instance of FQRP-G. As a minor result, they provide an algorithm (called CaR) to solve any instance of FQRP-G on a grid graph $n \times m^*$ with time complexity $\mathcal{O}(n^3)$, thus showing that m^* horizontal lanes are also sufficient.

An important byproduct of the research in [7], which will be relevant for the analysis proposed in this paper, is the definition of the *C-conflict directed graph* $F = (V, A)$, where V is the set of vehicles and A is the set of arcs, defined as follows: given two vehicles k_1 and k_2, $(k_1, k_2) \in A$ if and only if k_2 has a C-conflict with k_1. In other words, arcs are associated with C-conflicts: arc (k_1, k_2) means that the route of k_1 must be strictly below the route of k_2 in column $\omega(k_2)$ of graph G. Notice that the definition of C-conflict directed graph given above slightly differs from the one in [7], as the arc orientation is opposite. This allows us to restate one of the results in [7] as follows, and to provide a formal proof (recall that an arborescence is a directed rooted tree such that the path from the root to any other node is unique).

Proposition 1 ([7]). *The C-conflict directed graph F is a forest of arborescences.*

Proof. Suppose that the directed graph F contains a cycle k_1, k_2, ..., k_c, $k_{c+1} = k_1$. For any pair of consecutive vehicles in the cycle, k_i and k_{i+1}, vehicle k_{i+1} has a C-conflict with vehicle k_i by the definition of arc in F. It follows that the number of horizontal steps in the route of k_i (equal to $|\omega(k_i) - \alpha(k_i)|$) is strictly greater than the number of horizontal steps in the route of k_{i+1}, for any $i = 1, \ldots, c$. However, this contradicts the fact that $k_1 = k_{c+1}$. Therefore, the directed graph F does not contain cycles. Furthermore, as each vehicle can have a C-conflict with at most one other vehicle, each node of F has at most one entering arc, and thus the path from the root to any node is unique. It follows that F is a forest of arborescences. $\square$

3.3. A Mathematical Programming Formulation

In this section, we propose a mathematical programming formulation of FQRP-G. Mathematical programming is a well-known operations research tool to model and solve optimization problems. A mathematical programming model defines numerical decision variables and, based on these variables, an objective function, and a system of equations and inequalities (constraints): the objective function is the quantity to be maximized or minimized, whereas the constraints define the set of feasible solutions. Solving a mathematical programming model means finding a solution that satisfies all the constraints and optimizes the value of the objective function. Integer linear programming formulations are mathematical programming models where the objective, as well as the constraints, are linear functions of the decision variables, and (some of) the variables are restricted to assume integer values only. There is no known polynomial-time algorithm to solve general integer linear programming models (indeed, this is an NP-hard problem [41]), but standard techniques are available, like, e.g., branch and bound or cutting planes algorithms and further improvements (see, e.g., [42,43]), whose running time is expected to grow exponentially with the size (number of variables and constraints) of the formulation. However, these techniques are implemented by state-of-the-art solvers, which provide effective off-the-shelf tools to solve optimization problems formulated as integer linear programming models in a wide range of applications, including collision-free network design and routing (e.g., [25,37]), at least for moderate-size instances.

The integer linear programming formulation we propose for FQRP-G is based on network flow models (see, e.g., [26]). For each vehicle $k \in R$, let us introduce the following binary variables x_{ijk}^v and x_{ijk}^h, representing decisions about vertical and, respectively, horizontal moves:

- variable x_{ijk}^v is equal to 1 if the edge joining nodes (i, j) and $(i, j+1)$ belongs to the chosen shortest path of vehicle k (and 0 otherwise); these variables are defined for every triplet i, j, k such that $\alpha(k) \leq i \leq \omega(k)$ and $1 \leq j \leq m$;
- variable x_{ijk}^h is equal to 1 if the edge joining nodes (i, j) and $(i+1, j)$ belongs to the chosen shortest path of vehicle k (and 0 otherwise); these variables are defined for every triplet i, j, k such that $\alpha(k) \leq i \leq \omega(k) - 1$ and $1 \leq j \leq m$.

For each vehicle $k \in L$, let us introduce the following binary variables y_{ijk}^{v} and y_{ijk}^{h}, which are the homologous of x variables above:

- variable y_{ijk}^{v} is equal to 1 if the edge joining nodes (i,j) and $(i, j+1)$ belongs to the chosen shortest path of vehicle k (and 0 otherwise); these variables are defined for every triplet i,j,k such that $\omega(k) \leq i \leq \alpha(k)$ and $1 \leq j \leq m$;
- variable y_{ijk}^{h} is equal to 1 if the edge joining nodes (i,j) and $(i-1,j)$ belongs to the chosen shortest path of vehicle k (and 0 otherwise); these variables are defined for every triplet i,j,k such that $\omega(k) + 1 \leq i \leq \alpha(k)$ and $1 \leq j \leq m$.

We remind that the dummy level $m+1$ represents the vehicle exit points from the grid. Therefore, variables x_{imk}^{v} or y_{imk}^{v}, just above defined, are equal to 1 if vehicle k from the top of column i moves out of the grid.

Finally, let us introduce a variable z, whose meaning is the highest level where a horizontal move takes place.

The proposed integer linear programming formulation of FQRP-G (ILP) is reported in Figure 2. As from the objective function (5), we are interested in minimizing z, i.e., we want to find the minimum number of levels necessary for all the vehicles to complete all horizontal moves before reaching their final destinations.

Overall, constraints (6)–(11) guarantee that, for each vehicle, the edges associated with variables that take value 1 provide a (shortest) Manhattan path: constraints (6)–(8) are devoted to vehicles in R whereas constraints (9)–(11) to vehicles in L. Constraints (6) and (9) require that the route of vehicle k starts at position $(\alpha(k), 1)$ with either a horizontal step or a vertical one. Then, equalities (7) and (10) state flow conservation, that is: if vehicle k reaches node (i,j) (either with a vertical or a horizontal move, see the left-hand side), then k must perform either a vertical or a horizontal move starting from the same node (see the right-hand side). For the sake of clarity, notice that constraints are stated regardless of the fact that, for some boundary values of indexes i and j, some of the variables involved in (7) and (10) are not defined and must be replaced by 0. In the definition of constraint (7) for $k \in R$, this happens for the following variables: (i) $x_{i-1,j,k}^{h}$, if $i = \alpha(k)$; (ii) $x_{i,j,k}^{h}$, if $i = \omega(k)$ and (iii) $x_{i,j-1,k}^{v}$, if $j = 1$. With similar arguments, in the definition of constraint (10) for $k \in L$, the following variables must be replaced by 0: (i) $y_{i+1,j,k}^{h}$, if $i = \alpha(k)$; (ii) $y_{i,j,k}^{h}$, if $i = \omega(k)$ and (iii) $y_{i,j-1,k}^{v}$, if $j = 1$. Equalities (8) and (11) require that the route of vehicle k reaches the top of the destination column with a vertical step.

After the observation that defines sufficient conditions to avoid collisions related to node conflicts (see Section 3.1), such collisions are avoided by constraints (12): they state that at most one of the two vehicles involved in a given conflict can reach, with either a vertical or a horizontal move, the potential collision position, i.e., the same row in the conflict column. Even for these constraints, boundary index values are solved by replacing $x_{i,j-1,k_1}^{v} = y_{i,j-1,k_2}^{v} = 0$ in case $j = 1$. Even according to the sufficient conditions stated in Section 3.1, constraints (13) prevent collisions related to edge conflicts, since they exclude routes where two vehicles in such a conflict move between the interested columns at the same level.

Variable z is linked to variables x and y through (14) and (15), stating that at least j levels are required if at least one horizontal move takes place at row j.

$$
\text{(ILP)} \quad \min \quad z \tag{5}
$$

$$
\text{s.t.}
$$

$$
x^v_{\alpha(k),1,k} + x^h_{\alpha(k),1,k} = 1 \qquad k \in R \tag{6}
$$

$$
x^v_{i,j-1,k} + x^h_{i-1,j,k} = x^v_{i,j,k} + x^h_{i,j,k}
$$
$$
k \in R,\ \alpha(k) \le i \le \omega(k),\ 1 \le j \le m \tag{7}
$$

$$
x^v_{\omega(k),m,k} = 1 \qquad k \in R \tag{8}
$$

$$
y^v_{\alpha(k),1,k} + y^h_{\alpha(k),1,k} = 1 \qquad k \in L \tag{9}
$$

$$
y^v_{i,j-1,k} + y^h_{i+1,j,k} = y^v_{i,j,k} + y^h_{i,j,k}
$$
$$
k \in L,\ \omega(k) \le i \le \alpha(k),\ 1 \le j \le m \tag{10}
$$

$$
y^v_{\omega(k),m,k} = 1 \qquad k \in L \tag{11}
$$

$$
x^v_{i,j-1,k_1} + x^h_{i-1,j,k_1} + y^v_{i,j-1,k_2} + y^h_{i+1,j,k_2} \le 1
$$
$$
(k_1,k_2) \in C_{even},\ i = c(k_1,k_2),\ 1 \le j \le m \tag{12}
$$

$$
x^h_{i,j,k_1} + y^h_{i+1,j,k_2} \le 1 \qquad (k_1,k_2) \in C_{odd},\ i = c(k_1,k_2),\ 1 \le j \le m \tag{13}
$$

$$
z \ge j \cdot x^h_{i,j,k} \qquad k \in R,\ \alpha(k) \le i \le \omega(k)-1,\ 1 \le j \le m \tag{14}
$$

$$
z \ge j \cdot y^h_{i,j,k} \qquad k \in L,\ \omega(k)+1 \le i \le \alpha(k),\ 1 \le j \le m \tag{15}
$$

$$
\begin{aligned}
x^v_{ijk} &\in \{0,1\} \qquad k \in R,\ \alpha(k) \le i \le \omega(k),\ 1 \le j \le m \\
x^h_{ijk} &\in \{0,1\} \qquad k \in R,\ \alpha(k) \le i \le \omega(k)-1,\ 1 \le j \le m \\
y^v_{ijk} &\in \{0,1\} \qquad k \in L,\ \omega(k) \le i \le \alpha(k),\ 1 \le j \le m \\
y^h_{ijk} &\in \{0,1\} \qquad k \in L,\ \omega(k)+1 \le i \le \alpha(k),\ 1 \le j \le m \\
z &\in \mathbb{R}
\end{aligned} \tag{16}
$$

Figure 2. The integer linear programming formulation of FQRP-G (ILP).

The objective is to find the minimum of z. Finally, constraints (16) set variables x and y as binary and variable z real. Notice that z integrality follows, by (14), (15) and the objective function (5), from integrality of x^h and y^h. Moreover, even x^v and y^v could be defined as continuous, since their integrality follows from the one of x^h and y^h by (6), (7), (9) and (10).

We remark that the proposed ILP model describes, for each vehicle, a static flow on the grid network, since no time component is required to define both the decision variables and the constraints, in view of the conditions devised in Section 3.1 to bound the set of possible collision points. This is different from the mathematical programming formulations presented in, e.g., [25], where flows are defined on a time-expanded network, or [37], where the impact of the flow on different rhythmic routing intervals has to be considered. As a consequence, the size of ILP, in terms of the number of both variables and constraints, is considerably smaller than the corresponding formulations presented in previous literature, with benefits for the required solution time.

4. Heuristics

The mathematical model presented in Section 3 can be solved through off-the-shelf solvers for mixed-integer linear programming to obtain an optimal solution for a given FQRP-G instance, i.e., the minimum number of rows that allows collision-free nonstop Manhattan paths to route vehicles on, together with the paths themselves. However, due to the computational complexity of integer linear programming, we expect that the efficiency of the model, in terms of time to obtain the optimal solution, degrades with the size of the instance to handle, as in fact our computational experiments, presented in Section 5, ascertain. We thus propose two heuristics to solve FQRP-G. The first one, called Heuristic

A, is a reinterpretation of the CaR algorithm given by Cenci et al. in [7], but it is much simpler and improves the computational complexity, as will be stated in Proposition 4. It uses the C-conflict directed graph to generate vehicle routes that are simple Manhattan paths and it is able to always provide a feasible solution to FQRP-G. The second one, called Heuristic B, is more flexible in choosing Manhattan paths and attempts to determine the vehicle routes by giving priority to the horizontal moves of vehicles ranked on the basis of measures obtained from the C-conflict directed graph.

4.1. Heuristic A

Heuristic A is based on the C-conflict directed graph F. As from Proposition 1, each connected component of F is an oriented arborescence. Any such arborescence has a root, and its nodes can be partitioned according to their depth. The root has zero depth. The depth of any node is equal to the length of the unique path in F from the root to that node. For each node, we also define its height as the length of the longest path in F from that node to any of the leaves. The height of a connected component is equal to the length of the longest path from its root, i.e., the height of the root itself. Vehicles in S, as well as any vehicle that is not involved in C-conflicts, are isolated nodes in F, and have both height and depth equal to 0. The root of any non-trivial arborescence is either in L or in R.

We now state Heuristic A and, then, we discuss its correctness and properties. Given an instance of FQRP-G, in terms of the number of columns n, set of vehicles and related origins α and destinations ω, Heuristic A runs through the following steps:

1. Partition the set of vehicles into S, R and L and build the C-conflict directed graph F. Assume, without loss of generality, that a connected component with maximal height has the root in R (the case in L is similar).
2. Let p be any vehicle in a connected component rooted in R, and let l_p be its depth in F. The route of vehicle p is as follows: move vehicle p vertically on column $\alpha(p)$ to reach level $l_p + 1$, and then horizontally on level $l_p + 1$ until column $\omega(p)$; then move it vertically to its final destination.
3. Let q be any vehicle in a connected component rooted in L, and let l_q be its depth in F. The route of vehicle q is as follows: move vehicle q vertically on column $\alpha(q)$ to reach level $l_q + 2$, and then horizontally on level $l_q + 2$ until column $\omega(q)$; then move it vertically to its final destination.
4. The route of vehicles in S contains vertical steps only.

The following proposition shows that Heuristic A always provides a feasible solution.

Proposition 2. *The vehicle routes given by Heuristic A are nonstop collision-free simple Manhattan paths.*

Proof. In the output of Heuristic A, all the horizontal moves performed at any level have the same direction. Indeed, each path in F is a C-conflict path and, hence, it alternates vehicles in R and in L. If follows that, under the assumption that the maximum height is related to an arborescence rooted in R (the case in L is similar) all vehicles in R move horizontally on an odd row, and all vehicles in L on an even row. This corresponds to having one-way horizontal lanes, which prevents collisions related to edge conflicts from occurring. We observe that, trivially, Heuristic A outputs simple Manhattan paths, as each vehicle performs consecutively all its horizontal moves on the same level. This, together with one-way lanes, avoids collisions related to B-conflicts.

For each pair of vehicles p and q such that q has a C-conflict with p, the directed graph F contains the arc (p, q), and $l_q = l_p + 1$ holds. Therefore, vehicle p performs its horizontal moves on a lower level than q does, and the route of p is below the one of q in column $\omega(q)$, as required to avoid collisions related to C-conflicts. $\square$

Notice that, if Heuristic A is applied to a single connected component of F, then the number of grid levels used by the output solution is equal to one plus the height of that component. Therefore, given any instance, the number of grid levels needed by Heuristic A is equal the height of the highest connected component of F, added by 2. The term "+2" comes from the case in which the forest F contains two (or more) highest components, of which, one rooted in R and another in L. This proves the following

Proposition 3. *Given an instance of FQRP-G and its C-conflict direct graph, let $\bar{m}_R$ and $\bar{m}_L$ be the maximum height of an arborescence rooted in R and, respectively L. The number of levels required by Heuristic A is $\max\{\bar{m}_R, \bar{m}_L\} + a$, where $a = 1$ if $\bar{m}_R \neq \bar{m}_L$, $a = 2$ otherwise.*

The number of required levels is equal to the one stated for algorithm CaR proposed by Cenci et al. in [7]: as shown in [7], CaR optimally solves FQRP-G if the set of vehicle routes is restricted to simple Manhattan paths and under the hypothesis of one-way horizontal lanes. We thus have the following

Corollary 1. *Heuristic A finds the optimal solution of FQRP-G restricted to simple Manhattan paths and one-way horizontal lanes.*

In fact, as already stated above, Heuristic A is a reinterpretation of the CaR algorithm that improves its computational complexity (we recall that CaR runs in $\mathcal{O}(n)^3$).

Proposition 4. *Given an instance of FQRP-G on a grid network with n columns, the computational complexity of Heuristic A is $\mathcal{O}(n)$.*

Proof. In order to detect all C-conflicts, $\mathcal{O}(n)$ calculations are sufficient. Indeed, for any vehicle $k \in R$ (resp. in L), we only have to check if there is another vehicle moving from position $(2\omega(k) - \alpha(k), 1)$ and having its destination on the left (resp. on the right) of column $\omega(k)$; in such case, vehicle k has a C-conflict with the other vehicle. All the arcs of F can be thus detected in at most n (a bound on the number of vehicles) operations, and F built in $\mathcal{O}(n)$. All the data required by Heuristic A can be collected during a depth-first visit of F, which allows computing the depth and the height of any node in $\mathcal{O}(n)$. This shows that Step 1 takes $\mathcal{O}(n)$ operations. Concerning Steps 2 to 4, they simply assign the horizontal level to each vehicle, which can still be done in $\mathcal{O}(n)$. $\square$

4.2. Heuristic B

Heuristic B aims at calling non-simple Manhattan paths conveniently into play. The underlying idea is to find an appropriate order of the vehicles and, then, to sequentially route each vehicle on the "lowest" Manhattan path possible, i.e., a Manhattan path obtained by choosing a horizontal step whenever this is compatible with previously assigned paths.

Vehicles are sorted according to a measure of how critical it is to route them. For example, an order of the vehicles could provide a feasible set of routes only if, for any pair of vehicles k_1, k_2 such that k_2 has a C-conflict with k_1 on column $\omega(k_2)$, vehicle k_1 precedes k_2 in the order, since otherwise k_2 would have precedence in the horizontal move to reach the conflict column and stay below k_1 on it, which means that the C-conflict cannot be resolved (see Section 3.1). It follows that vehicles belonging to a C-conflict path should be sorted in the increasing order of their depth in the C-conflict directed graph F, which again plays an important role in prioritizing vehicles. We also observe that, in general, the assigned Manhattan paths are not simple and each level can be run in opposite directions; therefore, both edge conflicts and B-conflicts may actually generate collisions and have to be taken into account.

For each vehicle k, the following measures are considered:

- l_k: the depth of k in F;
- γ_k: length of the longest C-conflict path k belongs to. Notice that, if l_k and h_k are, respectively, the depth and the height of k in F, $\gamma_k = l_k + h_k$, and, in particular, $\gamma_k = 0$ if k is not involved in any C-conflict;

- δ_k: overall number of conflicts k is involved in;
- ρ_k: number of edge conflicts k is involved in.

Given an instance of FQRP-G, in terms of number of columns n, set of vehicles and related origins α and destinations ω, Heuristic B runs through the following steps:

1. Compute an upper bound $\overline{m}$ on the number of required levels (it can be simply equal to the number of vehicles, or it can be obtained by running Heuristic A).
2. Build the C-conflict directed graph F and, for each vehicle k, compute γ_k, l_k, δ_k and ρ_k;
3. Sort vehicles according to any order such that they appear by non-decreasing l_k;
4. For each vehicle k in the determined order, assign k to the "lowest" available Manhattan path, as recursively defined by the following rule (given for the case $k \in R$, the case $k \in L \cup S$ is similar):
   (a) let (i, j) be the actual position of vehicle k in the grid (initially set to $(\alpha(k), 1)$);
   (b) if $i = \omega(k)$ and $j = \overline{m}$, then output "feasible path for k found" and consider the next vehicle;
   (c) if $i = \omega(k)$ and all vehicles up to k in a C-conflict path are not involved in further conflicts, then k performs a vertical move;
   (d) otherwise, if $i \neq \omega(k)$, then check if the horizontal move to node $(i + 1, j)$ involves any collision with previously assigned paths (this could be related to a node-conflict if, after a unit of time, another vehicle will be in node $(i + 1, j)$, or an edge conflict if another vehicle is performing the opposite move from $(i + 1, j)$ to (i, j) at the same time); if the answer is "no conflict", then k performs the horizontal move to node $(i + 1, j)$;
   (e) otherwise, check if the vertical move to $(i, j + 1)$ involves any conflict with previously assigned paths (this could be a node-conflict if, after a unit of time, another vehicle will be in node $(i, j + 1)$); if the answer is "no conflict", then k vertically moves to node $(i, j + 1)$;
   (f) otherwise, output "no feasible path for k found" and stop.

With reference to Step 4c, we remark that, since vehicles are sorted by non-decreasing l_k, collisions related to C-conflicts are avoided, as for any arc (k_1, k_2) of F, the path of vehicle k_1 is set before the path of k_2. These are the only collisions associated with vertical moves on the destination columns, so that, in the case specified by Step 4c, checking their occurrence is redundant.

While, in the above case, C-conflicts are solved by appropriately ordering the vehicles in the first phase of the algorithm, remaining node-conflicts and edge conflicts are tentatively solved during Steps 4d–4e. However, we have no guarantee to avoid related collisions and, indeed, Heuristic B may get stuck if both horizontal and vertical moves of a vehicle at a given node are forbidden. Nevertheless, if Heuristic B is successful, the required number of levels is not bounded from below by the length of the longest C-conflict path, as it is the case for Heuristic A: we thus aim to empirically evaluate the probability of getting stuck and, if this is not the case, the ability of Heuristic B to provide better results than Heuristic A.

The actual performance of Heuristic B depends on the specific sorting adopted by Step 3. We propose two alternatives giving rise to:

- Heuristic B1: vehicles are sorted in lexicographic order by decreasing γ_k, increasing l_k, decreasing δ_k and decreasing ρ_k;
- Heuristic B2: vehicles are sorted in lexicographic order by increasing l_k, decreasing γ_k, decreasing δ_k and decreasing ρ_k.

We now discuss the computational complexity of Heuristic B.

Proposition 5. *Given an instance of FQRP-G on a grid graph with n columns, the computational complexity of Heuristic B is $\mathcal{O}(n^2 \overline{m})$.*

Proof. Step 1 to determine $\overline{m}$ can be done in $\mathcal{O}(n)$. The measures required by the sorting step can be computed by building and depth-first visiting the C-conflict directed graph F, which can be done in $\mathcal{O}(n)$ (as discussed in proof of Proposition 4). The sorting Step 3 takes $\mathcal{O}(n \log n)$. Since the number of moves in a Manhattan path is bounded by $n + \overline{m}$, and the number of vehicles by n, the complexity of Step 4, and of overall Heuristic B, is $\mathcal{O}(n^2 \overline{m})$. $\square$

5. Results

In the previous sections, we propose the integer linear programming (ILP) formulation and three heuristics (A, B1, and B2) to solve FQRP-G. Computational experiments have been conducted with the following purposes:

- determine to what extent, in terms of instance size and required running time, ILP is able to solve FQRP-G;
- assess the quality of the solutions output by Heuristic A (which, we recall, is optimal under one-way horizontal lanes and simple Manhattan paths hypothesis) in terms of additional required levels with respect to the (unrestricted) optimal solution provided by ILP;
- estimate the success rate of Heuristics B1 and B2 and their ability to find better solutions than Heuristic A.

We recall that, as discussed in Section 2, previous literature approaches to collision-free routing problems present limitations in their application to FQRP-G, since they do not consider grid-size minimization and, moreover, they allow for space-time deviations from nonstop Manhattan paths. The heuristic algorithm DA presented in [38] is able to solve FQRP-G, however it is dominated by Heuristic A for both efficiency since DA is $\mathcal{O}(n^2)$ whereas Heuristic A is $\mathcal{O}(n)$, and effectiveness. Indeed, as observed in [38], DA returns routing schedules made of simple Manhattan paths on one-way horizontal lanes and, hence, compliant with the hypothesis of Corollary 1: as a consequence, DA cannot provide better solutions than Heuristic A, which is optimal under such restrictions.

In our experiments, we consider two benchmarks. The first one is made of random instances with 10 up to 300 columns and vehicles: in particular, 20 instances are generated for each $n \in \{10, 25, 50, 75, 100, 150, 200, 300\}$ by randomly choosing the origin and the destination columns of each vehicle. The second benchmark includes 11 ad hoc instances with 105 up to 233 columns and vehicles, created on purpose as to contain long C-conflict paths, and more than one arborescences in the related C-conflict directed graph. ILP has also been run on a third benchmark of large random instances with $n \in \{350, 400, 500\}$, to determine the larger size instances ILP can solve in practice.

All the tests were run on a workstation equipped with an Intel Xeon E-2176G processor with 6 cores at 3.7 GHz, and 16 GB RAM.

ILP has been solved using the Cplex 12.9.0 engine [44] with a time limit of 30 min. In order to take the number of variables and constraints of ILP, hence running times, as small as possible, we run Heuristic A (whose running time, as we will see, is negligible) and set the parameter m in the ILP model equal to the number of levels output by Heuristic A.

Table 1 reports the computational results given by ILP and Heuristic A on the first random benchmark. Statistics involving ILP refer to tests on 10 out of 20 instances available per size. The first column specifies the instance size. The average, minimum and maximum number of levels used by ILP are reported in Columns 2 and 3. ILP running times, whose average (in seconds) appears in Column 4, are below the time limit in every instance and, therefore, data in Columns 2 and 3 refer to proven optimal values. Columns 5–8 are related to Heuristic A and give respectively: the average number of levels used by its solutions, the relative percentage error with respect to the optimal ILP value, the minimum and the maximum number of levels required by all the obtained solutions, and the maximum absolute gap between the number of levels used by the solutions of Heuristic A and the corresponding optimal values. Running times of Heuristic A are not specified as they are negligible (always fairly less than 1 ms).

Table 1. Experimental results of ILP and Heuristic A on random instances.

Instance	ILP			Heur A			
n	Avg	Min–Max	Time	Avg	Err%	Min–Max	Δ_{max}
10	2.2	2–3	0.02	2.25	0.0	2–3	0
25	2.4	2–3	0.09	2.95	15.0	2–4	1
50	3.0	3–3	0.79	3.50	20.0	3–5	1
75	2.9	2–3	1.75	3.45	15.0	3–4	1
100	3.1	3–4	5.90	3.75	23.3	3–5	2
150	3.0	3–3	42.54	4.05	30.0	3–5	2
200	3.0	3–3	151.33	4.25	40.0	3–5	2
300	3.0	3–3	1099.91	4.30	43.3	4–5	2

The computational results given by Heuristics B1 and B2 on the first random benchmark appear in Table 2. The percentage of instances where Heuristic B1 has been able to find a feasible solution (success rate) is reported in Column 2. Columns 3 to 6 refer to these successful instances and report: the average, minimum and maximum number of levels required by B1 (Columns 3 and 5 respectively); the average per cent error in the number of levels used by B1 with respect to the optimal value output by ILP (Column 4); the maximum absolute gap between the number of levels used by B1 and the optimal values (Column 6). Always referring to successful instances for B1, Column 7 compares the performances of Heuristic B1 versus Heuristic A, reporting the percentages of successful instances in which B1 uses less (win) or more (lose) levels than A. Columns 8 to 13 report the same information for Heuristic B2. Again, statistics involving ILP refer to tests on 10 out of 20 instances available per size.

Table 2. Experimental results of Heuristics B1 and B2 on random instances.

	Heur B1					Heur B2					
n	Succ%	Avg	Err%	Min–Max	Δ_{max}	Win-Lose%	Succ%	Avg	Err%	Min–Max	Δ_{max}	Win-Lose%
10	90	2.72	50.00	2–4	2	5–40	90	2.72	50.00	2–4	2	5–40
25	40	3.75	56.25	2–5	3	5–30	55	3.64	55.56	2–5	3	10–35
50	5	4.00	33.33	4–4	1	0–0	5	3.00	0.00	3–3	0	100–0
$\geq$75	0	–	–	–	–	–	0	–	–	–	–	–

ILP was able to find the optimal solution of all the instances within the time limit, and running times are consistently less than a few seconds up to 100 vehicles. For larger sizes, running time grows almost exponentially, as expected. Indeed, we performed a further test of ILP on the third benchmark, observing that only four out of ten cases with $n = 350$ (and no other larger instances) are solved to optimality. In the remaining cases with $n = 350$, ILP always finds feasible solutions whose difference with respect to the best available lower bound (optimality absolute gap) is 2.5 levels on average (maximum 4). The success rate on 400 columns instances is 90%, i.e., ILP finds feasible (even if not provably optimal) solutions for 9 out of 10 instances, with an optimality absolute gap of 3 levels on average (maximum 4). For $n = 500$, the success rate is 60%, with optimality absolute gap of 3.5 levels (maximum 4). We also observe that, as far as the third benchmark is concerned, the number of required horizontal lanes never exceeds 6 in the proposed feasible solutions.

Heuristic A is extremely fast, and, as from Table 1, it finds solutions that, even for larger random instances, take no more than 5 levels and at most 2 additional horizontal lanes with respect to the optimal values.

Running times of Heuristics B1 and B2 are negligible as well (always less than 10^{-2} s), however, their performance is poor. Both B1 and B2 get stuck in all the instances with 75 or more vehicles. The success rate is acceptable only for very small instances and just, in a few cases, B1 and B2 are able to improve over Heuristic A (with B2 showing slightly better results than B1). Summarizing the overall performance of Heuristic A on random instances is by far better than B1 and B2.

As observed above, the number of levels required by Heuristic A is very small in random instances. However, we recall that it is strictly connected to the length of the longest C-conflict path in the instance, so that, according to Equations (3) and (4), instances exist where the collision-free paths outputted by Heuristic A need more than a few horizontal lanes to be seamlessly operated. Therefore, we consider the second benchmark of ad hoc generated instances containing long C-conflict paths, more than one arborescence in the C-conflict directed graph, and further edge conflicts and B-conflicts between vehicles in the same or different arborescences. Table 3 reports the related computational results, showing a row for each instance. The number of vehicles and the length of the longest observed C-conflict path appears in Columns 1 and 2. Columns 3 and 4 give the number of levels required by the solution of Heuristic A and by the optimal solution of the ILP model, respectively. The ILP model running times, in seconds, are listed in the last column (the table does not show Heuristic A running times, since they are always less than 10^{-3} s). Results for Heuristics B1 and B2 are not reported, since they always fail in providing feasible routes.

Table 3. Experimental results of ILP and Heuristic A on ad hoc instances.

Instance		Heur A	ILP	
n	Longest C-Path	Used Levels	Used Levels	Time
105	27	28	4	2.88
117	30	31	4	2.63
129	33	34	4	4.11
141	36	37	3	4.44
153	39	40	4	10.50
161	41	42	4	7.49
173	44	45	4	10.63
189	48	49	4	16.92
201	51	52	4	13.61
221	56	57	4	22.11
233	59	60	4	38.05

ILP solves all the instances of the second benchmark to optimality, still providing routes that can be operated on a few (at most four) horizontal lanes. It is thus self-evident that Heuristic A is not appropriate to solve FQRP-G on these ad hoc instances, as it needs many more levels with respect to the optimal solution. Indeed, performing horizontal steps on more than one level is crucial, in presence of long chains of C-conflicts, to save levels. However, the ad hoc instances in the third benchmark do not appear much harder to be solved with ILP in terms of computational time.

6. Discussion

The methods presented in the previous sections allow us to find provably optimal or heuristic solutions to FQRP-G. The problem is relevant for the design and the operation of automated transportation systems where the routing network consists of intersecting horizontal and vertical lanes, vehicles move between opposite sides (e.g., from bottom to top) and a network of sensors supports safe and efficient operations: port container terminals, automated warehouses, train terminals, etc., are some significant examples that can be approximated by such routing networks. By solving FQRP-G, the number of horizontal lanes and a set of routes is determined that can be seamlessly operated without intermediate stops nor deviations from static shortest paths (efficient routes) and without any collision (safe routes). In real-time, the sensing network and related logic monitor the operations and manage further conflicts just in case they arise due to unpredicted events (vehicle breakdowns, network interruptions, etc.), making the overall routing system robust against disruptions, and further reducing the risk of collisions.

While the number of vertical lanes is often determined by the facility layout, the number of required horizontal lanes should be carefully dimensioned at the design phase, in order to minimize the cost of the underlying transportation and sensing infrastructure. In this work, we have devised and tested four possible approaches to solve FQRP-G and determine the minimum number of levels, given vehicles' initial positions and final destinations: an exact method (ILP) based on solving an integer linear programming formulation of the problem by standard solvers, and three heuristics (A, B1 and B2) that prioritize vehicles based on the properties of a graph summarizing C-conflicts between vehicles.

Experiments on benchmarks of random and ad hoc instances show that, from a computational point of view, ILP is able to find the proven minimum number of horizontal lanes (with related vehicle routes) for instances of up to 300 vehicles, even if running times seem to be suitable for real-time operations of up to about 100 vehicles. For larger random instances, Heuristic A always provides, in negligible running time, feasible routes with at most two additional horizontal lanes, if compared to the optimal solutions, while heuristics B1 and B2 often fail in finding a set of non-conflicting vehicle paths.

From a network design perspective, it is interesting to notice that the optimal solution for the tested instances (up to 300 vehicles) always requires no more than 3 levels (4 in 2 out of 80 cases), thus suggesting that the size of the transportation network can be set to a relatively small number of horizontal lanes. Even more interestingly, our experimental results show that, at the cost of a few additional horizontal lanes, Heuristic A can be run to produce feasible seamless routes for the case where, due to limited computational resources, solving ILP is unpractical. Moreover, Heuristic A has the advantage of providing simple Manhattan paths that can be run on a network with one-way lanes and leads to a simpler network to design, monitor and maintain, as well as to smoother, safer and simpler routes to operate. The drawback is that the number of levels required by Heuristic A may be very large with respect to the optimal one, as our experiments on ad hoc instances show: however, such instances (with long chains of vehicles in C-conflict paths) seem to be extreme cases and, in fact, they never occurred in random experiments. Moreover, they get solved by ILP in less than 40 s, even for the larger 233 vehicles instance, with optimal solutions requiring, as for random instances, no more than four horizontal lanes. It follows that an automated grid transportation network can be conveniently designed with a relatively small number of horizontal lanes and operated through ILP or Heuristic A (depending on instance size and available computational resources), leaving to the sensor network and to the run-time collision detection and avoidance system (based, e.g., on more general methods for collision-free routing presented in literature) the rare cases where the proposed methods do not find feasible solutions to FQRP-G.

Our experiments with ILP show that a grid routing network with four horizontal lanes has always been able to accommodate routing paths according to the requirements of FQRP-G. In case an exact solution method (like ILP) is not conveniently available, the proposed heuristic would require at most five horizontal lanes in almost all of the FQRP-G instances. For the residual cases, a grid network with five horizontal lanes may not guarantee nonstop routing on Manhattan paths for all vehicles: in such events, the envisioned routing system can be integrated with state-of-the-art algorithms for multi-agent pathfinding, like the ones presented in the literature review, in order optimize any required space-time deviations from nominal shortest routes.

7. Conclusions

In this work, we addressed FQRP-G, where a set of vehicles has to be routed on a grid network according to a set of nonstop collision-free Manhattan paths that minimizes the overall number of required horizontal lanes. Such paths can be seamlessly operated with no further control logic for collision and deadlock detection and avoidance, leaving a sensor network as the only task to guarantee safe operations against an unexpected vehicle or infrastructure disruptions during the real-time execution.

We have presented an integer linear programming formulation of FQRP-G, called ILP, and three heuristics, showing that:

- a theoretical analysis provides properties of conflicting paths that have been exploited to devise improved mathematical models and solution algorithms for FQRP-G. In particular: ILP formulates the problem on a static graph model, whereas the formulations proposed by literature for problems related to FQRP-G rely on a dynamic (time-expanded) graph, thus requiring a larger number of variables and constraints; Heuristic A fairly improves the computational complexity with respect to the implementation proposed by [7];
- ILP, by means of state-of-the-art off-the-shelf mathematical optimization software, can solve instances of up to 100 vehicles in a few seconds at most, providing the minimum number of horizontal lanes and related routes. ILP can even solve larger instances of up to hundreds of vehicles, at the cost of longer running times, which may be not compliant with real route-execution environments;
- one of the proposed heuristics, called Heuristic A, is very efficient and effective, even for instances with hundreds of vehicles. It always runs in negligible time, and, with only rare exceptions that never showed up in random benchmarks, it finds routing paths requiring just a few horizontal lanes (one or two) more than the optimal solution;
- from a routing network design perspective, our empirical study shows that a grid with four or five horizontal lanes normally allows for finding collision-free nonstop Manhattan paths for all the vehicles of FQRP-G. With such sizing, the needing to integrate the routing system with further state-of-the-art algorithms for multi-agent pathfinding (as to optimize possible delays and deviations from shortest routes) is rare and limited to some infrequent exceptions where the methods proposed in this paper would require higher grids.

Further research is needed towards heuristic algorithms that, like B1 or B2, do not rely on one-way lanes and on simple Manhattan paths, which, according to the theoretical results reviewed in this work, is mandatory to enable a smaller number of required levels for the instances that are critical for Heuristic A. Possible lines for future studies could also involve exact solution methods for FQRP-G, based on either the model proposed in this work or alternative mathematical programming formulations, and the extension of FQRP-G and related solution approach to more and more realistic settings, e.g., considering arbitrary vehicle origins and destinations or more general grids.

Author Contributions: Conceptualization, G.A., C.D.F. and L.D.G.; methodology, G.A., C.D.F. and L.D.G.; software, C.D.F. and L.D.G.; validation, G.A., C.D.F. and L.D.G.; formal analysis, G.A., C.D.F. and L.D.G.; data curation, G.A., C.D.F. and L.D.G.; writing—original draft preparation, G.A., C.D.F. and L.D.G.; writing—review and editing, G.A., C.D.F. and L.D.G. All authors have read and agreed to the published version of the manuscript.

Funding: This research received no external funding.

Data Availability Statement: Data are available from the corresponding authors.

Conflicts of Interest: The authors declare no conflict of interest.

References

1. Andreatta, G.; De Giovanni, L.; Monaci, M. A fast heuristic for airport ground-service equipment-and-staff allocation. *Procedia Soc. Behav. Sci.* **2014**, 108, 26–36. [CrossRef]
2. Möhring, R.H.; Köhler, E.; Gawrilow, E.; Stenzel, B. Conflict-free real-time AGV routing. In *Operations Research Proceedings 2004, Proceedings of the Operations Research 2004 Conference, Tilburg, The Netherlands, 1–3 September 2004*; Fleuren, H., Hertog, D., Kort, P., Eds.; Springer: Berlin/Heidelberg, Germany, 2005; pp. 18–24.
3. Chung, S.H. Applications of smart technologies in logistics and transport: A review. *Transp. Res. Part E Logist. Transp. Rev.* **2021**, *153*, 102455. [CrossRef]
4. Chocholáč, J.; Boháčová, L.; Kučera, T.; Sommerauerová, D. Innovation of the process of inventorying of the selected transport units: Case study in the automotive industry. *LOGI Sci. J. Transp. Logist.* **2017**, *8*, 48–55. [CrossRef]
5. Maghazei, O.; Netland, T.H.; Frauenberger, D.; Thalmann, T. Automatic drones for factory inspection: The role of virtual simulation. In *Advances in Production Management Systems. Artificial Intelligence for Sustainable and Resilient Production Systems*; Dolgui, A., Bernard, A., Lemoine, D., von Cieminski, G., Romero, D., Eds.; Springer: Cham, Switzerland, 2021; Volume 633, pp. 457–464.
6. Stopka, O.; Kampf, R. Determining the most suitable layout of space for the loading units' handling in the maritime port. *Transport* **2018**, *33*, 280–290. [CrossRef]
7. Cenci, M.; Di Giacomo, M.; Mason, F. A note on a mixed routing and scheduling problem on a grid graph. *J. Oper. Res. Soc.* **2017**, *68*, 1363–1376. [CrossRef]
8. Gao, J.; Zhang, L. Trade-offs between stretch factor and load-balancing ratio in routing on growth-restricted graphs. *IEEE Trans. Parallel Distrib. Syst.* **2009**, *20*, 171–179.
9. Vis, I.F.A. Survey of research in the design and control of automated guided vehicle systems. *Eur. J. Oper. Res.* **2006**, *170*, 677–709. [CrossRef]
10. Kim, K.H.; Jeon, S.M.; Ryu, K.R. Deadlock prevention for automated guided vehicles in automated container terminals. In *Container Terminals and Cargo Systems*; Springer: Berlin/Heidelberg, Germany, 2007; pp. 243–263.
11. Moorthy, R.; Hock-Guan, W.; Wing-Cheong, N.; Chung-Piaw, T. Cyclic deadlock prediction and avoidance for zone-controlled AGV system. *Int. J. Prod. Econ.* **2003**, *83*, 309–324. [CrossRef]
12. Zheng, K.; Tang, D.; Gu, W.; Dai, M. Distributed control of multi-AGV system based on regional control model. *Prod. Eng.* **2013**, *7*, 433–441. [CrossRef]
13. Wu, N.Q.; Zhou, M.C. Resource-oriented Petri nets for deadlock avoidance in automated manufacturing. In Proceedings of the 2000 IEEE International Conference on Robotics and Automation, San Francisco, CA, USA, 24–28 April 2000; pp. 3377–3382.
14. Wu, N.Q.; Zhou, M.C. Modeling and deadlock avoidance of automated manufacturing systems with multiple automated guided vehicles. *IEEE Trans. Syst. Man Cybern. Part B* **2005**, *35*, 1193–1202. [CrossRef]
15. Cho, H.B.; Kumaran, T.K.; Wysk, R.A. Graph theoretic deadlock detection and resolution for flexible manufacturing systems. *IEEE Trans. Robot. Autom.* **1995**, *11*, 413–421.
16. Zhai, W.; Tong, X.; Miao, S.; Cheng, C.; Ren, F. Collision detection for UAVs based on GeoSOT-3D grids. *ISPRS Int. J. Geo-Inf.* **2019**, *8*, 299. [CrossRef]
17. Koszelew, J.; Karbowska-Chilinska, J.; Ostrowski, K.; Kuczyński, P.; Kulbiej, E.; Wołejsza, P. Beam search algorithm for anti-collision trajectory planning for many-to-many encounter situations with autonomous surface vehicles. *Sensors* **2020**, *20*, 4115. [CrossRef]
18. Zhang, Z.; Guo, Q.; Chen, J.; Yuan, P. Collision-free route planning for multiple AGVs in an automated warehouse based on collision classification. *IEEE Access* **2018**, *6*, 26022–26035. [CrossRef]
19. Gawrilow, E.; Klimm, M.; Möhring, R.H.; Stenzel, B. Conflict-free vehicle routing. *EURO J. Transp. Logist.* **2012**, *1*, 87–111. [CrossRef]
20. Alon, N.; Yuster, R.; Zwick, U. Color-coding. *J. Assoc. Comput. Mach.* **1995**, *42*, 844–856. [CrossRef]
21. Krishnamurthy, N.; Batta, R.; Karwan, M. Developing conflict-free routes for automated guided vehicles. *Oper. Res.* **1993**, *41*, 1077–1090. [CrossRef]
22. Qiu, L.; Hsu, W.J. A bi-directional path layout for conflict-free routing of AGVs. *Int. J. Prod. Res.* **2001**, *39*, 2177–20195. [CrossRef]
23. Davoodi, M.; Abedinb, M.; Banyassady, B.; Khanteimouri, P.; Mohadesb, A. An optimal algorithm for two robots path planning problem on the grid. *Robot. Auton. Syst.* **2013**, *61*, 1406–1414. [CrossRef]
24. Gawrilow, E.; Köhler, E.; Möhring, R.H.; Stenzel, B. Dynamic routing of automated guided vehicles in real-time. In *Mathematics: Key Technology for the Future. Joint Projects between Universities and Industry 2004–2007*; Jäger, W., Krebs, H.J., Eds.; Springer: Berlin, Germany, 2008; pp. 165–178.
25. Yu, J.; La Valle, S.M. Planning optimal paths for multiple robots on graphs. In Proceedings of the IEEE International Conference on Robotics and Automation (ICRA), Karlsruhe, Germany, 6–10 May 2013.
26. Ahuja, R.K.; Magnanti, T.L.; Orlin, J.B. *Network Flows: Theory, Algorithms, and Applications*; Prentice Hall: Upper Saddle River, NJ, USA, 1993.
27. Zafar, M.N.; Mohanta, J.C. Methodology for path planning and optimization of mobile robots: A review. *Procedia Comput. Sci.* **2018**, *133*, 141–152. [CrossRef]

28. Yuan, R.; Dong, T.; Li, J. Research on the collision-free path planning of multi-AGVs system based on improved A* algorithm. *Am. J. Oper. Res.* **2016**, *6*, 442–449. [CrossRef]
29. Mugarza, I.; Mugarza, J.C. A coloured Petri net- and D* Lite-based traffic controller for Automated Guided Vehicles. *Electronics* **2021**, *10*, 2235. [CrossRef]
30. Santos, J.; Rebelo, P.M.; Rocha, L.F.; Costa, P.; Veiga, G. A* based routing and scheduling modules for multiple AGVs in an industrial scenario. *Robotics* **2021**, *10*, 72. [CrossRef]
31. Sharon, G.; Stern, R.; Felner, R.; Sturtevant, N.R. Conflict-based search for optimal multi-agent pathfinding. *Artif. Intell.* **2015**, *219*, 40–66. [CrossRef]
32. Atzmon, D.; Stern, R.; Felner, A.; Wagner, G.; Barták, R.; Zhou, N. Robust multi-agent path finding and executing. *J. Artif. Intell. Res.* **2020**, *67*, 549–579. [CrossRef]
33. Lübbecke, E.; Lübbecke, M.E.; Möhring, R.H. Ship traffic optimization for the Kiel canal. *Oper. Res.* **2019**, *67*, 791–812. [CrossRef]
34. Eichler, D.; Bar-Gera, H.; Blachman, M. Vortex-based zero-conflict design of urban road networks. *Netw. Spat. Econ.* **2013**, *13*, 229–254. [CrossRef]
35. Boyles, S.D.; Rambha, T.; Xie, C. Equilibrium analysis of low-conflict network designs. *Transp. Res. Rec.* **2014**, *2467*, 129–139. [CrossRef]
36. Liyanage, S.P.; Pravinvongvuth, S. Obtaining the optimum block length of the Chet network: An at-grade transportation network without signalized intersections, roundabouts, or stop signs. *Engineer* **2018**, *50*, 37–46. [CrossRef]
37. Lin, X.; Li, M.; Shen, Z.M.; Yin, Y.; He, F. Rhythmic control of automated traffic—Part II: Grid network rhythm and online routing. *Transp. Sci.* **2021**, *55*, 988–1009. [CrossRef]
38. Andreatta, G.; De Giovanni, L.; Salmaso, G. Fleet quickest routing on grids: A polynomial algorithm. *Int. J. Pure Appl. Math.* **2010**, *62*, 419–432.
39. Andreatta, G.; De Francesco, C.; De Giovanni, L.; Salmaso, G. A Note on FQRP-G. Available online: http://www.math.unipd.it/~luigi/manuscripts/FQRP-G/fqrp.pdf (accessed on 2 December 2021).
40. Di Giacomo, M.; Mason, F.; Cenci, M. A note on solving the Fleet Quickest Routing Problem on a grid graph. *Cent. Eur. J. Oper. Res.* **2020**, *28*, 1069–1090. [CrossRef]
41. Garey, M.R.; Johnson, D.S. *Computers and Intractability: A Guide to the Theory of NP-Completeness*; W. H. Freeman & Co.: New York, NY, USA, 1979.
42. Achterberg, T.; Bixby, R.E.; Gu, Z.; Rothberg, E.; Weninger, D. Presolve reductions in Mixed Integer Programming. *INFORMS J. Comput.* **2020**, *32*, 473–506. [CrossRef]
43. Wolsey, L.A. *Integer Programming*, 2nd ed.; Wiley: Hoboken, NJ, USA, 2020.
44. IBM CPLEX Optimizer. Available online: https://www.ibm.com/it-it/analytics/cplex-optimizer (accessed on 15 November 2021).

Article

Responsive Dashboard as a Component of Learning Analytics System for Evaluation in Emergency Remote Teaching Situations

Emilia Corina Corbu [1] and Eduard Edelhauser [2,*]

[1] Department of Mathematics and Informatics, University of Petrosani, 332003 Petrosani, Romania; corinacorbu@upet.ro

[2] Department of Management and Industrial Engineering, University of Petrosani, 332003 Petrosani, Romania

* Correspondence: eduardedelhauser@upet.ro; Tel.: +40-722-562-167

Abstract: The pandemic crisis has forced the development of teaching and evaluation activities exclusively online. In this context, the emergency remote teaching (ERT) process, which raised a multitude of problems for institutions, teachers, and students, led the authors to consider it important to design a model for evaluating teaching and evaluation processes. The study objective presented in this paper was to develop a model for the evaluation system called the learning analytics and evaluation model (LAEM). We also validated a software instrument we designed called the EvalMathI system, which is to be used in the evaluation system and was developed and tested during the pandemic. The optimization of the evaluation process was accomplished by including and integrating the dashboard model in a responsive panel. With the dashboard from EvalMathI, six online courses were monitored in the 2019/2020 and 2020/2021 academic years, and for each of the six monitored courses, the evaluation of the curricula was performed through the analyzed parameters by highlighting the percentage achieved by each course on various components, such as content, adaptability, skills, and involvement. In addition, after collecting the data through interview guides, the authors were able to determine the extent to which online education during the COVID 19 pandemic has influenced the educational process. Through the developed model, the authors also found software tools to solve some of the problems raised by teaching and evaluation in the ERT environment.

Keywords: online education; learning analytics; emergency remote teaching; responsive dashboard; eLearning and digital transformation of education; IT for education

Citation: Corbu, E.C.; Edelhauser, E. Responsive Dashboard as a Component of Learning Analytics System for Evaluation in Emergency Remote Teaching Situations. *Sensors* **2021**, *21*, 7998. https://doi.org/10.3390/s21237998

Academic Editor: Joaquin Ordieres Meré

Received: 29 October 2021
Accepted: 26 November 2021
Published: 30 November 2021

Publisher's Note: MDPI stays neutral with regard to jurisdictional claims in published maps and institutional affiliations.

1. Introduction

During the pandemic crisis, governments from various countries decided to force the closure of educational institutions and universities by implementing new learning models that could help the education sector to continue its work exclusively online. Several types of teaching were adopted, including mobile learning and blended learning. According to UNESCO, only 20% of countries globally were equipped with online teaching devices and programs before the pandemic [1].

In the academic year of 2020/2021, a pandemic year, the world's universities were forced, in most cases, to cancel face-to-face courses and to use online education. The responses of European countries differed, and online education took place to varying degrees—25%, 50%, 75%, or 100%, depending on local conditions. Alternating distance learning with present or face-to-face learning (F2F) has been the subject of several studies [2,3].

The drastic changes imposed by the pandemic have had a multitude of effects on learning environments, including open ones. Learning environments have undergone a changing trend from learning management systems to personal learning environments [4], and in terms of learning infrastructures, they have provided learning services from open educational resources to a classroom framework [5]. In the UK, according to the YouGov study in February 2021, it was shown that the adaptation of educational tools was achieved by using video conferencing applications such as Zoom, Google Meet, and Microsoft Teams, and for communication with students, Microsoft Teams and Google Classroom were used as collaborative platforms.

Teaching exclusively online has raised a multitude of issues, namely video transmission problems, slow access to platforms, untimely responses to questions in class, students' hopes to get corresponding guidance after class, and teachers' feedback communicated in a timely manner so as to improve the students' learning. Considering the resources used in multiple studies, proposals were made to use online resources for as many courses as possible, as well as to add as many activities as possible to online platforms to satisfy as many students as possible [6].

The learning analytics system through embedded dashboards was used by the authors to monitor the activities that generate the learning process. It was also used in the monitored activities, by analyzing algorithms using educational data mining techniques [7] and viewing information, to monitor the degrees of involvement and reflection.

The study was based on the proposed model and the designed and implemented application, and it was conducted in regard to the educational process in crisis situations or ERT situations, that is, online teaching and evaluation, representing an element of novelty in the field of online educational resources. Training carried out under the pressure of time with minimal resources is called emergency remote teaching (ERT). Through the application developed, the authors tried to solve some of the problems that this type of teaching raises. There is still the fear that by the end of 2021, or after a certain period of time, the pandemic experience will be repeated, which is why the authors focused on developing a package of software tools. Regarding the evaluation, in the second semester of the 2019/2020 academic year, which was carried out exclusively online, the authors thought about the need to develop tools that help teachers in teaching, both in the process of evaluating courses and evaluating the students. For this purpose, a model was designed for the evaluation, a system called the EvalMathI system. The EvalMathI system is software that supports teachers in their teaching activities, and it can be used for two evaluation processes, namely course evaluation and student evaluation. From the designed software, the dashboard tools necessary for monitoring and evaluating several disciplines were implemented and tested. In addition, an optimization module was added by introducing and integrating the dashboard into a responsive panel to facilitate and streamline the evaluation process.

The results of the study are based on the answers of the students involved in an investigation conducted by both authors. The sample was composed of 157 students in the 2019/2020 academic year and 143 students in the 2020/2021 academic year. Questionnaires were distributed, and 190 and 339 answered the survey each year, respectively. The students who responded can be considered representative from the perspective of the tested model and application.

The process of monitoring and the evaluation of the six courses through the EvalMathI dashboard were also attended by nine teachers from the three faculties of the University of Petrosani, the university where the study took place, namely four from the Faculty of Mining, three from the Faculty of Sciences, and two from the Faculty of Mechanical and Electrical Engineering. For each of the six monitored courses, tools were developed for the evaluation of the discipline through the parameters analyzed, highlighting the percentage achieved by each course on various components.

By elaborating the learning analytics and evaluation model (LAEM) and the proposed application—the EvalMathI system—the authors managed to coordinate the monitoring and evaluation activity of the six optional courses carried out with 300 students. Throughout this study, the authors aimed to answer the following questions regarding the central objective by evaluating the utility of the proposed learning analytics and evaluation model (LAEM) as well as the efficiency of the EvalMathI system software application:

Q1—How did EvalMathI affect the evaluation process of the courses in an ERT situation?
Q2—What are the best EvalMathI dashboard tools regarding content evaluation?
Q3—How is the content of each course assessed through EvalMathI in terms of relevance and applied scientific content, coherence, and consistency?
Q4—What skills were obtained by the students in completing the six courses?

2. Related Work

2.1. Emergency Remote Teaching (ERT) and Learning System Process

Lately, the teaching–learning process has been the subject of many studies; previous research has referred to different processes as blended learning, face-to-face learning, or eLearning. Because of the COVID-19 threat, universities and colleges have had to decide how they can continue teaching under conditions of acute uncertainty. Many institutions have opted to suspend face-to-face classes, including the operation of laboratories, opting for online courses.

In a normal situation, the planning, preparation, and elaboration of a completely online university course requires an elaboration time between six and nine months, time that the teachers did not have during the pandemic [8]. It was also impossible for every university professor to suddenly become an expert in online teaching and learning. The process of preparing and implementing online resources was prolonged over time, increasing stress and pressure, and it was sincerely acknowledged that there were many online teaching experiences in which instructors failed to fully prepare materials, leading to a large probability of suboptimal implementation [9].

In this time of crisis, teachers and students have made the best use of the available resources, but it must be acknowledged that there is a big difference between vocational training on online platforms and education carried out under the pressure of time with minimum resources—emergency remote teaching (ERT) [8].

In contrast to activities that are planned and designed from the beginning to be online, emergency remote teaching is a temporary form of education that is used to carry out the training process in an alternative way in crisis circumstances. This mode of training involves the use of entirely remote teaching solutions. The main objective is not to create a robust educational ecosystem but rather to ensure temporary access to training in a rapid manner in an emergency or in a period of emergency crisis. This is the main difference between ERT and classic online learning [8].

During the pandemic, some educational institutions supplemented the hardware support made available to teachers for conducting online courses. The technical staff of universities were also involved to facilitate the educational process. Carrying out all activities exclusively in an online environment meant flexibility in the teaching and learning processes [10,11], but the problems raised by teaching exclusively online were multiple because many of the systems that provided the resources were overworked and even exceeded capacity [9].

In ERT, speed and "just get it online" are detrimental to the quality of the course, so the authors of [9–11] believe that courses created in an ERT situation should be a temporary solution to an immediate problem. Further, the principles of the universal design for learning (UDL) should focus on creating flexible, inclusive, student-centered learning environments that ensure that all students have access to courses and activities [12].

2.2. Learning Analytics Dashboards and Learning Analytic Evaluation Models

Siemens (2010) defines learning analytics as "the use of intelligent data, learner-produced data, and analysis models to discover information and social connections, and to predict and advise on learning". In recent years, a series of tools have been developed to monitor and/or to evaluate learning activities and then to visualize them in learning dashboards. Learning dashboards, depending on how the teaching process is carried out, are divided into three categories—dashboards based on face-to-face courses, dashboards for face-to-face work groups, and dashboards that work in online learning or blending learning [13,14].

Dashboards that collect data from traditional face-to-face courses have been integrated into multiple applications, such as Backstage [15], which is a dashboard that displays Twitter activity during the course activities [16], Classroom Salon [17] which allows teachers to create, manage, and analyze social networks to view the contribution of each member, and in which the dashboard allows for viewing the contribution of each member to view the correct answers received, and finally, Slice 2.0. [18], which is a system that interconnects the teacher's slides with the students' devices, and the teacher can view the students' annotations.

Several dashboards for working in face-to-face groups and a classroom orchestration design [19] have been developed, including TinkerBoard [20], Collaid [21], and Class-on [22–24], and some of them were developed by taking data from OpenSocial [25].

A learning analytic evaluation model usually contains the following groups of parameters: the relevant student actions, captured data on relevant actions, awareness, reflection, sense-making, and impact, and effectiveness, efficiency, and usefulness.

1. Relevant student actions. In face-to-face teaching, time and date, location, who and what kind of device they use, and even the background sounds were analyzed, and each item has a greater or smaller significance. Students find that social interaction is somewhat useful, especially for blended or face-to-face courses [26]. Teachers consider the visualization of social interaction more important because it is useful in identifying students who do not collaborate with others or those who collaborate excessively. Teachers consider that the effort of students is very important. However, the usefulness is seen differently by teachers and students. Students often believe that the data collected does not reflect the effort made, while teachers perceive the data to be useful for an in-depth view and possibly for the identification of students potentially at risk. All the data mentioned above refer to quantitative data, which is why it was investigated whether it was possible to increase the qualitative data, such as the number of re-tweets or comments on a blog post, and whether that can indicate the relevance of such communication. This idea was already explored by the authors of the Backstage dashboard [16].

2. Capture data on relevant actions. Many learning analytic dashboards take data from virtual sensors through certain tools and resources, such as laptop or desktop user interactions and social media, through hashtags or blog comments. One of the main problems tracking learning activities, an aspect called automated tracking. Previous studies have shown that students rate the usefulness of dashboards as low when some of the activities covered take place outside of the learning environment [27]. Regarding personal learning environments, which include a wide variety of tools and services that aggregate data from various sources—Twitter, blog posts, comments, software environments, and so on—the feedback is positive. The use of sensors such as cameras or microphones to capture student data for monitoring and counseling activities is the subject of computer-supported collaborative learning (CSCL) research. They are present to a small extent in learning dashboard applications because they provide real-time feedback to students or teachers [28].

3. Awareness, reflection, sense-making, and impact. Bakker et al. [29] presented research on the use of sensors to capture physiological responses and to both estimate stress levels and provide feedback to employees on their current work schedule. Such research on the level of awareness and reflection in learning environments and the impact of such awareness have been presented at the Learning Analytics and Knowledge (LAK) conference [30,31].

4. Effectiveness, efficiency and, utility. Efficacy has been measured in terms of better engagement [24,32], higher grades [26,33,34], post-test results [34], lower retention rates [26], and improved self-assessment [35]. The results of a long-term experiment regarding course signal [26] indicated that there is an impact on retention rates and grades. There was also a significant difference in improving self-assessment in an evaluation of the

CALMS system [36]. An evaluation of TADV [20] indicated that the overall satisfaction with the course for students using the dashboard was higher, satisfaction that was measured in terms of self-esteem and the recommendation of the course to other students. In other experiments, student involvement has been measured to obtain information on the potential impact. The results of Morris et al. [24] indicated that there was no increase in involvement when learners used the scoreboard. The results of iTree [35] indicated that the dashboard does not encourage learners to post messages on a forum, but there is an increase in the reading of posts. Efficiency was measured in a Class-on assessment experiment [23], which assessed whether using a dashboard during class sessions helped to distribute a teacher's time more accurately. Usage and utility evaluations were performed—either by teachers or students, or both. The perceived usefulness of the Student Inspector [37] and LOCO-Analyst [21] dashboards was, for example, evaluated by teachers and was high for both categories of dashboard users. The results of the evaluations for SAM and StepUp indicated that the perceived utility is often greater for teachers than for students [38]. The results of the LOCO-Analyst assessment [21] also indicated that the perceived utility was significantly higher in a case study in which several data points were used to provide a perspective on the learning activity.

2.3. Dashboards for Online Learning

For retrieving data from an online environment, dashboards for online learning have been created for online learning and blended learning. Course signals [26] predict and visualize learning outcomes based on three data sources—grades in the course so far, time spent on a task, and past performance.

The dashboard developed by Carnegie Mellon University [39] is highly detailed, and concepts and how they are carried out on different course activities may need additional attention from the student. Displaying various parameters differentiates dashboards for online learning. A student activity meter (SAM) [38] displays the progress of a particular course and illustrates the time spent by students in different study environments. LOCO-Analyst [40], the Moodle dashboard [41], and GLASS [42] are tools that visualize student feedback and different levels of performance in different ways. Student Inspector [43] visualizes the use of data in the Active Math environment. Tell Me More [44] provides visualizations of exercise results. The CALM system [45] visualizes comparative levels of knowledge through self-assessments [33].

There are models with components dedicated especially to students, such as Teacher Advisor [46], which is based on manual interventions to automatically generate tips for the student, or StepUp [47], which is a model designed for mobile devices for students who apply learning analytics techniques for awareness and self-reflection.

In their latest work, Vieira et al. [48] analyzed visual learning analytics and concluded that there are few studies that have simultaneously deepened complex visualizations and educational theories, a statement also supported by Jivet et al. [49], who analyzed learning dashboards from the students' point of view.

From the category of learning dashboards used on cloud platforms [50], Amazon Web Services offers two solutions for monitoring—AWS CloudTrail resources [51], a managed service to track user activity and API usage, and Amazon CloudWatch, a monitoring service of cloud resources and applications. There are several workarounds on the market that offer more powerful dashboards for cloud monitoring, including Opsview Monitor [52], Spectrum [53], SignalFx [54], and AWS Cloud Monitoring [55]. However, these are all expensive enterprise solutions that are difficult to use in academic or education fields in countries like Romania.

Lonn, in one of the most recent studies focused on learning analytics tools that use dashboards that measure performance, stated that they may decrease learner mastery orientation and the students' exposure to graphics of their academic performance may negatively affect the students' interpretations of their own data as well as their subsequent academic success [56]. Thus, Sedrakyan et al. focused on feedback and its speed in certain activities during laboratory hours to improve the feedback given to students, and they did not focus on academic performance [57]. Irons et al. [58] stated that there is a direct proportionality between student feedback and impact, so the faster the feedback, the more substantial the impact in learning; thus, the ability to give timely feedback is very important.

From the analysis, it can be stated that it is a challenge to retrieve data on the evaluation process from emergency remote teaching, so one of the contributions of this study is that it proposes a tool that is actually a system that comes both in support of teachers as well as students working in an ERT situation. This tool was partially validated in the period of 2020/2021 at a university in Romania.

3. Materials and Methods

From the analysis performed and from the evaluation of the difficulties encountered in the teaching and evaluation process in ERT, the authors proposed a method of collecting data on the results of different teaching and evaluation activities, as well as their processing and subsequent visualization in a certain form of the activities monitored, minimizing the time required for corrections.

The proposed learning analytics and evaluation model (LAEM) was developed by the authors and is intended to be a support for teachers conducting ERT by collecting, integrating, and analyzing data from various sources through a semi-automated process. Subsequently, the designed software tool, the EvalMathI system is intended to be a support for professors at the University of Petrosani in their attempt to automate and streamline certain components of the evaluation process. EvalMathI proposes a solution for teachers and students to access information about courses by completing questionnaires regarding the evaluation process. EvalMathI was also tested according to the evaluation process, and the results indicate that it is a responsive dashboard.

In order to establish the component elements of the input data for the learning analytics block, data collected from the learning activities and data that evaluate the activities carried out on various platforms were analyzed on the basis of the project POCU 12596, implemented by the authors as project managers at the University of Petrosani, and financed by the European Commission through the Romanian Operational Program Human Capital (OPHC) [59]. The virtual class platform for the courses was analyzed, monitored, and evaluated [60,61].

3.1. The Learning Analytics and Evaluation Model (LAEM) Design, Based on Virtual Sensors

In developing the model for the proposed assessment system, the learning analytics process model (LAPM) proposed by Verbert, K. [62] was adapted by the authors, and the modified model is detailed in Figure 1, model in which the project's groups of parameters are represented.

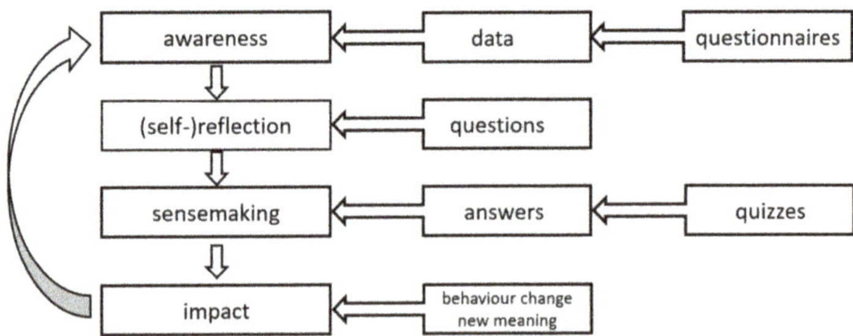

Figure 1. The learning analytics process model adaptation of [62].

This model was customized by the authors for the conditions in which the teaching–learning process took place by ERT at the University of Petrosani in the two pandemic years. In developing the LAEM model, the authors analyzed the following groups of parameters: the relevant student actions, captured data on relevant actions, awareness, reflection, sense-making, and impact, and effectiveness, efficiency, and usefulness, all of which were captured by virtual sensors. Then, the groups of parameters analyzed in the LAPM were customized for LAEM as well as the EvalMathI system.

In developing the LAEM model for the design of the EvalMathI system, the time and date, the location, who and what kind of device they use, the background noise, the use of resources, and the results of tests and exercises were taken into account. These results are the main data taken from the EvalMathI system dashboard to be a support in the evaluation process performed by teachers. The LAEM model that generated the EvalMathI system dashboard retrieves data from virtual sensors through laptop or desktop interactions. In applying LAEM and EvalMathI system customization for this parameter, we tried to assess awareness by measuring the feedback received in the virtual classroom, but this developed system will be the subject of further research.

This study answered questions related to the evaluation of courses conducted in an ERT situation through our own design, EvalMathI, software that displays the status of indicators monitored for the evaluated courses in a responsive dashboard.

The tool does not evaluate the students' activity at this moment, but the authors intend to further develop that part as well. Thus, regarding the analysis performed on existing solutions, the authors considered the possibilities of collecting data from the virtual classes and implementing a virtual sensor for them.

Customizing the LAEM and EvalMathI system for this parameter measured the effectiveness from the perspective of the students because they obtained better results in the tests measured in the second series of courses, demonstrating the efficiency and usefulness of EvalMathI.

The conceptual scheme of the learning analytics and evaluation model presented in Figure 2 applied in ERT was elaborated by the authors. In this model, the learning activities included reading, lectures, quizzes, projects, media, tutoring, homework, research, assessments collaboration, social media, and discussions, resulting as input data for learning analytics.

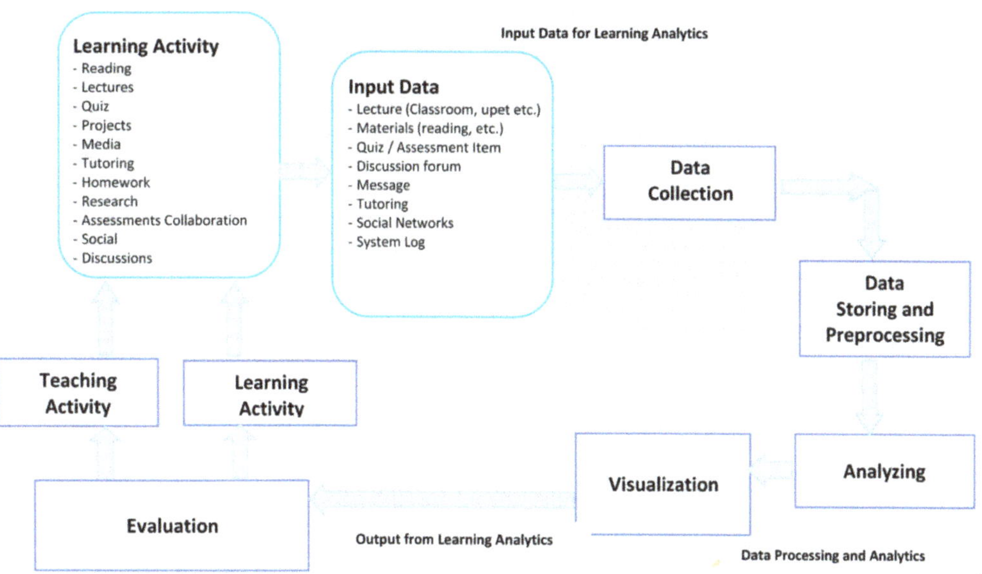

Figure 2. The learning analytics and evaluation model proposed by the authors for designing the EvalMathI system in ERT cases, based on virtual sensors.

Following the research carried out on various learning models, starting with analysis, design, development, implementation, and evaluation (ADDIE) [43], the authors established the main stages of the process. In the next stage, the authors analyzed the proposed CIPP evaluation model [12] and the proposed adaptive learning model [37]. The authors proposed the model from Figure 2 as the best for exclusively online courses in ERT.

The learning analytics and evaluation model proposed for the EvalMathI system, and schematically presented in Figure 2, shows the way in which data taken from learning activities constitute input data in the data collection. The main learning activities that the projected model considers are reading, lectures, quizzes, projects, media, tutoring, homework, research, assessments collaboration, social media, and discussions. From these activities, the input data block consists of lectures, materials, quiz/assessment items, discussion forms, messages, tutoring, social networks, and data from the system log. The data processing and analytics block incorporates the data collection and storing, and the preprocessing, analyzing, and visualization. The output of this block constitutes the input data for the evaluation block. The evaluation block analyzes the data from the learning analytics and sends an indicator of the situation at a certain course to the teaching activity as well as one indicator to the learning activity. Each of the two activities should be rethought or corrected in order to increase the indicator's value.

3.2. Design and Development of the EvalMathI System Software

The proposed LAEM model was validated by the EvalMathI system, a component of which helps teachers in evaluation processes, which is why the LAEM model was used and the application was designed and developed. The EvalMathI system is a web-based application that can be used by students and teachers. The application also allows for testing as well as disciplinary evaluation and monitoring.

For the database, in the 2020/2021 academic year, the authors used the MySQL database for teachers, students, and the courses created; it will be extended later to the administrative staff. For the development of the EvalMathI system, technologies were used for indexing, searching, and analyzing the data available under the Apache server, Apache 3.2.4 Open Source License [63].

The EvalMathI system admin panel and the main responsive are presented in Figure 3, representing two solutions. Each of the two solutions performs different sections in the evaluation process. EvalMathI was developed in PHP and collects data from teachers and students using pages created for each type of user. The evaluations made by the experts are collected using Elasticsearch and are then processed with Kibana.

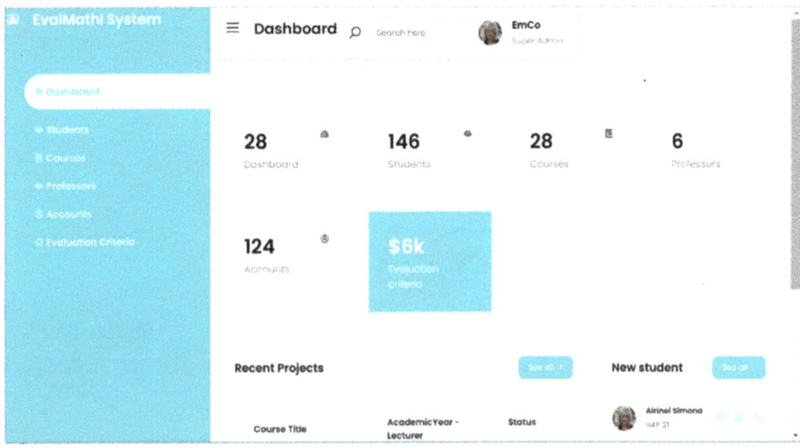

Figure 3. EvalMathI system admin panel and main page.

From a structural point of view, the EvalMathI system has access to the MySQL databases of students, teachers, and courses. It provides aggregated information regarding the use of resources at a given time in a given framework by a particular user, detailed information about a specific activity carried out by a specific student in a specific setting, and the percentage of progress made for a certain course in a certain moment of time, on certain parameters studied, evaluated by a group of experts with well-defined roles in the activity.

Users with a teacher profile can view each student's progress in a particular course. The scores obtained for tests taken and the degree of completion of each session can be tracked. The benefits of the dashboard for this profile include the monitoring of several students at a time, providing automatic feedback from students as well as obtaining useful parameters for evaluation.

Users with a student profile can view the evaluation of their progress for each test taken, the tests they still have to complete if the instructor agrees, and a comparison of their progress with the rest of the class. The benefits of the dashboard to this profile include self-regulated learning, planned learning, thinking and evaluating tasks and contexts [64], motivation [65], and an overview of the class.

The EvalMathI system offers users with the system administrator profile the ability to view detailed or aggregated resource consumption, to view history, use resources, and monitor current resource consumption. Elasticsearch, a popular and powerful distributed search and analytics engine based on Apache Lucene and designed for horizontal scalability, reliability, and easy management, was designed to further connect the system to other databases. It combines the speed of searching with the power of analytics via a sophisticated, developer-friendly query language, covering structured, unstructured, and time series data [66]. For the visualization and presentation, Chart.js was used, a visualization platform that allows for the interaction between the graphs, and the Kibana tool [67] was used for the subsequent developments of the application that will include histograms and geo maps.

3.3. Data Collection through the Activity Analysis Sheet (AAS) and its Validation

Nine teachers were used as evaluators from the three faculties of the University of Petrosani. They participated in the monitoring and evaluation processes of the six courses using the EvalMathI dashboard. Four were from the Faculty of Mines, three from the Faculty of Sciences, and two from the Faculty of Mechanical and Electrical Engineering. Six optional courses were monitored, namely Sustainable Development, Creativity and Innovation, Environmental Management, Renewable Energy, Cyber Security, and Web Programming. These optional courses were attended by students and master students, with 157 students in the 2019/2020 academic year and 143 students in the 2020/2021 academic year.

3.3.1. Instruments and Investigation Tools

An online questionnaire was the investigation tool selected for this study, as an online survey was the most significant and reliable method for conducting this study. The collected data were processed by elaborating the structure of the data matrix and encoding the answers of the questionnaire applied. The results of the questionnaires were processed using the IBM SPSS Statistics application, version 23, with which the variables were also verified.

The study was based on three online questionnaires. Two of them were given to the students, the "Questionnaire for evaluating the teaching activity", composed of 17 questions, and "Course observation", composed of 31 questions. One was distributed to the teachers—"Activity product analysis", composed of 31 questions.

The two surveys completed by the students took place over two stages. The first one was completed by 157 students in May 2020, and the second one was completed by 143 students in March–May 2021. The 300 students involved in the POCU 122596 project were monitored during six online courses carried out within the project. The students chose one or more of these six courses, and the total number of students involved in our survey was 529.

The survey completed by the teachers who presented the six courses took place during the same period and a group of nine experts, teachers—also called evaluators—participated during the same POCU 122596 project. Each of the nine experts evaluated the curricula, the teaching, and the evaluation system of the three courses in six different stages of the project, for a total of 162 responses.

3.3.2. Questionnaire Validation

By using the three proposed questionnaires, the authors' intention was to evaluate the curricula of six courses used in an online teaching and evaluation system through seven parameters—Content Design Intra- and Interdisciplinary Relationship (CDDR), Content Design Intra and Intercurricular Relationship (CDCR), Relevance for Life and Applied Scientific Content (RLASC), Degree of Structuring (DS), Degree of Systematization (DSY), Coherence (CH), and Consistency (CS). The percentages achieved by each course were determined regarding components such as content, adaptability, skills, or involvement.

The construct validity of the questionnaire was tested using the Pearson correlation matrix of major variables related to the online level. In addition, the fidelity and internal consistency of the questionnaire were tested using the Cronbach alpha for the multiple Likert questions.

After receiving the data from the respondents, they were processed accordingly using inferential statistics such as the Cronbach alpha coefficient to assess reliability. These statistical analysis tools were used to process the questionnaires given to the target group. To transform the information gathered with the questionnaires, the authors used the variables in SPSS—nominal, ordinal variables, which are qualitative variables, and the range and ratio variables, which are quantitative. To assess the reliability, the Cronbach coefficient was used; as indicated by Sekaran, a Cronbach alpha coefficient of 0.70 or higher is considered reliable and acceptable.

The authors assessed the reliability using the Cronbach alpha coefficient for the main variables. A value of 0.829 indicates a high level of internal consistency in this study with this specific sample. The results for the seven parameters that evaluated the curricula and the correlation matrix show the strength of the association between the variables, as demonstrated in Table 1. The Cronbach alpha coefficients for all variables were well above the threshold of 0.70, and it can be deduced that the results meet the reliability hypothesis and that the reflective constructs have sufficient reliability.

Table 1. Summary of processed cases of variables. Reliability statistics.

Variable	Cronbach's Alpha
Content Design Intra- and Interdisciplinary Relationship (CDDR)	0.964
Content Design Intra- and Intercurricular Relationship (CDCR)	0.829
Relevance for Life and Applied Scientific Content (RLASC)	0.835
Degree of Structuring (DS)	0.872
Degree of Systematization (DSY)	0.856
Coherence (CH)	0.834
Consistency (CS)	0.829

3.3.3. Population and Sample—Respondents

This study was based on the responses of the students of two faculties of the University of Petrosani, the Science and the Mining Faculties. The total number of students of these two faculties was 2153. It is important to mention that the University of Petroșani has only three faculties; the authors selected the Science and the Mining Faculties because their teaching activities involve students from only these two faculties, so the students could be easily contacted, and as a result, their responses were representative. The 529 answers represent 25% of the total number of students, as can be seen in Table 2. The students of these two faculties who answered the questionnaire are students in the bachelor's or master's degree programs.

Table 2. The number of students who answered the questionnaire from the Mining and Science Faculties of the University of Petroşani.

Online Monitored Course	Academic Year	Number of Students/Course	% of total Number of Students/Academic Year
Sustainable Development	2019/2020	74	47%
Creativity and Innovation	2019/2020	41	26%
Environmental Management	2019/2020	18	11%
Renewable Energy	2019/2020	31	20%
Cyber Security	2019/2020	12	8%
Web Programming	2019/2020	14	9%
Sustainable Development	2020/2021	69	48%
Creativity and Innovation	2020/2021	49	34%
Environmental Management	2020/2021	58	41%
Renewable Energy	2020/2021	60	42%
Cyber Security	2020/2021	63	44%
Web Programming	2020/2021	40	28%
Total		529	

The study was based on the University of Petrosani, a small university that can be considered representative, not for the whole Romanian education system, but for the small Romanian universities. This assumption can be based on the fact that small universities are similar in the field of online education because, unfortunately, they started the massive implementation and use of eLearning only after 15 March 2020. The Romanian National Council for the Financing of Higher Education statistics provide the total number of Romanian students for the 2019/2020 academic year—459,899. The same statistics show that out of the total 49 Romanian State Universities, 32 are small universities with around 5000 students, representing 30% of Romanian students. On the other hand, these small universities are very important and representative because most of them are comprehensive universities with the most fundamental areas of study, such as engineering, social sciences, and humanities. The University of Petrosani is a very small university, with 3565 students and engineering and social science areas of study.

In the field of eLearning, all these small universities are similar; they did not previously have distance learning, so they had to adjust very quickly to the new conditions of education generated by the COVID-19 Pandemic. Additionally, these small universities could not afford to buy an eLearning platform produced by one of the major world players in this field. These small universities have adopted low-priced eLearning platforms, such as LMS, developed by Moodle, or free collaborative educational platforms.

This study was also based on the responses of the evaluators from the same university, teachers that evaluated other teachers during the development of the POCU 122596 project in the 2019/2020 and 2020/2021 academic years. Each of the nine experts evaluated the curricula, the teaching, and the evaluation system through seven main parameters—CDDR, CDCR, RLASC, DS, DSY, CH, and CS—for three courses at six different stages of the project, for a total of 162 results. This number of evaluation results could be approximated as 10% of the total number of courses at the University of Petrosani that are evaluated by one evaluator per course at one time.

3.3.4. Questionnaire Description

To monitor the activity, we used a questionnaire called the "Activity Products Analysis Sheet" (APAS). The APAS was carried out in APAS version S1 and APAS version S2 for each series of students who participated in the courses. For the monitoring of the courses, the Analysis Sheet had the following components: (1) Analysis of the course content; (2) Adaptation to the requirements of the project; (3) Competences brought to the students; (4) Determination of the students' activity; (5) Other items.

1. Regarding the content analysis, the analyzed parameters were whether the content design supports the intra- and interdisciplinary relationship, whether the content design supports the intra- and intercurricular relationships, whether the course design supports the lifelong relevance of the content, the applied scientific content, the degree of structuring, the degree of systematization, course coherence, and course consistency.

2. Regarding the adaptation to the project requirements, we analyzed the following parameters: to what extent it offers intellectual activity skills; to what extent it offers skills of applicative activity; it is correlated with the project objectives.

3. Regarding the competences taught to the students, the following parameters were monitored: to what extent the courses offer professional, social, or other competences.

4. In order to determine the students' activity, the following parameters were monitored: the attitude towards learning; attendance at the course; the attitude and responsibility of the students towards solving the work tasks; collaboration in the learning process; the degree of use of knowledge, skills, and attitudes in new learning contexts; the progress made by the students during the course.

These parameters were not included in this course analysis and evaluation study because they were input data for the student evaluation. These parameters influenced the learning activities (LA) in the LAEM model and not the teaching activity. These parameters are described as the parameters of the evaluation block of the LAEM.

5. Other elements taken into account were the educational environment, the place where the course took place and the platform used; funding, material resources, and curricular auxiliaries used; presentation of materials, student work, the general atmosphere during the course, and other observations

In the questionnaires used at the first stage of the study, there were some questions that refer to these parameters, but they were not used in the present study. During the first stage of the study, the 2019/2020 academic year, the authors recorded data for these parameters, but the authors did not interpret them, as they were not considered representative.

4. Case Study Results: Dashboard for Courses Evaluation

4.1. EvalMathI System Dashboard for Courses Evaluation—Beta Version

The dashboard designed by the authors for the course evaluation was developed in the first stage in a beta version in the 2019/2020 academic year and tested in the first semester of the 2020/2021 academic year when the University of Petrosani transitioned all teaching activities to online in an emergency remote teaching (ERT) situation. In ERT, the courses were held in a video conference system, and the main platforms used were Zoom, Google Meet, and Microsoft Teams. The activities carried out during laboratory classes were developed in virtual classes created for each course. In the virtual classes, the activities carried out by the groups of students for each discipline were monitored. The student assessments were also conducted online by developing tests and quizzes in Microsoft Forms. The exams at the end of the first semester were taken by participating in a video conferencing system and completing quizzes.

The beta version of EvalMathI presented in Figure 4 was designed to evaluate six courses in the 2019/2020 academic year and contains filters that are suitable for different categories or fields. It can also offer information for a selected course.

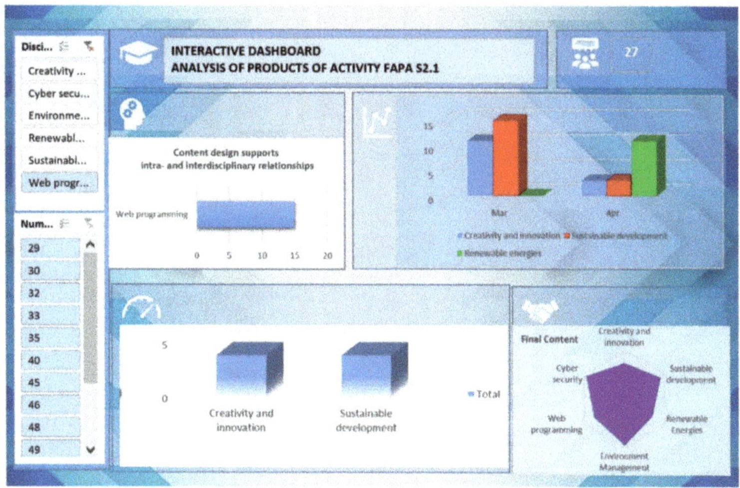

Figure 4. EvalMathI dashboard for courses panel—beta version.

4.2. EvalMathI System for Course Evaluation

The answers to the four questions, Q1, Q2, Q3, and Q4, which support the central objective of the study—the design and testing of an evaluation model under ERT condition— are presented as follows:

Q1. How did EvalMathI affect the evaluation process of the courses in an ERT situation?

For the interactive communication between teachers and students, the authors designed a window in the application with which users can interact to share information and communicate. Regarding the optimization use mode, an attempt was made to improve the stability, security, and compatibility of the platforms. Because the method of archiving information was scattered among a number of applications, and their unification required consistent effort, the authors thought it would be useful for the application to gather the information needed for an evaluation in a single window, and based on this information, the evaluation can be carried out. The optimization of the usage method was solved by inserting and integrating the dashboard in a responsive panel in order to facilitate and make the evaluation process more efficient.

4.2.1. EvalMathI System Responsive Application Programing Interface (API) Optimization

The responsive dashboard of the EvalMathI presented in Figure 5 was developed for the course evaluation and tested during the pandemic. By introducing and integrating the dashboard in a responsive panel, the functionalities of the applications were optimized, and at the same time, the evaluation process was facilitated and streamlined. After collecting the data through the interview guides, the authors were able to evaluate the level of influence on online education induced by the COVID-19 pandemic, and its influence on the courses throughout the studied period.

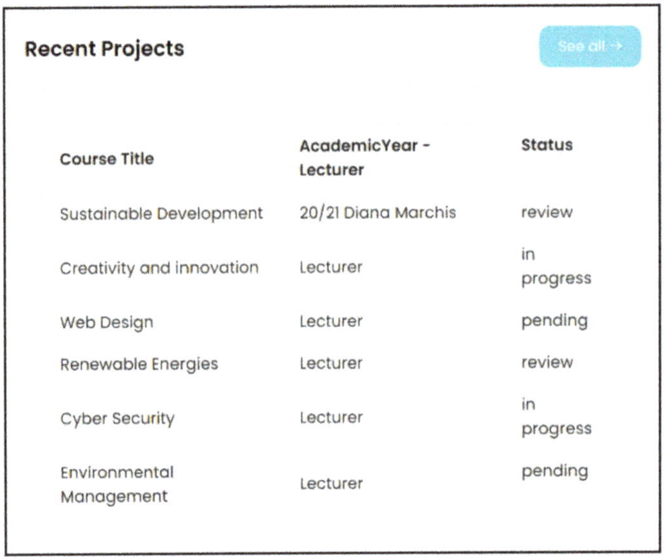

Recent Projects

See all →

Course Title	AcademicYear – Lecturer	Status
Sustainable Development	20/21 Diana Marchis	review
Creativity and innovation	Lecturer	in progress
Web Design	Lecturer	pending
Renewable Energies	Lecturer	review
Cyber Security	Lecturer	in progress
Environmental Management	Lecturer	pending

Figure 5. Responsive API EvalMathI—courses panel.

4.2.2. Results of Monitoring and Evaluation of the Courses with EvalMathI Dashboard Software

Six online courses were monitored in the 2020/2021 academic year with the EvalMathI dashboard for this study. For each of the six monitored courses, the evaluation of the course was carried out by the analyzed parameters highlighting the percentage achieved by each course on various components. With the help of the dashboard in the EvalMathI system, the activity carried out for the six optional courses was monitored, following the evolution of several parameters, such as content, adaptability, skills, and involvement.

Q1. How did EvalMathI affect the evaluation process of the courses in an ERT situation?

The final version of the responsive dashboard and EvalMathI, presented in Figure 6, represent two solutions, each of which performs different parts in the evaluation process. EvalMathI was developed in PHP and collects data from teachers and students using pages created for each type of user. The evaluations made by the experts are collected using Elasticsearch and then processed with Kibana. EvalMathI uses dynamic pages for creating dynamic content. The contents of the pages integrated with Elasticsearch are easy to analyze, being very intuitive and interactive. Field values can be easily seen with various filters. The responsive dashboard displays four panels, each of them answering questions regarding the evaluated courses—Content Design Intra- and Interdisciplinary Relationship (CDDR), Relevance for Life and Applied Scientific Content (RLASC), Profession Student Skills (PSS), and Degree of Structuring (DS).

Q2. What are the best EvalMathI dashboard tools regarding content evaluation? What does this response mean for the intra- and interdisciplinary relationship indicator (CDDR)?

In order to determine the content according to the CDDR indicator, the first panel, presented in Figure 7 from the responsive dashboard indicates the time interval of the courses' development between March and June 2021. The results of the indicator have a median value between 3.5 and 4, which indicates that the course content sections at different times of evaluation corresponded by an 87.5% proportion.

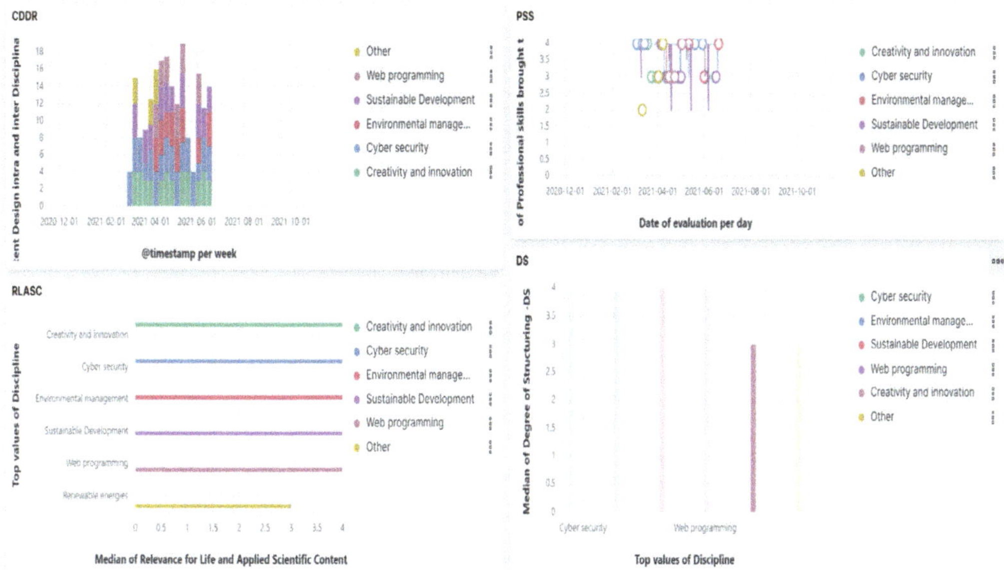

Figure 6. Final version of the responsive dashboard presenting Content Design Intra- and Interdisciplinary Relationship (CDDR), Relevance for Life and Applied Scientific Content (RLASC), Profession Student Skills (PSS), and Degree of Structuring (DS).

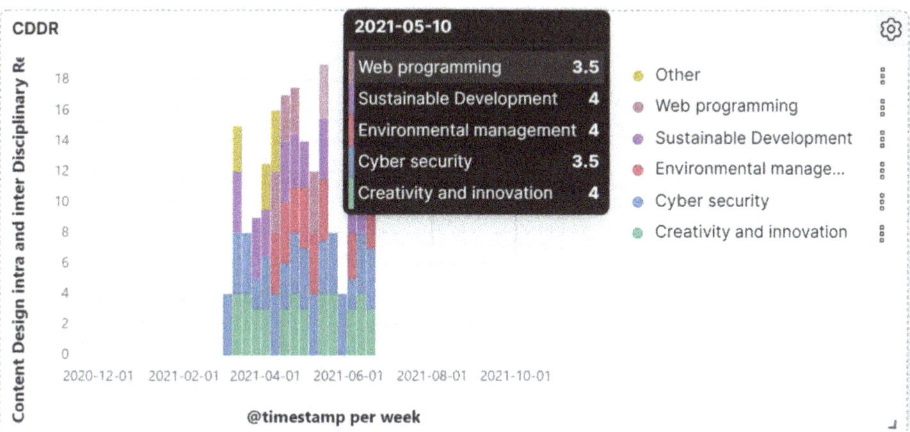

Figure 7. Responsive dashboard presenting Content Design Intra- and Interdisciplinary Relationship (CDDR).

Q3. How is the content of each course assessed through EvalMathI in terms of relevance and applied scientific content, coherence, and consistency?—What response does the responsive dashboard give regarding the relevance for life of each evaluated course, the RLASC indicator?

The result presented in Figure 8 can be interpreted with the assumption that there were chapters or sections of this course for which the applied scientific content made up 75% or less of this indicator. The presence of the "Other" category for this discipline indicates a lack of completion of some fields. When fields are not completed, Elasticsearch interprets the results in this manner.

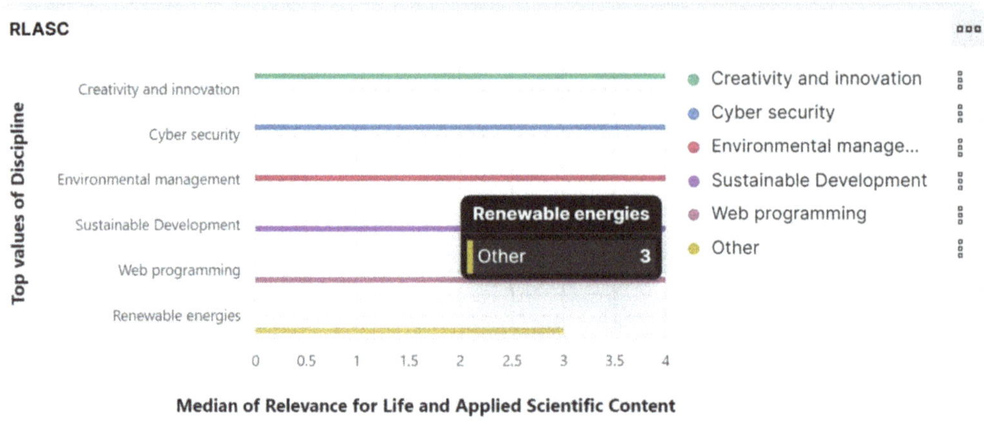

Figure 8. Responsive dashboard presenting Relevance for Life and Applied Scientific Content (RLASC).

Q3. How is the content of each course assessed through EvalMathI in terms of relevance and applied scientific content, coherence, and consistency?—What response does the responsive dashboard give regarding the degree of structuring (DS)?

Regarding the Degree of Structuring (DS) and the Degree of Systematization (DSY), the indicators presented in Figure 9, the results extracted from Kibana are relative and investigations should be continued.

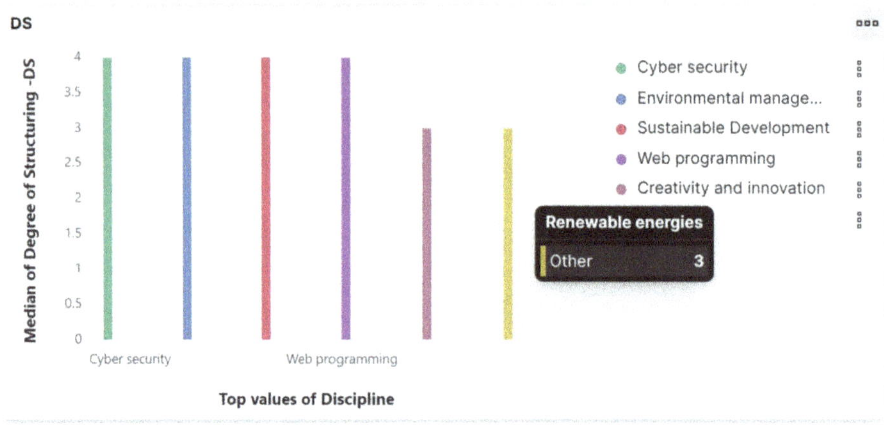

Figure 9. Responsive dashboard presenting Degree of Structuring (DS).

Q4. What skills were obtained by the students in completing the six courses? What response does the responsive dashboard give regarding the PSS indicator?

Interactively, by positioning the cursor on each position, as it is presented in Figure 10, the authors found the responsive dashboard response for each course as the average value of the PSS indicator on the date it was completed. During the first stage, in the beta version, when data were collected with FAPA S1, the authors used Chart.js for representing the data in a responsive zone using responsive objects such as panels of courses, canvases, and so forth. In the second stage, the authors redefined the responsive dashboard concept because the application contained responsive charts. With Chart.js, the authors represented the responsive graphs and chose Chart.js because it set a certain value for that container.

All the data were resized and the graph was responsive, responding at the width of the container, which was very useful in that stage. During the second stage, the authors added the Elastic Search component and the meaning of responsive was expanded.

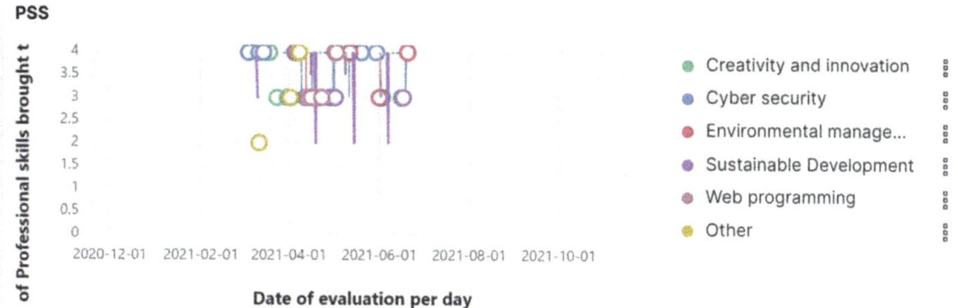

Figure 10. Responsive dashboard presenting Professional Student Skills (PSS).

5. Discussion

In the process of developing the LAEM model, the authors of the current study analyzed other learning models, such as the ADDIE model [43], the CIPP evaluation model [12], and the adaptive learning model [66]. Based on the presumption that the time factor is essential in ERT situations, in the present study, the authors proposed a model, called LAEM, which was adapted to the requirements of ERT for the academic level in which the evaluation block would give a direct answer to the teaching and learning activities.

The COVID-19 pandemic, perhaps one of the longest disruptive periods, except for times of wars, has irrevocably affected education and all related activities. Researchers around the world, whether they work in education or not, have studied the effects on all stakeholders involved in the education process. Many of them have also tried to look to the future and to estimate how the new face of education will look. A search in important databases such as the Web of Science Core Collection, Elsevier's Scopus, The Directory of Open Access Journals, and Springer, and using the keywords "learning analytics" or "emergency remote teaching" or "responsive dashboard" led the authors to almost 600 results. A more refined search made within the open-access database journals published by MDPI, with the words "learning analytics" or "emergency remote teaching" or "responsive dashboard", led the authors to 15 results, important papers published mainly in the Sustainability, Mathematics, Education Science, and Sensors journals during the COVID-19 period.

The result of the present study can be compared with 10 of these relevant papers [68–77]. While other researchers have analyzed the effects of ERT on high school teachers [68], state universities [69], and the challenges faced by educational institutions [70], for the proposed model, the developed EvalMathI system was tested to be able to answer questions Q1–Q4, questions that support the development of the model for the evaluation system (LAEM), and also validate the software instrument called EvalMathI. Other previous studies have shown that the teaching process in ERT can be improved mainly by improving the method of interactive communication and by optimizing the use of resources. Thus, for interactive communication between teachers and students, first, a new window was designed in EvalMathI, with which users can interact so that they can share information and communicate. The authors answered Q1—How useful is EvalMathI in evaluating courses in an ERT situation?—by introducing and integrating the dashboard in a responsive panel to facilitate and streamline the evaluation process. In addition, other researchers have previously analyzed students' performance in an ERT situation [71], the challenges faced by math teachers in an ERT situation [72], the level of emotions in the learning process [75], the factors influencing home learning [78], and students' emotions and the perception of teachers in ERT [76,77]. In this context, the present study analyzed the methodology of evaluation in ERT conditions and proposed a tool called EvalMathI, which was tested in a case study of six courses conducted in ERT at our university.

In another study [73], the authors dealt with the evaluation of the quality of courses in ERT situations, proposing a dashboard of indicators for decision making. However, the present study found results in the content of each course evaluated through EvalMathI in terms of the intra- and interdisciplinary relationship and the intra- and intercurricular relationship. In the present study, considering the relevance and the application of the content, EvalMathI highlighted two parameters for each discipline. Considering the degree of structuring and the degree of systematization of the content, the response to Q3 shows that EvalMathI evaluates each course. Thus, the authors could determine which courses had high scores, based on two indicators, the coherence and consistency of the contents. It was determined that the Cyber Security course obtained good scores, while the Web Programming course needs to be rethought to improve these indicators.

6. Conclusions

The sudden transition from face-to-face (F2F) education to online made in the education field in March 2020, without prior teacher training and without minimal knowledge of the tools used in online education by the students, has generated an educational flaw. Very few Romanian universities developed dedicated online platforms before the pandemic, which is why switching to online education actually meant switching to an ERT education system.

Based on their higher level of management, countries from western Europe have managed to overcome the pandemic waves, but at this time, Eastern Europe still faces major problems generated by the pandemic, problems that also affect education at all levels. Suddenly moving to an exclusively online system and using unprepared online teaching and evaluation modules at the academic level in the entire Romanian education system in the last year and a half has generated many unsolved problems.

According to the authors, future education will probably become more of a hybrid model in the academic field, while online secondary education will probably remain only an adjuvant for the teachers. This conclusion is based on the fact that online learning is second nature for students, and now, after a year and a half of online education, teachers are more prepared and trained. In addition, the transition to distance learning is now easier because different software companies have created resources that could help both teachers and students to migrate more easily to the online system. This conclusion is also supported by the results of another five papers published by the authors on the same subject during the 2020/2021 academic year [34,61,79–81], as well as by their expertise as members of the Romanian Agency for Quality Assurance in Higher Education (ARACIS) and of the Executive Agency for Higher Education, Research, Development and Innovation Funding (UEFISCDI), and as managers in the top management level of the university.

Another opinion of the authors of this study is that the future hybrid education system, composed of F2F and online education, will be more of an ERT system than a standard online education system based on eLearning platforms, which is why instruments, such as those presented in this paper, will become more necessary and useful in the future.

In the context of a future hybrid education system, the model and the application developed and proposed by the authors could solve some of the problems caused by this type of teaching, including online education in ERT conditions, because there is fear that by the end of 2021, the educational system will have to pass again through such an experience. This is one of the reasons the authors have tried to develop software tools for teachers, software that could solve some of the problems raised by teaching and evaluations exclusively online in ERT situations.

Author Contributions: The contribution of the two authors has been balanced in all phases of the development of this study, both in the methodological part and in the writing part of this manuscript and of its various parts. The literature review was contributed by E.E. and the results are mostly

E.C.C.'s contribution. The writing of the discussion and the conclusions were produced through debate between both contributors of the work, which allowed for enriching the arguments based on the different opinions. Conceptualization, E.C.C. and E.E.; methodology, E.C.C. and E.E.; validation, E.C.C. and E.E.; formal analysis, E.C.C. and E.E.; investigation, E.C.C.; resources, E.C.C. and E.E.; data curation, E.C.C. and E.E.; writing—original draft preparation E.C.C. and E.E.; writing—review and editing, E.C.C. and E.E; visualization, E.C.C.; supervision, E.E project administration, E.E.; funding acquisition, E.E All authors have read and agreed to the published version of the manuscript.

Funding: This research received no external funding.

Institutional Review Board Statement: Regarding the Institutional Review Board Statement, the ethical review and approval of this study were non-interventional, and the confidentiality of the respondents was maintained by the responses being completely anonymous, and only aggregated data are presented. The study was conducted according to the guidelines of the Declaration of Helsinki.

Informed Consent Statement: Informed consent was obtained from all subjects involved in the study.

Data Availability Statement: The data presented in this study are available upon request from the first author.

Conflicts of Interest: The authors declare no conflict of interest.

References

1. Data.europa.eu. Education during COVID-19; Moving towards e-Learning. Available online: https://data.europa.eu/en/impact-studies/covid-19/education-during-covid-19-moving-towards-e-learning (accessed on 15 November 2021).
2. Clark, D. EU: Share of Individuals Doing an Online Course 2020. Published by D. Clark. 2020. Available online: https://www.statista.com/statistics/1099445/internet-use-in-schools-in-the-european-union/ (accessed on 4 October 2021).
3. Online Courses in Europe 2020 | Statista Available Online. Available online: https://www.statista.com/ (accessed on 4 October 2021).
4. Verbert, K.; Govaerts, S.; Duval, E.; Santos, J.L.; Van Assche, F.; Parra, G.; Klerkx, J. Learning dashboards: An overview and future research opportunities. *Pers. Ubiquitous Comput.* **2013**, *18*, 1499–1514. [CrossRef]
5. Duval, E.; Wiley, D. Guest Editorial: Open Educational Resources. *IEEE Trans. Learn. Technol.* **2010**, *3*, 83–84. [CrossRef]
6. Chen, G.D.; Chang, C.K.; Wang, C.Y. Ubiquitous learning website: Scaffold learners by mobile devices with information-aware techniques. *Comput. Educ.* **2008**, *50*, 77–90. [CrossRef]
7. Cen, H.; Koedinger, K.; Junker, B. Automating Cognitive Model Improvement by A* Search and Logistic Regression. In Proceedings of the AAAI 2005 Educational Data Mining Workshop, Carnegie Mellon University 5000 Forbes, Pittsburgh, PA, USA, 10 July 2005.
8. Hodges, C.; Moore, S.; Lockee, B.; Trust, T.; Bond, A. The Difference between Emergency Remote Teaching and Online Learning EDUCAUSE Review 2020. Available online: https://vtechworks.lib.vt.edu/handle/10919/104648 (accessed on 25 November 2021).
9. Information for Ohio State University Students, Faculty and Staff, the Ohio State University, Wexner Medical Center. Available online: https://safeandhealthy.osu.edu/ (accessed on 20 March 2021).
10. Coronavirus and Higher Education Resources Bryan Alexander Blog. Available online: https://bryanalexander.org/coronavirus/coronavirus-and-higher-education-resources (accessed on 17 March 2021).
11. Zimmerman, J. Coronavirus and the Great Online-Learning Experiment. The Chronicle of Higher Education, March 10. Available online: https://www.chronicle.com/article/Coronavirusthe-Great/248216 (accessed on 17 March 2021).
12. Stufflebeam, D.L.; Zhang, G. *Reporting Evaluation Findings in the Cipp Evaluation Model: How to Evaluate for Improvement and Accountability*; Guilford Publications: New York, NY, USA, 2017; pp. 273–280, Chapter 11.
13. Leitner, P.; Ebner, M. Development of a Dashboard for Learning Analytics Development of a Dashboard for Learning Analytics in Higher Education. In Proceedings of the Part II of Learning and Collaboration Technologies, Technology in Education: 4th International Conference, LCT 2017, Vancouver, BC, Canada, 9–14 July 2017; pp. 293–301.
14. Chen, T.; Peng, L.; Jing, B.; Wu, C.; Yang, J.; Cong, G. The Impact of the COVID-19 Pandemic on User Experience with Online Education Platforms in China. *Sustainability* **2020**, *12*, 7329. [CrossRef]
15. Yu, Y.-C.; You, S.-C.D.; Tsai, D.-R. Social Interaction Feedback System for the Smart Classroom. In Proceedings of the 2012 IEEE International Conference on Consumer Electronics (ICCE), Las Vegas, NV, USA, 13–16 January 2012; pp. 500–501.
16. Pohl, A.; Bry, F.; Schwarz, J.; Gottstein, M. Sensing the Classroom: Improving Awareness and Self-Awareness of Students with Backstage. In Proceedings of the International Conference on Interactive and Collaborative Learning (ICL), Villach, Austria, 26–28 September 2012.
17. Barr, J.; Gunawardena, A. Classroom Salon: A Tool for Social Collaboration. In Proceedings of the 43rd ACM Technical Symposium on Computer Science Education (Sigcse'12), Raleigh, NC, USA, 29 February–3 March 2012; ACM: New York, NY, USA; pp. 197–202.

18. Fagen, W.; Kamin, S. Developing Device-independent Applications for Active and Collaborative Learning with the SLICE Framework. In *EdMedia+ Innovate Learning*; Association for the Advancement of Computing in Education (AACE): Chesapeake, VA, USA, 2012; pp. 1565–1572.
19. Cuendet, S.; Bonnard, Q.; Kaplan, F.; Dillenbourg, P. Paper interface design for classroom orchestration. In Proceedings of the CHI'11 Extended Abstracts on Human Factors in Computing Systems (CHI EA'11), Vancouver, BC, Canada, 7–12 May 2011; pp. 1993–1998.
20. Son, L.H. Supporting Reflection and Classroom Orchestration with Tangible Tabletops. Ph.D Thesis, École Polytechnique Fédérale De Lausanne, Lausanne, Switzerland, 2012. Available online: https://infoscience.epfl.ch/record/174680 (accessed on 17 March 2021).
21. Martinez Maldonado, R.; Kay, J.; Yacef, K.; Schwendimann, B. An interactive teacher's dashboard for monitoring groups in a multi-tabletop learning environment. In Proceedings of the 11th International Conference on Intelligent Tutoring, Chania, Greece, 14–18 June 2012; Springer: Berlin/Heidelberg, Germany; pp. 482–492.
22. Gutiérrez Rojas, I.; Crespo García, R.M.; Delgado Kloos, C. Orchestration and feedback in lab sessions: Improvements in quick feedback provision. In *European Conference on Technology Enhanced Learning*; Springer: Berlin/Heidelberg, Germany, 2011; Volume 6964, pp. 424–429.
23. Gutiérrez Rojas, I.; Crespo García, R.M. Towards Efficient Provision of Feedback Supported by Learning Analytics. In Proceedings of the 12th International Conference on Advanced Learning Technologies (ICALT), Rome, Italy, 4–6 July 2012; pp. 599–603.
24. Morris, M.R.; Piper, A.M.; Cassanego, T.; Winograd, T. Supporting Cooperative Language Learning: Issues in Interface Design for an Interactive Table, Stanford University Technical Report. Available online: https://hci.stanford.edu/cstr/reports/2005-08.pdf (accessed on 25 November 2021).
25. Santos, J.L.; Govaerts, S.; Verbert, K.; Duval, E. Goal-oriented visualizations of activity tracking: A case study with engineering students. In Proceedings of the 2nd International Conference on Learning Analytics and Knowledge (LAK'12), Vancouver, BC, Canada, 29 April–2 May 2012; pp. 143–152.
26. Arnold, K.E.; Pistilli, M.D. Course signals at Purdue: Using learning analytics to increase student success. In Proceedings of the 2nd International Conference on Learning Analytics and Knowledge (LAK'12), Vancouver, BC, Canada, 29 April–2 May 2012; pp. 267–270.
27. Leony, D.; Crespo, R.M.; Perez-Sanagustin, M.; Parada, G.H.A.; de la Fuente Valentin, L.; Pardo, A. Coverage Metrics for Learning-Event Datasets Based on Client-Side Monitoring. In Proceedings of the IEEE 12th International Conference on Advanced Learning Technologies (ICALT), Rome, Italy, 4–6 July 2012; pp. 652–653.
28. Dillenbourg, P.; Zufferey, G.; Alavi, H.; Jermann, P.; Do-Lenh, S.; Bonnard, Q.; Cuendet, S.; Kaplan, F. *Classroom Orchestration: The Third Circle of Usability*; CSCL 2011 Proceedings; International Society of the Learning Sciences: Hong Kong, China, 2011; Volume 1, pp. 510–517.
29. Bakker, J.; Holenderski, L.; Kocielnik, R.; Pechenizkiy, M.; Sidorova, N.J. Stess@Work: From measuring stress to its understanding, prediction and handling with personalized coaching. In Proceedings of the 2nd ACM Sighit International Health Informatics Symposium, Miami, FL, USA, 28–30 January 2012; pp. 673–678.
30. Ferguson, R. Learning Analytics: Fattori trainanti, sviluppi e storie. *Ital. J. Educ. Technol.* **2014**, *22*, 138–147.
31. Suthers, D. From contingencies to network-level phenomena: Multilevel analysis of activity and actors in heterogeneous networked learning environments. In Proceedings of the Fifth International Conference on Learning Analytics and Knowledge, Poughkeepsie, NY, USA, 16–20 March 2015.
32. Nakahara, J.; Hisamatsu, S.; Yaegashi, K.; Yamauchi, Y. Itree: Does the Mobile Phone Encourage Learners to Be More Involved in Collaborative Learning? In Proceedings of the 2005 Conference on Computer Support for Collaborative Learning, Taipei, Taiwan, 30 May–4 June 2005; pp. 470–478.
33. Kim, I.; Kim, R.; Kim, H.; Kim, D.; Han, K.; Lee, P.H.; Mark, G.; Lee, U. Understanding smartphone usage in college classrooms: A long-term measurement study. *Comput. Educ.* **2019**, *141*, 103611. [CrossRef]
34. Edelhauser, E.; Lupu-Dima, L. Is Romania Prepared for eLearning during the COVID-19 Pandemic? *Sustainability* **2020**, *12*, 5438. [CrossRef]
35. Kim, S.; Yun, K.; Park, J.; Choi, J.Y. Skeleton-Based Action Recognition of People Handling Objects. In Proceedings of the IEEE Winter Conference on Applications of Computer Vision (WACV), Waikoloa, HI, USA, 7–11 January 2019; IEEE: Piscataway, NJ, USA, 2019; pp. 61–70.
36. Pandea, C.; Witschela, H.F.; Montecchiaria, A.M.; Montecchiaria, D. Hybrid Conversational AI for Intelligent Tutoring Systems. In Proceedings of the AAAI Spring Symposium: Combining Machine Learning with Knowledge Engineering, Stanford University, Palo Alto, CA, USA, 21–23 March 2021.
37. Allen, M. Designing Online Asynchronous Information Literacy Instruction Using the ADDIE Model. In *Distributed Learning Chandos*; Maddison, T., Kumaran, M., Eds.; Elsevier: Amsterdam, The Netherlands, 2017; pp. 69–91.
38. Govaerts, S.; Verbert, K.; Duval, E.; Pardo, A. The student activity meter for awareness and self-reflection. In *Proceedings of the Chi'12 ACM Annual Conference Extended Abstracts on Human Factors in Computing Systems Extended Abstracts*; Austin, TX, USA, 5–10 May 2012, pp. 869–884.
39. Dollár, A.; Steif, P.S. Web-based Statics Course with Learning Dashboard for Instructors. In Proceedings of the Computers and Advanced Technology in Education (CATE 2012), Napoli, Italy, 25–27 June 2012.

40. Ali, L.; Hatala, M.; Gašević, D.; Jovanović, J. A qualitative evaluation of evolution of a learning analytics tool. *Comput. Educ.* **2012**, *58*, 470–489. [CrossRef]

41. Podgorelec, V.; Kuhar, S. Taking Advantage of Education Data: Advanced Data Analysis and Reporting in Virtual Learning Environments. *Electron. Electr. Eng.* **2011**, *114*, 111–116. [CrossRef]

42. Leony, D.; Pardo, A.; de la Fuente Valentín, L.; Sánchez de Castro, D.; Delgado Kloos, C. GLASS: A learning analytics visualization tool. In Proceedings of the 2nd International Conference on Learning Analytics and Knowledge LAK'12, Vancouver, BC, Canada, 29 April–2 May 2012; pp. 162–163.

43. Scheuer, O.; Zinn, C. How did the e-learning session go? The Student Inspector. In Proceedings of the 13th International Conference on Artificial Intelligence and Education (AIED 2007), Los Angeles, CA, USA, 9 July 2007; pp. 487–494.

44. Lafford, B.A. Review of Tell Me More Spanish. *J. Lang. Learn. Technol.* **2004**, *8*, 21–34.

45. Kerly, A.; Ellis, R.; Bull, S. CALMsystem: A Conversational Agent for Learner Modelling. In *Applications and Innovations in Intelligent Systems XV*; Springer: Berlin/Heidelberg, Germany, 2007; pp. 89–102.

46. Kobsa, E.; Dimitrova, V.; Boyle, R. Using student and group models to support teachers in web-based distance education. In Proceedings of the 10th International Conference on User Modeling, Edinburgh, Scotland, UK, 24–29 July 2005; pp. 124–133.

47. Santos, J.L.; Verbert, K.; Duval, E. Empowering students to reflect on their activity with StepUp!: Two case studies with engineering students. In Proceedings of the EFEPLE11 2nd Workshop on Awareness and Reflection in Technology-Enhanced Learning, Saarbrücken, Germany, 18 September 2012.

48. Vieira, C.; Parsons, P.; Byrd, V. Visual learning analytics of educational data: A systematic literature review and research agenda. *Comput. Educ.* **2018**, *122*, 119–135. [CrossRef]

49. Jivet, I.; Scheffel, M.; Specht, M.; Drachsler, H. License to Evaluate: Preparing Learning Analytics Dashboards for Educational Practice. In Proceedings of the 8th International Conference on Learning Analytics and Knowledge, Sydney, NSW, Australia, 7–9 March 2018; ACM: New York, NY, USA; pp. 31–40.

50. Naranjo, D.N.; Prieto, J.R.; Moltó, G.; Calatrava, A. A Visual Dashboard to Track Learning Analytics for Educational Cloud Computing. *Sensors* **2019**, *19*, 2952. [CrossRef]

51. Amazon. AWS Cloudtrail. Available online: https://aws.amazon.com/cloudtrail/ (accessed on 6 September 2021).

52. Opsview. Opsview Monitor. Available online: https://www.opsview.com/ (accessed on 6 September 2021).

53. Spectrum. Available online: https://spectrumapp.io/ (accessed on 6 September 2021).

54. SignalFx. Available online: https://signalfx.com/ (accessed on 6 September 2021).

55. SolarWinds. AWS Cloud Monitoring. Available online: https://www.solarwinds.com/topics/aws-monitoring (accessed on 6 September 2021).

56. Lonn, S.; Aguilar, S.J.; Teasley, S.D. Investigating student motivation in the context of a learning analytics intervention during a summer bridge program. *Comput. Hum. Behav.* **2015**, *47*, 90–97. [CrossRef]

57. Sedrakyan, G.; Malmberg, J.; Verbert, K.; Järvelä, S.; Kirschner, P.A. Linking learning behavior analytics and learning science concepts: Designing a learning analytics dashboard for feedback to support learning regulation. *Comput. Hum. Behav.* **2020**, *107*, 105512. [CrossRef]

58. Irons, A. *Enhancing Learning through Formative Assessment and Feedback*; Routledge: Abingdon, UK, 2007.

59. Educația, o Șansă Pentru Valea Jiului!—POCU 122596. 2020. Available online: https://www.upet.ro/proiecte/122596/ (accessed on 13 September 2020).

60. Educația, o Șansă Pentru Valea Jiului!—POCU 122596. 2021. Available online: https://elearning-upet.ro/index.php#welcome (accessed on 17 November 2021).

61. Edelhauser, E.; Lupu-Dima, L. One Year of Online Education in COVID-19 Age, a Challenge for the Romanian Education System. *Int. J. Environ. Res. Public Health* **2021**, *18*, 8129. [CrossRef] [PubMed]

62. Verbert, K.; Duval, E.; Klerkx, J.; Govaerts, S.; Santos, J.L. Learning Analytics Dashboard Applications. *Am. Behav. Sci.* **2013**, *57*, 1500–1509. [CrossRef]

63. Cho, Y.S. Prospect for Learning Analytics to Achieve Adaptive Learning Model 2015, Seoul, Korea. Available online: https://www.slideshare.net/zzosang/prospect-for-learning-analytics-to-achieve-adaptive-learning-model (accessed on 20 September 2020).

64. Community-led The Apache Way. Available online: https://www.apache.org/licenses/LICENSE-2.0 (accessed on 1 February 2021).

65. Pintrich, P.R. A Conceptual Framework for Assessing Motivation and Self-Regulated Learning in College Students. *Educ. Psychol. Rev.* **2004**, *16*, 385–407. [CrossRef]

66. Corrin, L.; de Barba, P. Exploring Students Interpretation of Feedback Delivered through Learning Analytics Dashboards. In Proceedings of the Annual Conference of the Australian Society for Computers in Tertiary Education (Ascilite 2014), Rhetoric and Reality: Critical Perspectives on Educational Technology, Dunedin, New Zealand, 23–26 November 2014.

67. Gormley, C.; Tong, Z. *Elasticsearch: The Definitive Guide*; O'Reilly Media, Inc.: Newton, MA, USA.

68. Gupta, Y. *Kibana Essentials*; Packt Publishing Ltd.: Birmingham, UK.

69. Seabra, F.; Teixeira, A.; Abelha, M.; Aires, L. Emergency Remote Teaching and Learning in Portugal: Preschool to Secondary School Teachers' Perceptions. *Educ. Sci.* **2021**, *11*, 349. [CrossRef]

70. Chierichetti, M.; Backer, P. Exploring Faculty Perspectives during Emergency Remote Teaching in Engineering at a Large Public University. *Educ. Sci.* **2021**, *11*, 419. [CrossRef]
71. Colclasure, B.C.; Marlier, A.; Durham, M.F.; Brooks, T.D.; Kerr, M. Identified Challenges from Faculty Teaching at Predominantly Undergraduate Institutions after Abrupt Transition to Emergency Remote Teaching during the COVID-19 Pandemic. *Educ. Sci.* **2021**, *11*, 556. [CrossRef]
72. Hidalgo, G.I.; Sánchez-Carracedo, F.; Romero-Portillo, D. COVID-19 Emergency Remote Teaching Opinions and Academic Performance of Undergraduate Students: Analysis of 4 Students' Profiles. A Case Study. *Mathematics* **2021**, *9*, 2147. [CrossRef]
73. Ní Fhloinn, E.; Fitzmaurice, O. Challenges and Opportunities: Experiences of Mathematics Lecturers Engaged in Emergency Remote Teaching during the COVID-19 Pandemic. *Mathematics* **2021**, *9*, 2303. [CrossRef]
74. Mejía-Madrid, G.; Llorens-Largo, F.; Molina-Carmona, R. Dashboard for Evaluating the Quality of Open Learning Courses. *Sustainability* **2020**, *12*, 3941. [CrossRef]
75. Nandi, A.; Xhafa, F.; Subirats, L.; Fort, S. Real-Time Emotion Classification Using EEG Data Stream in E-Learning Contexts. *Sensors* **2021**, *21*, 1589. [CrossRef] [PubMed]
76. Ciordas-Hertel, G.P.; Rödling, S.; Schneider, J.; Di Mitri, D.; Weidlich, J.; Drachsler, H. Mobile Sensing with Smart Wearables of the Physical Context of Distance Learning Students to Consider Its Effects on Learning. *Sensors* **2021**, *21*, 6649.
77. Kohnke, L.; Zou, D.; Zhang, R. Pre-Service Teachers' Perceptions of Emotions and Self-Regulatory Learning in Emergency Remote Learning. *Sustainability* **2021**, *13*, 7111. [CrossRef]
78. Sosa Díaz, M.J. Emergency Remote Education, Family Support and the Digital Divide in the Context of the COVID-19 Lockdown. *Int. J. Environ. Res. Public Health* **2021**, *18*, 7956. [CrossRef]
79. Edelhauser, E.; Lupu-Dima, L. Managerial Research of Online Education in the Primary and Secondary Romanian Education System During COVID 19 Crisis. In Proceedings of the MATEC Web of Conferences, Sibiu, Romania, 2–4 June 2021; volume 343, p. 07006. [CrossRef]
80. Edelhauser, E.; Lupu-Dima, L.; Grigoras, G. A Managerial Perspective over One Year of Online Education in Romania. A Case Study at the University of Petrosani. In Proceedings of the MATEC Web of Conferences, Petroani, Romania, 27–27 May 2021; volume 342, p. 09001. [CrossRef]
81. Edelhauser, E.; Lupu-Dima, L. A Comparative Study between Two Different Academic Years in the Online Education System for the Engineering and Economic Field. In Proceedings of the 16th International Symposium in Management, Management, Innovation and Entrepreneurship in Challenging Global Times, Timisoara, Romania, 22–23 October 2021.

Article

Effects of the Uncertainty of Interpersonal Communications on Behavioral Responses of the Participants in an Immersive Virtual Reality Experience: A Usability Study

Shirin Hajahmadi [1] and Gustavo Marfia [2],*

[1] Department of Computer Science and Engineering, University of Bologna, 40126 Bologna, Italy
[2] Department of the Arts, University of Bologna, 40126 Bologna, Italy
* Correspondence: gustavo.marfia@unibo.it

Abstract: Two common difficulties which people face in their daily lives are managing effective communication with others and dealing with what makes them feel uncertain. Past research highlights that the result of not being able to handle these difficulties influences people's performance in the task at hand substantially, especially in the context of a social environment such as a workplace. Perceived uncertainty of information is a key influential factor in this regard, with effects on the quality of the information transfer between sender and receiver. Uncertainty of information can be induced into the communication system in three ways: when there is any kind of information deficit that makes the target message unclear for the receiver, when there are some requested changes that could not be predicted by the receiver, and when the content of the message is so interconnected and complex that it limits understanding. Since uncertainty is an inseparable feature of our lives, studying the effects that different levels of it have on individuals and how individuals nevertheless accomplish the tasks of daily living is of high importance. Modern technologies such as immersive virtual reality (VR) have been successful in providing effective platforms to support human behavioral and social well-being studies. In this paper, we suggest the design, development, and evaluation of an immersive VR serious game platform to study behavioral responses to the uncertain features of interpersonal communications. In addition, we report the result of a within-subject user study with 17 participants aged between 20 and 35 and their behavioral responses to two levels of uncertainty with subjective and objective measures. The results convey that the application successfully and meaningfully measured some behavioral responses related to exposure to different levels of uncertainty and overall, the participants were satisfied with the experience.

Keywords: human-computer interaction; virtual reality; serious game; interpersonal communications; uncertainty; social well-being; behavioral responses

Citation: Hajahmadi, S.; Marfia, G. Effects of the Uncertainty of Interpersonal Communications on Behavioral Responses of the Participants in an Immersive Virtual Reality Experience: A Usability Study. *Sensors* **2023**, *23*, 2148. https://doi.org/10.3390/s23042148

Academic Editors: Pietro Manzoni, Claudio Palazzi and Ombretta Gaggi

Received: 29 December 2022
Revised: 2 February 2023
Accepted: 6 February 2023
Published: 14 February 2023

1. Introduction

Most of us would agree that uncertainty, in many circumstances, is not something we like to experience. For example, we do not wish to be uncertain about our ability to pay bills at the end of each month, our work and educational prospects, and our health [1]. Some studies in neuroscience also support this claim by providing evidence that the human brain is hardwired to interpret uncertainty as a danger and respond to it with fear and stress [2,3]. A human brain under uncertainty tends to overestimate and dramatize danger [4], jump to conclusions [5], and underestimate its ability to handle it [6–8].

Following this approach to uncertainty, the goal has been to reduce it [9,10]. For example, people are encouraged to reduce the uncertainty of loss of income in old age or of possible unemployment with saving money, paying taxes, and buying insurance policies [1]. In education, traditionally, uncertainty is often seen as a threat and removed by exposing students to clearly defined problems, following predefined methods of solving them, to reach expected outcomes [11]. The reality is that we live in an uncertain and complex world [1]. Despite our best efforts, things do not always go as planned, and unexpected events may happen. Hence, one should strive to accept uncertainty, performing tasks aware of its existence instead of amplifying its fear with the risk of arguing with life rather than living it. The recent experience with COVID-19 supports such an idea [12]. This is why many educators have recently sought the best ways to provide a structured and supportive learning environment to prepare young students to respond productively to the challenges originating from dealing with uncertainty [13,14]. As described by Beghetto [15], novel learning environments should structurally offer uncertainty, engaging students with it, teaching them how to sit with its difficulty, how to explore, how to generate and evaluate new possibilities, and, most importantly, take action based on them [16]. In this way, uncertainty may act as a catalyst for creative answers rather than an unbeatable barrier. This approach motivates the idea of designing and implementing platforms to support the study of the behavioral responses that the uncertainty may trigger [12,17,18].

The broad concept of uncertainty is, in fact, closely connected with that of information which, in turn, is at the core of interpersonal communications [19,20]. Interpersonal communication concerns the study of social interaction between people and tries to understand how verbal and written dialogues, as well as nonverbal actions, are used to achieve communication goals [21]. Studies show individuals facing different levels of uncertainty have different behavioral responses, from negative to positive [22–24]. The ways a human being may deal with an uncertain situation may differ based on individual differences [25], culture [26], and the level of expertise [27]. Hillen et al presented a conceptualization of an individual's experience of uncertainty based on a categorization of potential responses [28]. In such a model, ambiguities or/and complexities generate(s) stimuli to the information system. Uncertainties appear when individuals perceive (consciously become aware of) their existence. Cognitive, emotional, and behavioral responses then follow such a perception.

Virtual Reality (VR) systems may act as feasible platforms to assist in understanding behavioral responses to the uncertainty of interpersonal communications, as they may provide 3D spaces involving the same kind of navigational and communication challenges experienced in the real world [29]. With VR, it is possible to create structured environments where the ability of people to cope with challenges can be observed, behavioral data gathered, eventual achievements and feedback engineered, and strategies for skill improvement applied in a top-down fashion [30–33]. In VR, people can express their ideas, feel in control, and accomplish tasks and communicate with others [34–37]. This raises the potential to enjoy and engage in activities in the digital space and then apply them to the real world to improve one's social well-being [38]. In addition, creating such an experience in the context of a serious game can support situated cognition by contextualizing a player's experience in an engaging and realistic environment [39]. In addition, it can benefit from those game design techniques that support the idea that uncertainty could potentially maintain a user's attention and engagement, providing the motivation to continue even in challenging moments [1].

Considering this domain, we propose the design and development of an immersive virtual reality experience whose scope is to support the investigation of how people manage uncertainty while performing tasks in a workplace scenario. This experience, implemented as a serious game, aims at simulating a workplace scenario, a social environment where success in managing effective interpersonal communication appears very important [40–44].

With this work, we aim to contribute to the research community by providing answers to the research questions below:

- RQ1: How do the participants rate their experience with different tasks in terms of perceived uncertainty?
- RQ2: How do different degrees of uncertainty affect users' behavior and performance in this immersive virtual workplace scenario?
- RQ3: How are the users' subjective responses to uncertainty related to the objective responses?
- RQ4: How does the user evaluate the quality of his/her experience?

This paper is structured as follows. Section 2 discusses related work. Section 3 describes the interaction techniques, environment, and task design process of this immersive VR serious game. Section 4 describes the result of a usability study that evaluates the user experience of the proposed VR system as well as reports some behavioral responses. Section 5 discusses the main findings of the experiment. Section 6 concludes the paper and discusses future opportunities for research.

2. Related Work

In this section, we present and discuss the works that fall closest to our contribution. A good body of research has focused on the study of "Navigational uncertainty" and its effect on the user's spatial navigation performance and behavior [45–48]. In this area of research, uncertainty has been mostly introduced into the system by creating a perception of disorientation [49] and curing conflict [50] for the user, resulting in an increase in his/her information-seeking behavior. In their recent review, Keller et al. [51] proposed that collecting and analyzing continuous navigational data obtained from the participants in virtual reality experiences that create navigational uncertainty can potentially provide important insight into their information-seeking behavior. For example, in this research [46], the authors focused on the "Looking around behavior" as a common type of information-seeking behavior of participants when experiencing navigational uncertainty. They recorded continuously the heading direction and tried to find its relation to navigational success measures. From this body of literature, we could conclude the potential and importance of the data that could be captured from VR experiences to provide insights into the behavioral responses of people, especially in the study of the effects of a variable, such as uncertainty, on behavior.

Another area in which the study of uncertainty has received a lot of attention is gaming. As Costikyan et al. [1] claim, games could improve by purposefully applying the concept of uncertainty in their designs. Uncertainty could act as a catalyst to hold users' attention and interest; mastering it may help pursue a game's goal in an efficient and non-threatening way [52]. In addition, Costikyan et al. [1] support these claims by citing the sociologist Roger Callios [53] "Play is... uncertain activity. Doubt must remain until the end, and hinges upon the denouement... every game of skill, by definition, involves the risk for the player of missing his stroke and the threat of defeat, without which the game would no longer be pleasing. The game is no longer pleasing to one who, because he is too well trained or skillful, wins effortlessly and infallibly".

In the following, we review some examples of games that exploit uncertainty in their design and present a comparison of their features in Table 1:

- Gone Home [54] is a first-person exploration game designed to put players in unknown situations, engaging them to stay and accomplish some tasks, such as uncovering the narration by non-linear progression through searching the space. This game puts a player in the shoes of a young woman who returns home and finds that her family is absent. As Veale et al. [55] also discussed, Gone Home is a video game that uses effective storytelling to create empathy and a sense of responsibility in users by placing them within a recent historical moment. In this way, it exposes the user to the positive and negative elements of the past and encourages him/her to stay in the game and reflect on these elements [56]. While not strong on interactivity, the game through a careful visual, spatial, and audio design of the environment leads its users to explore the house along a twisting, uncertain path and find out what happened to the woman's family through an analysis of imperfect clues from the memorabilia, journals, and other items left around the various rooms. During the experience, there are notes, voices, and letters from or to her family that motivate and guide her in the exploration. These items of cues can be kept in the inventory and reviewed whenever desired [54]. Considering an interest in the study of navigational behaviors of users, Bonnie Ruberg [57] argues that with a deeper analysis of the interactive elements of the game, the player path is linear instead of meandering despite what it seems the game encourages players to do. The path is already set and the locked, or hidden doors prevent the user to have access to some areas unless they trigger an event or find an object that unlocks this barrier in a predefined order.

- Don't Starve [58] is a survival game that places the user in the role of a scientist who finds himself in a strange and unfamiliar world. The goal is to collect and effectively use survival tools. An uncertain scenario amounts to the interaction with the frogs in the game, as this creates ambiguity, as it is unclear whether they are hostile. For example, they can represent food, but different outcomes may result from eating them. The game successfully engages the user to accept this ambiguity till effectively able to develop higher-level strategies to interact with them. Farah et al. [59] studied the multiplayer expansion of the game to track cooperative features and teamwork behavioral markers.

- Wenge xu et al. [60] developed a motion-based survival game, GestureFit, that involves the user in a fight with a monster. They induced uncertainty in the system through three uncertain game elements: false attacks (creating the perception that there would be a chance that the system is tricking the player to waste a defense move by defending against a false attack), misses (creating the perception that there would be a chance that the actual hit will be interpreted as a miss), and critical hits (creating the perception that there would be a chance that an attack would be a critical attack and produce more damage than a normal one). In this way, they created two different levels of uncertainty, one with inducing these uncertain elements into the game and the other without. After, they conducted a study to measure the effects of levels of uncertainty (certain and uncertain), the display type (VR and LD), and age (young adults and middle-aged adults) on the game experience, performance, and exertion level. Their results showed that for the kind of game they designed, virtual reality could improve game performance. In addition, they found that the uncertain elements that they applied in their design might not help enhance the overall game experience, but could help increase the user's exertion.

- RelicVE [61] is a virtual reality (VR) game that gives the user a similar role to an archaeologist and engages him in an exploration process of an archaeological discovery experience. It exploits uncertainty in the design of their exploration process by placing the user in a situation where s/he does not know the shape and features of the target artifact and only can discover it by gradually and strategically using available tools and physical movements. In this way, when the user hits specific triggers, a new part of the information about the artifact will be uncovered. They also managed the complexity of the game by the complexity of the shape and volume of the artifact. They integrated VR interaction techniques in the design of their virtual system to create an experience close to the real-world experience of archaeologists and in this way increased the immersion and physical activity of the user during the experience. In addition, they used a timer and a health bar to add the element of time pressure to the experience. To evaluate the experience, the authors also conducted a usability study that found the experience to be innovative as it can improve players' learning and motivation by adding the elements of uncertainty into the design.

To the best of our knowledge, no previous work took full advantage of the available technologies, such as virtual reality, to induce structured uncertainty and investigate the influence of uncertainty levels on human behavior with a focus on interpersonal communications. Our study tries to take this step from within the design and development of such an application by applying some of the design techniques inspired by the previous games in this area and virtual reality techniques that improve the user experience and the study of behavior.

Table 1. Key features of the games reviewed in the related work section.

Name of the Game	Type of the Game	Type of the Experience	Applied Uncertainty Approach	Requested Task
Gone Home	Exploration game; Role-playing game	Video game	Put players in unknown situations; Encourage the user to explore scattered objects along a twisting, uncertain path	To find out what has happened to the character's sister and family by digging through scattered documents
Don't Starve	Exploration game; Role-playing game	Video game	Put the user in a strange and unfamiliar world; Put the user in a situation that is uncertain about the outcome of some of his/her choices	To find through trial and error the way that can survive from the threats coming from the environment
GestureFit	Motion-based survival game	Immersive virtual reality	Induce uncertainty in the system through three uncertain gameplay elements (false attacks, misses, and critical hits)	To stay alive and perform gestures to defeat a monster
RelicVE	Exploration game; Role-playing game	Immersive virtual reality	Place the user in a situation where s/he does not know the shape and features of the target artifact from the beginning	To take away earth from the artifact using the available tools and physical movement without damaging it

3. Experimental Setting

In this section, we describe the experiment we conducted to study the effects of different levels of uncertainty on behavioral responses, performance, and quality of the experience of the participants resorting to objective and subjective measures. Figure 1 visualizes the stages of this experimental design.

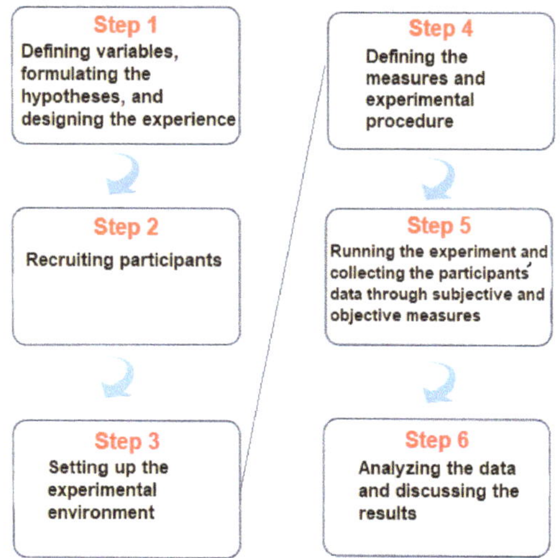

Figure 1. The stages of the experimental design.

3.1. Participants

We recruited 17 participants (3 female, 14 male, age: 20–35, M = 25.05, SD = 1.75) who are affiliated with our university to take part in our study. Detailed demographic information appears in Table 2. Before starting our experiment, we asked them to rate (1 = never, 5 = every day) any previous experience with virtual reality using a head-mounted display (M = 2.88, SD = 1.36) and their level (1 = low, 5 = high) of English proficiency (M = 3.47, SD = 1.01).

Table 2. Participants' demographic information.

Sex	Number	Percentage (%)
Female	3	17.65
Male	14	82.35
Age	**Number**	**Percentage (%)**
20–30	17	100
30–35	0	0
Nationality	**Number**	**Percentage (%)**
Italian	17	100
Others	0	0
Education level	**Number**	**Percentage (%)**
High school graduate, diploma or the equivalent	1	5.9
Bachelor's degree	12	70.59
Master's degree	3	17.65
Doctorate degree	1	5.9

3.2. Materials

We now proceed to present the design and implementation choices of the virtual environment.

3.2.1. Setup

In our experiment, participants navigated in a virtual office via an HTC Vive Pro HMD (refresh rate: 90 Hz, resolution: 1440 × 1600 pixels, FoV 110°) connected to a workstation (Intel(R) Core(TM) i7-6850K CPU @ 3.60 GHz, 3.60 GHz). The environment was developed using Unity 3D version 2019.4.35 f1. Unity 3D is a game engine developed by the Unity Technologies (San Francisco, CA, USA). It is a very famous platform that has been used by game developers across the world. The data analysis was performed using R version 4.2.2 and RStudio version 2022.07.2+576. R is a programming language and software environment for statistical computing and graphics, developed by the R Development Core Team and maintained by the R Foundation (Vienna, Austria). RStudio is an integrated development environment (IDE) for the R programming language, developed by RStudio, Inc. (Boston, MA, USA).

3.2.2. Experience Design

The experience was designed as a role-playing serious game where a user, in the role of a new employee, is exposed to two different levels of uncertainty in the context of interpersonal communication in a workplace scenario. To this aim, the story plot that develops within the experience takes inspiration from Amelia Bedelia, the protagonist and title character of the children's book series authored by Peggy Parish [62]. Amelia Bedelia is a housekeeper who takes her instructions literally because her boss could not be present in the house on the first day of her work. The instructions include lexical ambiguity coming from each sentence. Despite such ambiguity, Amelia stays positive and expresses her excitement to do her job well and make her boss happy, but she repeatedly misunderstands the guidance. Inspired by this story line, our application implements:

- An absence of guidance when a user is following and executing given instructions. This design limits the access to sources of information, asking the user to focus and rely on the already provided knowledge or information that may come after from sources such as panels and phone calls;
- Specific means of communicating instructions, which may be textual with the use of panels (and sometimes verbal, e.g., through phone calls) as a result of the absence of guidance;
- Specific instruction communication patterns in the form of sequences of sentences, such as what appears in step-by-step construction manuals. At the same time, it enables an experiment designer to purposely reduce the amount of available information and change the complexity of the sentences to control the amount of ambiguity and complexity in the system;
- The possibility of applying lexical ambiguity, as another potential source of confusion;
- A friendly environment supporting understanding and empathy in interpersonal communications.

Please see the Table 3 for a comparison between these features in the proposed platform and Amelia Bedelia story.

Table 3. A comparison of the elements in the proposed platform with those in the Amelia Bedelia story.

Name of the Element	Description of the Element in Amelia Bedelia Story	Description of the Element in Our platform	Reason for Use
Absence of guidance	The housekeeper is asked to accomplish tasks based on the given instructions while does not have access to anybody to communicate her doubts.	The same situation is true here for the user but also s/he can save in the system the type of problem which is facing (a problem with the interface and/or a problem with the instruction).	This design choice limits the access to the sources of information, asking to focus and rely on the already provided knowledge or may come after, from sources such as panels and phone calls.
Specific means of communicating instruction	Textual on a printed paper	Both textual and verbal that comes from the panels and voice calls	Communicating instructions in both verbal and textual forms would be in favor of cognitive load and managing the attention of the user during the experience [63].
Specific instruction communication patterns and sources of ambiguity	Instructions are presented in sequences of sentences, as appearing in step-by-step construction manuals. These instructions include lexical ambiguity coming from each sentence.	The same is followed here. In addition, the complexity of instructions also changes with increasing the degree of interconnectivity among parts of the sentences. In addition, available tasks and information change with receiving unpredictable phone calls coming from the boss.	This design choice provides the possibility for the experiment designer to purposely reduce the amount of available information and change the complexity of the sentences to control the amount of ambiguity and complexity in the system. In addition, the possibility of applying lexical ambiguity, as another potential source of confusion is provided.
A friendly environment supporting understanding and empathy in interpersonal communications	A nice house with friendly relationships between Amelia and the family	A nice virtual office and the friendly voice of the boss	Experiencing this environment potentially keeps the user's interest to stay till the end of the experiment and accept the challenges.

The application includes two phases: the "Familiarization" phase and the "Main" phase. The goal of the "familiarization" phase is to remove any uncertainty arising from unfamiliarity with VR interfaces and the related context. For this purpose, it provides information and a step-by-step tutorial with feedback to familiarize the user with the context and allow the user to feel confident with the interactions that will then be executed. The user in this phase will get to know the boss, his/her role, the space s/he will be working in, the means of communication, and the way s/he can accomplish the tasks indicated by the boss using the available interfaces. At the end of this phase and when the system confirms the user has successfully executed all steps, s/he will reach the virtual office by pressing the "Move to the office" button from the panel on the left hand (see Figure 2 for some screenshots taken from the familiarization phase).

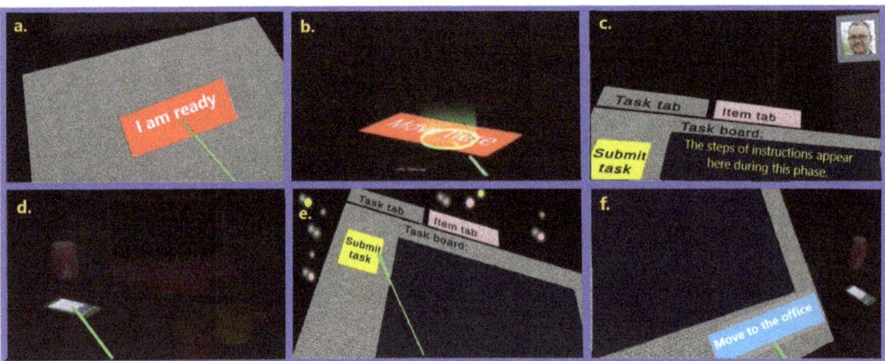

Figure 2. Screenshots showing some steps of the familiarization phase, from left to right: (**a**) selecting the "I am ready" button by the user; (**b**) teleporting to a destination; (**c**) reading instructions of the task from the panel on the user's left hand; (**d**) removing an interactive object; (**e**) submitting the current task; (**f**) selecting the "Move to the office" button to move there.

The main phase starts with the user finding himself/herself inside a virtual office in front of a door. After 10 s, a phone starts ringing, and s/he should answer. The boss is on the phone, welcoming and asking the player to follow some instructions, explaining three options that will be available during the experience: submitting the task, suppressing it, and requesting help using the buttons on the panel. The boss also says that if something important comes up he will call again. By pressing the "I am ready" button on the panel, a description of the first task appears. The user can now teleport to move within the office environment, removing objects based on the instructions. The removed objects then become visible in the "Item" tab of the panel. The user can cancel a previous removal by pressing the close button near each image. The user will be asked to complete a second task either by pressing the "Submit task" button or the "I do not do this task" button. After removing a specific number of objects, in the middle of the second task the phone will ring. The boss warns the user that it may be necessary to cancel previous removals to follow a new set of instructions. Task 2 finishes either by pressing the "Submit task" button or the "I will not do this task" button. The user can also decide to exit the game by pressing the "Exit the game" button.

The virtual office is hence furnished with interactive and non-interactive objects as well as two dynamic blackboards as two sources of information (See Figure 3 for some screenshots showing the virtual office environment). As described in Figure 4, there are five possible sources of information in the experience: 1. a small blackboard displays the name of the current task; 2. a big blackboard communicates the current status; 3. a small blackboard attached to the panel is a closeup of the big one; 4. a phone that blinks and rings when the boss calls; and 5. a task board showing the instructions for each task. The different parts of the panel are shown in Figure 5. An example of teleporting and removing interactive objects may be viewed in Figure 6. In addition, to increase the immersion, during the main phase an ambient sound is played, simulating the sounds coming from nearby offices to help reduce the confounding effects of noises coming from the real world.

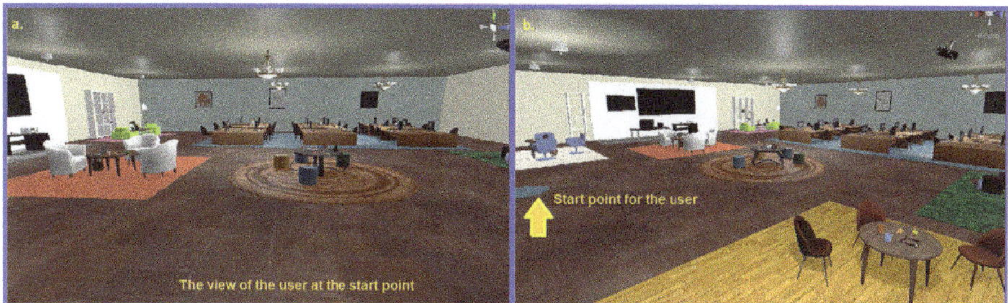

Figure 3. Screenshots showing two views of the office: (**a**) the view of the office from the perspective of the user at the beginning; (**b**) Another view of the office.

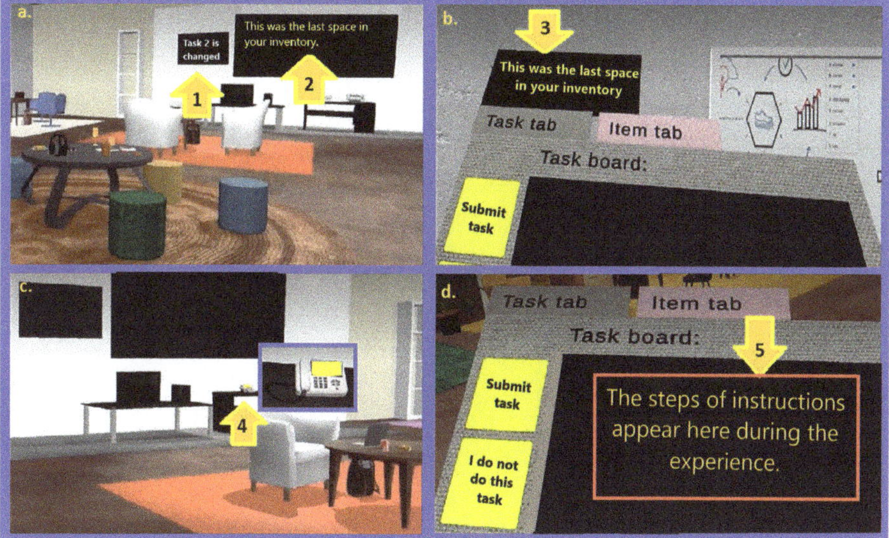

Figure 4. Screenshots showing five sources of information for the user during the experience: (**a**) Arrow 1 points to a small blackboard that displays the name of the current task. Arrow 2 points to a big blackboard that communicates the current status of the tasks during the experience; (**b**) Arrow 3 points to a small blackboard attached to the panel that is a closeup of the big one; (**c**) Arrow 4 points to a phone that blinks and rings when the boss calls; (**d**) Arrow 5 points to a task board that shows the instructions for each task.

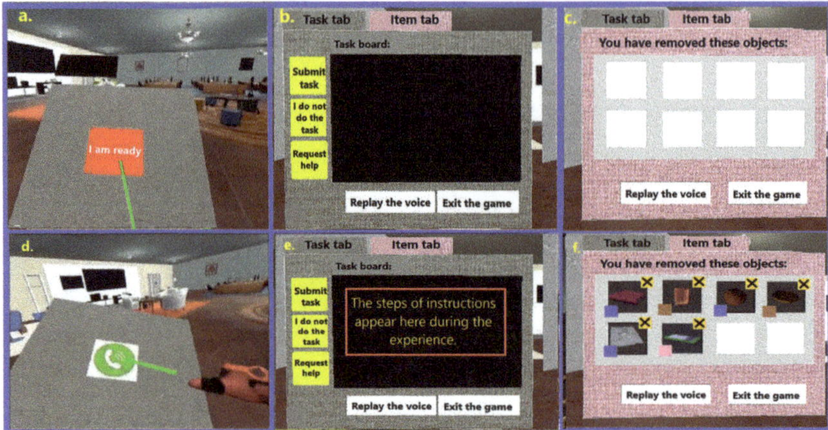

Figure 5. Screenshots showing different parts of the panel for interactions: (**a**,**d**) selecting the "I am ready" and "Answering the phone" buttons by the user; (**b**,**e**) visualization of the task tab; (**c**,**f**) visualization of the item tab.

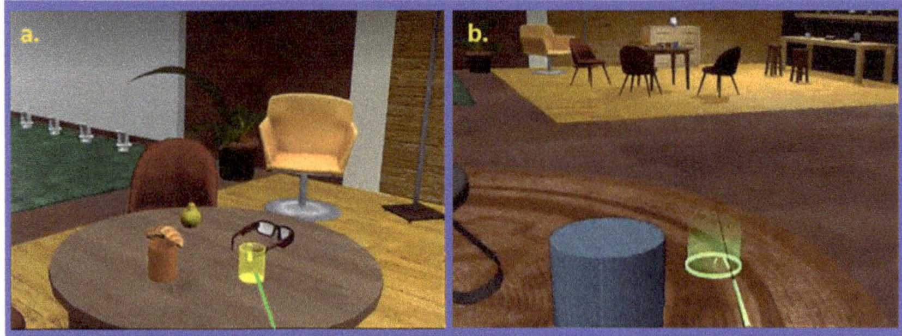

Figure 6. Screenshots showing the participant when: (**a**) removing interactive objects; (**b**) using the teleportation to move in the environment.

The tasks amount to sequences of instructions to search and remove objects expressed in written form or verbally at different moments in the experience. The tasks include two levels of uncertainty, as inspired by the definition of Hillen et al. [28]. As explained before, uncertainty appears in terms of ambiguity, probability, and complexity. Ambiguities result from incomplete guides and instructions. Probable situations appear as unexpected task changes, and complexity as a change in the number of causal factors in the instructions. Following this guide, we proposed these two tasks to represent two levels of uncertainty:

Task 1 (or base task): This includes a simple and clear set of instructions. For each step, the number, place, and color of objects that should be removed are clearly expressed.

Instructions for Task 1:

- Step 1: Go to the blue rug area. Remove the red glasses and green smartphone from the desk.
- Step 2: Go to the blue rug area. Remove the apple and orange from the tables.
- Step 3: Go to the brown rug area. Remove the blue cup from the table.
- Step 4: Go to the red rug area. Remove the green wallet and the sandwich from the table.
- Step 5: Go to the white rug area. Remove the pink book from the table.

Task 2 (or with an intermediate level of uncertainty): This is a task whose degree of uncertainty includes more complex instructions when compared to Task 1. This Task comprises lexical ambiguity in the instruction, instructions with missing information, and the possibility of instruction changes on the fly (See Figure 7 for the placements of objects in seven different areas of the environment with their associated rugs).

Instructions for Task 2:

- Step 1: Go to the brown rug area. Remove any food not positioned on a plate from the table.
- Step 2: Remove the glasses from the brown rug area.
- Step 3: If you find some mugs in the pink rug area, remove the orange one.
- Step 4: Go to the green rug area. Remove the pillows that are closest to the hat.
- Step 5: If you find calculators on the table and a bag under the table in the white rug area, remove the calculators.

Instructions for Task 2 (after change):

- Step 1: Go to the brown rug area. Remove any food on plates on the table.
- Step 2: Remove the glasses (if you find them there) from the brown rug area.
- Step 3: If you find any books in the blue rug area, remove any pencils near them.
- Step 4: Go to the red rug area. If a bag lies under the table, do not remove the smartphone.
- Step 5: If you see any mugs on the table in the yellow rug area, remove them.

Figure 7. Screenshots from seven different areas for the placements of objects characterized by the color of their associated rugs.

3.3. Methods

3.3.1. Procedure

After the participants read the consent form and provided their informed consent, we briefly explained that the experience would develop in a virtual office and that they would be asked to perform some tasks there. Then, a short introduction of the HTC Vive Pro headset, controllers, sensors, and their applications for this study was provided. Then, the users started with the familiarization phase and were guided to the main phase of the experiment. Afterward, the users answered demographic and evaluation questions that will be analyzed later. Figure 8 provides a diagram showing the steps that the participant experiences.

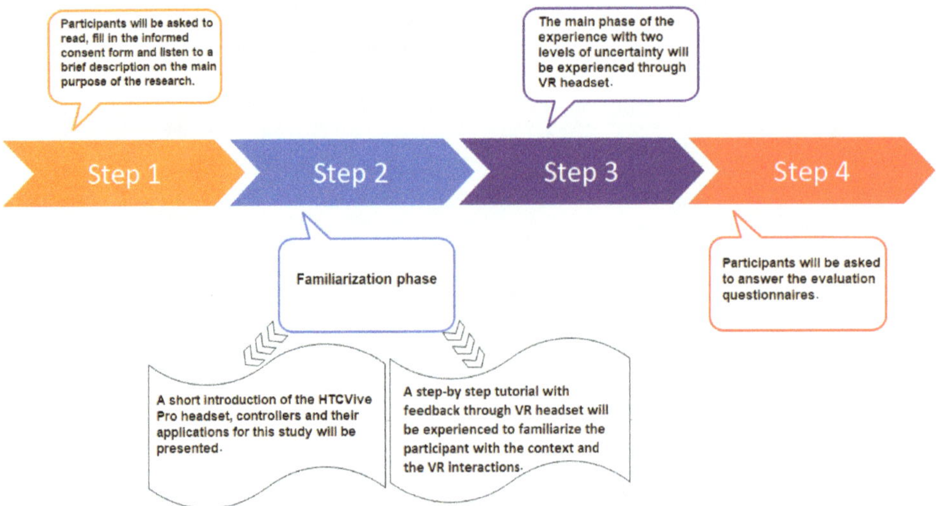

Figure 8. The steps that the participant experiences.

3.3.2. Measures

In this section, we describe the objective and subjective measures used to test our hypotheses.

Objective measures

The application records behavioral responses that could be inferred from the HTC Vive controllers and the headset log data. The following variables were measured:

- Variables related to the time:
  - Time to submit Task 1;
  - Time to submit Task 2;
  - Response time to new messages in Task 1;
  - Response time to new messages in Task 2.
- Variables related to the position: Position of the user in each moment.

Subjective measures

We utilized multiple questionnaires to evaluate participants' subjective experience with the application as detailed below.

- The Demographic questionnaire: This questionnaire asked participants about their nationality, sex, age, and education level.
- Level of English proficiency questionnaire: A five-point Likert scale was used to rate the level of English proficiency of participants (See Table 4).
- Previous Experience with immersive VR: A five-point Likert scale was used to rate the previous experience with immersive VR of participants (See Table 4).
- Perceived uncertainty questionnaire: In this questionnaire, using a five-point Likert scale, participants were asked to rate their level of perceived uncertainty for each task after the experiment (See Table 4).
- System Usability Scale (SUS) questionnaire: This questionnaire [64] consists of 10 items and utilizes a scale of 1 (Strongly disagree) to 5 (Strongly agree) to provide a "quick and dirty" reliable tool for measuring usability.
- Slater–Usoh–Steed presence questionnaire (SUS): This questionnaire [65] consists of five items and utilizes a scale of one to seven to assess participants' sense of being there in a virtual office.

- The immersive experience questionnaire (IEQ): This questionnaire [66] comprises 31 items and utilizes a scale of 1 (Not at all) to 7 (A lot) to measure the subjective experience of being immersed while playing a virtual serious game.
- Motion sickness questionnaire (MSAQ): This questionnaire [67] comprises 16 items, utilizes a scale of 1 to 9, and is a valid instrument for the assessment of motion sickness.
- Intolerance of Uncertainty Scale (IUS): This questionnaire [68] consists of 27 items and utilizes a scale of 1 (Not at all characteristic of me) to 5 (Entirely characteristic of me) that assesses emotional, cognitive, and behavioral reactions to ambiguous situations, implications of being uncertain, and attempts to control the future.

Table 4. Questionnaires used to rate the level of English proficiency, the previous experience with immersive VR, and the perceived uncertainty of Task 1 and Task 2.

Measuring Item	Question	Range
Level of English proficiency	How do you rate your level of English proficiency?	(1. Poor–5. Very good)
Previous Experience with immersive VR	Have you experienced virtual reality with a head-mounted display in the past?	(1. Never–5. Everyday)
Rating the perceived uncertainty of Tasks 1 and 2	From the definition of Hillen et al. [28], changes in three sources can provide uncertainty in an information system: probability, ambiguity, or/and complexity of information. You can perceive uncertainty if you can become aware of its existence. Based on this definition and your experiment with the virtual office, what score will you give to the perceived uncertainty of Task 1 (the same question for Task 2)?	1–5

4. Results

In this section, we present the objective and subjective results of our experiment concerning our research questions:

We compared the ratings that the participants gave to the perceived uncertainty of two tasks with the Wilcoxon signed-rank test. The result found a significant difference between them ($v = 0$, $p = 0.0003553 < 0.05$), suggesting that overall the participants rated Task 2 with higher perceived uncertainty than Task 1 (See also Figure 9 for a visual comparison of the ratings).

To find the effects of different degrees of induced uncertainty on the user's behavior, we first confirmed the normality of the data with the Shapiro–Wilk test at the 5% level. Then, we conducted the Paired t-test. The results did not yield a significant difference between the response time in the two tasks ($t(16) = 1.44$, $p = 0.084 > 0.05$). However, the box plot in Figure 10 visually shows a lower response time to pick up the phone in Task 2 when compared to Task 1.

Since the normality of the data was rejected by the Shapiro–Wilk test at the 5% level, using the Wilcoxon signed-rank test ($v = 0$, $p = 0.00001526 < 0.05$), we found a significant difference between the task completion time for Task 1 and Task 2 (See also Figure 11 to see a visual comparison between the amounts).

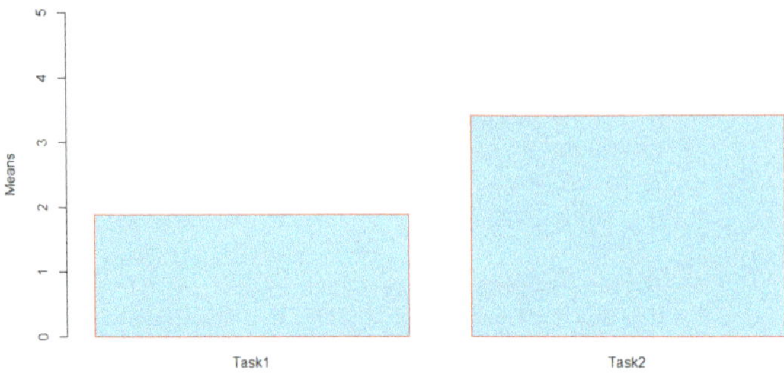

Figure 9. A comparison between the means of ratings obtained from the participants for their perceived uncertainty of Task 1 (M = 1.88, SD = 1.11) and Task 2 (M = 3.41, SD = 1.00); Range of answers = 1–5.

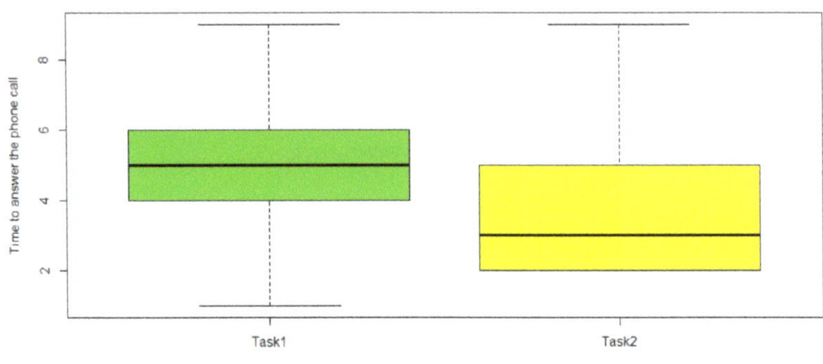

Figure 10. A comparison between the response time of the user to the source of information (the phone call) during Task 1 and Task 2, measured in [s].

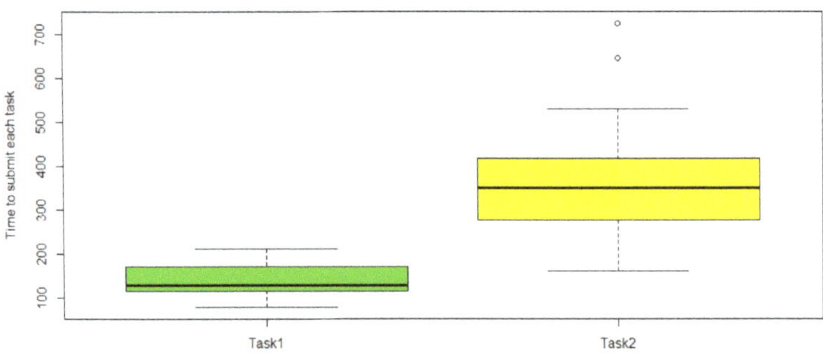

Figure 11. A comparison between the task time completion of the user for Task 1 and Task 2, measured in [s]. In these data, two completion time values identified as outliers which are shown in white circles in the figure.

To report the differences in the change of position in Task 1 in comparison to Task 2, Figures 12 and 13 present a visual comparison of participants' change of position.

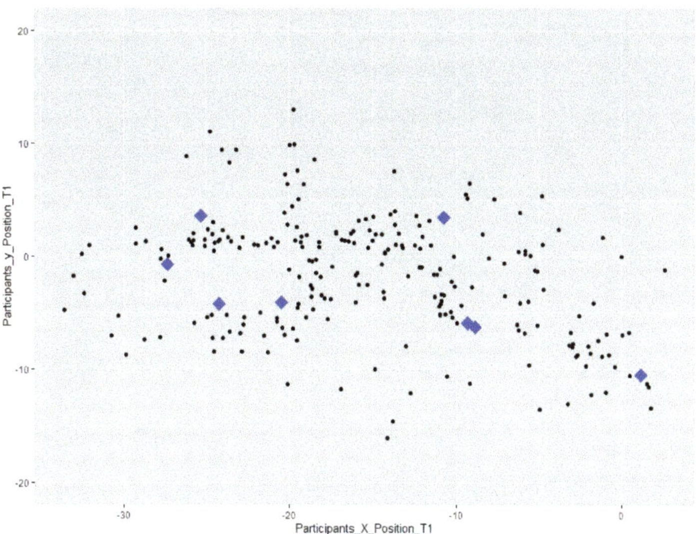

Figure 12. A visualization of participants' change of position in Task 1; the 8 task targets are shown in blue.

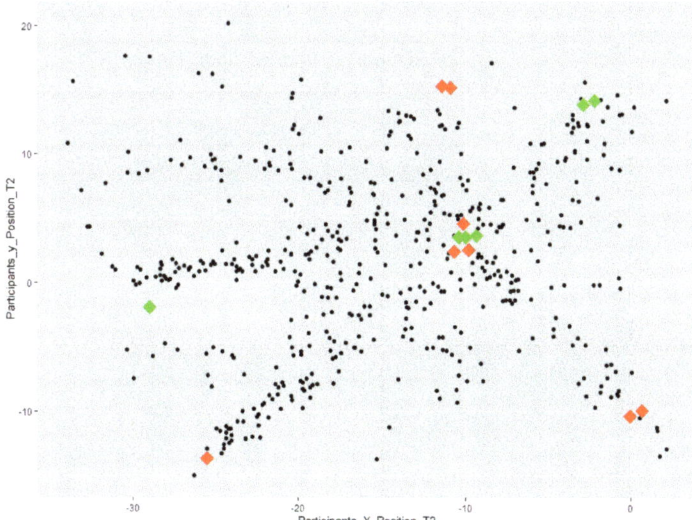

Figure 13. A visualization of participants' change of position in Task 2; the task targets related to before the change of the task are shown in red; the task targets related to after the change of the task are shown in green.

We used Pearson's r-test to measure the strength and direction of the possible linear relationship between the scores on system usability, immersion, presence, motion sickness, and intolerance of uncertainty questionnaires and the recorded time to answer the second call (i.e., response time to the source of information in Task 2). We also used this test for finding the possible relationships between the scores of these questionnaires and the time spent on the second task. See Table 5 for the results of these tests.

Table 5. Correlation values between subjective and objective measures.

Questionnaire Name	Time to Answer the Call in Task 2	Task Competition Time for Task 2
The System Usability Scale (SUS)	$r = -0.38$	$r = -0.15$
Immersive Experience Questionnaire (IEQ)	$r = 0.15$	$r = 0.4$
Slater–Usoh–Steed presence questionnaire (SUS)	$r = 0.24$	$r = 0.29$
Motion sickness questionnaire (MASQ)	$r = 0.21$	$r = 0.16$
Intolerance of Uncertainty Scale (IUS)	$r = -0.09$	$r = 0.07$

Table 6 reports the mean and standard deviation of scores obtained from the questionnaires about the quality of the participants' experience and their intolerance to uncertainty. Figures 14–20 present a visual comparison of the data obtained.

Table 6. Mean and standard deviation of scores received from participants' answers to the questionnaires.

Questionnaire Name	Mean	SD	Range
The System Usability Scale (SUS)	76.32	9.89	[1–100]
Immersive Experience Questionnaire (IEQ)	65.84	7.13	[1–100]
Slater–Usoh–Steed presence questionnaire (SUS)	64.37	14.50	[1–100]
Motion sickness questionnaire (MASQ)	22.18	12.48	[1–100]
Intolerance of Uncertainty Scale (IUS)	59.48	11.63	[1–100]

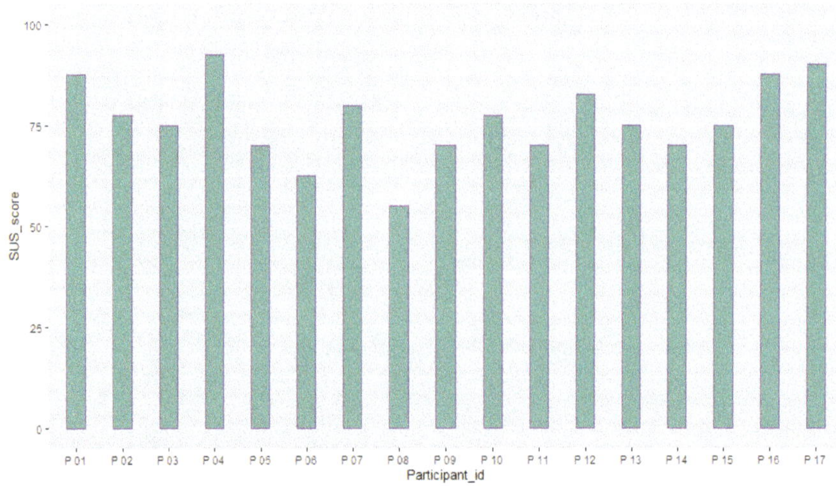

Figure 14. A comparison between the scores obtained from the participants from the System Usability Scale (SUS) questionnaire (M = 76.32, SD = 9.89).

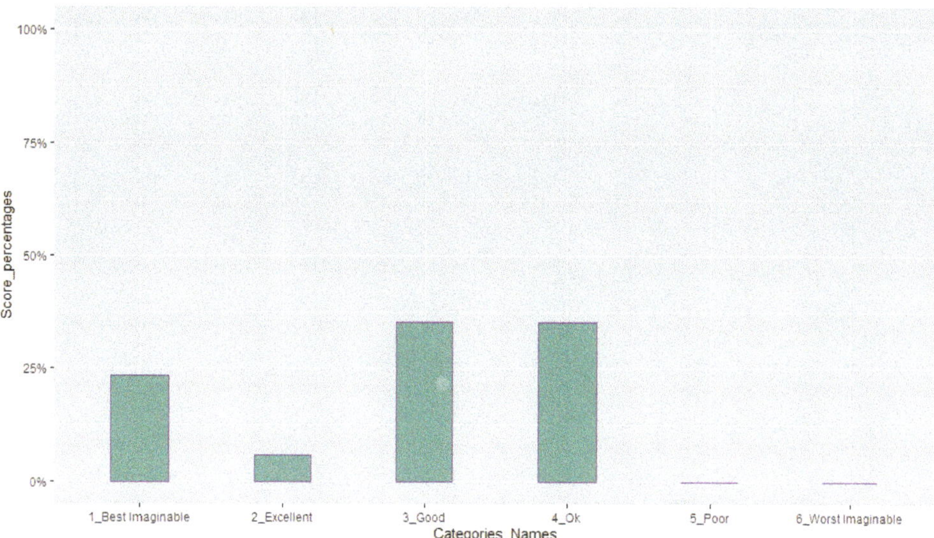

Figure 15. Percentages of ratings for each category from the System Usability Scale (SUS) questionnaire, categories from left to right are: best imaginable (score > 84.1); excellent (72.6 < score < 84.0); good (62.7 < score < 72.5); ok (51.7 < score < 62.6); poor (26 < score < 51.6); worst imaginable (score < 25).

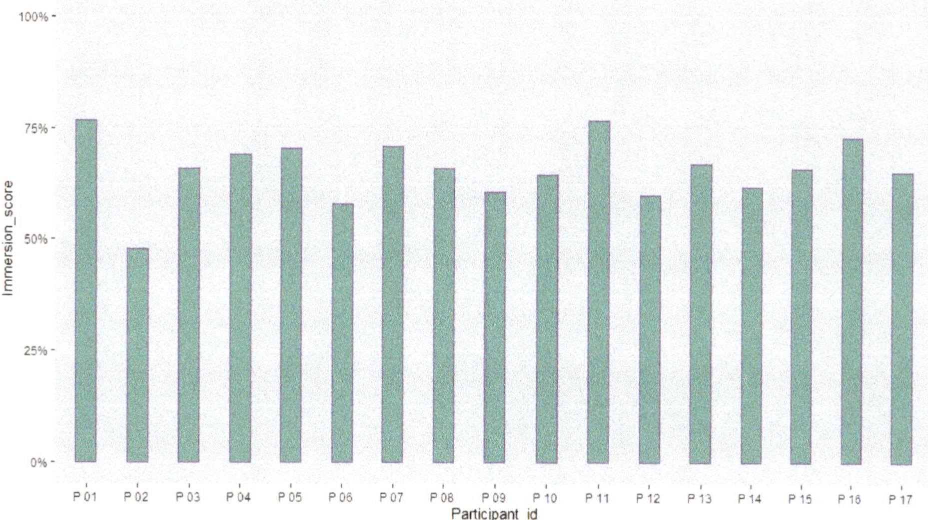

Figure 16. A comparison between the scores obtained from the participants from the Immersive Experience questionnaire (IEQ) (M = 65.84 (%), SD = 7.13 (%)).

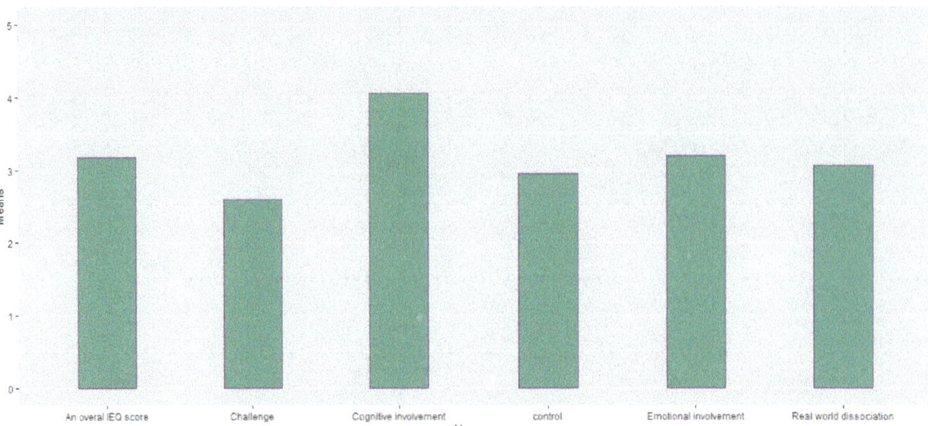

Figure 17. Range of (1: Not at all; 5: A lot) for a comparison between the means of scores for the components of the Immersive Experience questionnaire (IEQ) scale, from left to right: overall IEQ score (M = 3.17); challenge (M = 2.59); cognitive involvement (M = 4.06); control (M = 2.95); emotional involvement (M = 3.20); and real-world dissociation (M = 3.059).

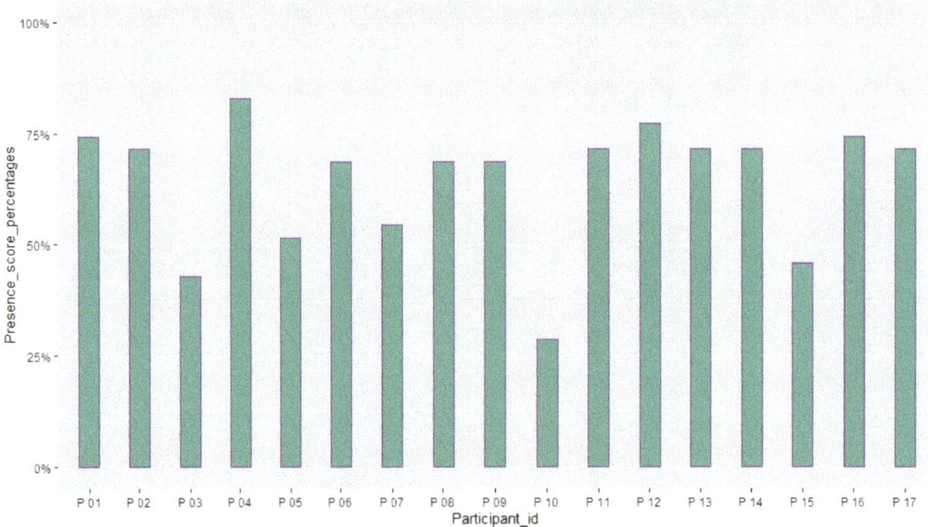

Figure 18. A comparison between the scores obtained from the participants from the Slater–Usoh–Steed presence (SUS) questionnaire (M = 64.37 (%), SD = 14.50 (%)).

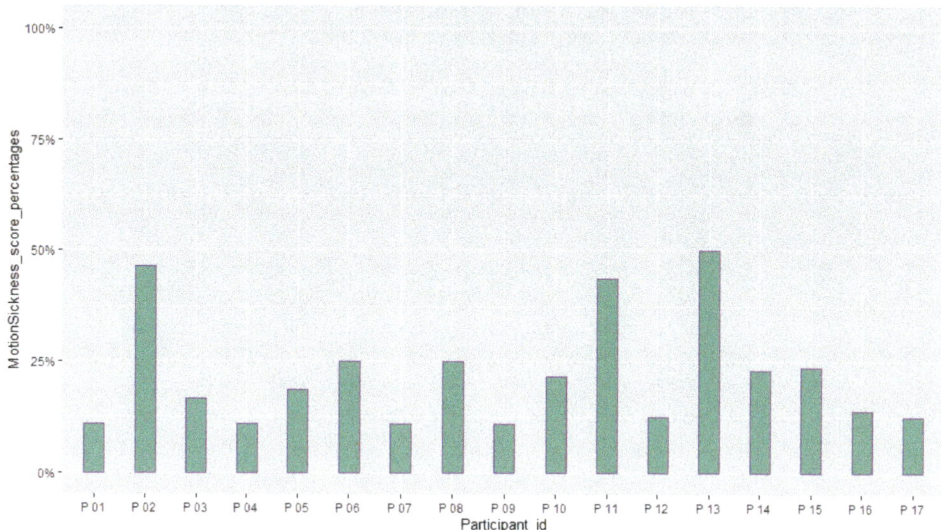

Figure 19. A comparison between the scores obtained from the participants from the Motion Sickness questionnaire (MASQ) (M = 22.18 (%), SD=12.84 (%)).

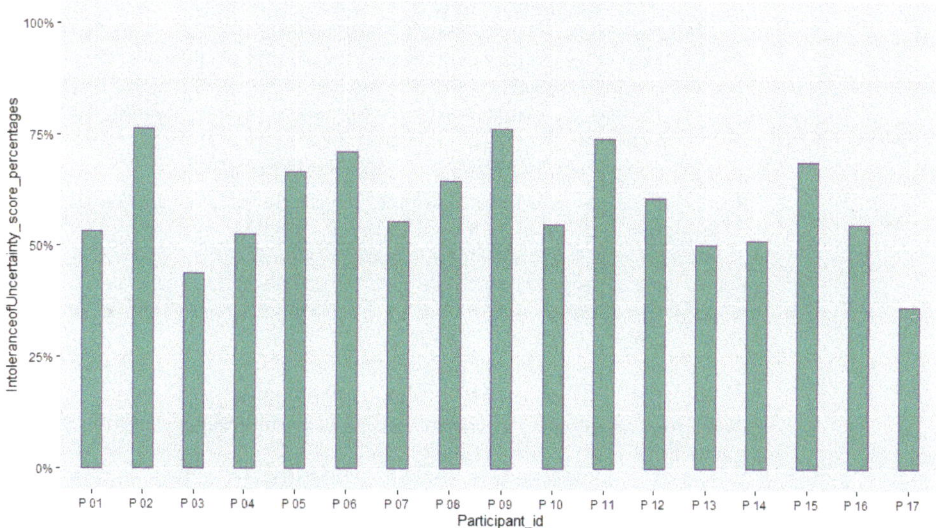

Figure 20. A comparison between the scores obtained from the participants from the Intolerance of Uncertainty Scale (IUS) (M = 59.48 (%), SD=11.63 (%)).

5. Discussion

In this section, we present and discuss the main findings of the experiment in more detail.

The main purpose of this study was to suggest the design and implementation of a VR platform that is able to create the experience of uncertainty of interpersonal communications on two levels and to record and report human behavioral responses to this exposition. In this paper, we addressed these research questions:

RQ1: Is there any significant difference between subjective ratings of participants for perceived uncertainty of Task 1 and Task 2?

Our findings from a comparison of the post-experiment ratings of the participants to the perceived uncertainty of two tasks indicate the potential of the proposed design to successfully produce at least two levels of uncertainty in the experience of the system.

RQ2: How do different degrees of induced uncertainty affect the users' behavior and performance in this immersive virtual workplace scenario?

- RQ2-1: Does the response time of the user to reach the source of information (the phone call) differ in the two tasks?
- RQ2-2: Does the task completion time for the user for each task differ?
- RQ2-3: How does the change of participants' position in Task 1 differ from Task 2?

In this paper, we targeted the study of the differences between behavioral responses to the experience of two levels of uncertainty. In particular, we focused on studying the real-time records of the time and position of the participants.

Related to time, we were interested in two variables:

1. Response time: In particular, we were interested in the participant's response time to new information coming from a highly influential source that was directly associated with the boss, a phone call. Our expectation was that by increasing the degree of induced uncertainty, the participant would show a different response time, but we did not find a significant difference in this comparison with the applied statistical test.
2. Task completion time: In particular, we were interested in the time that the participants persist in accomplishing each task as an objective measure of the user's tolerance to the change of uncertainty in the system. For this reason, we did not consider the correctness or wrongness of following the instructions, and the user was free to end tasks at any moment without experiencing any time pressure or encouraging or discouraging feedback. Related to this setting, by increasing the level of uncertainty of the task, our expectation was that the participant spends a different amount of time accomplishing the task. The results of the study were aligned with this expectation by reporting a significantly higher completion time for Task 2.

Related to position: The positions of the participants in Task 1 and Task 2 were recorded by the application every 4 s. The distribution of them along two axes of X and Y and in comparison with the targets for each individual task is visualized in Figures 12 and 13. From a visual comparison of the two plots, we can see that in total, participants in Task 2, a task with an increase in uncertainty level, have more changes of position. This finding is aligned with what we were expecting.

In sum, our experiment suggests that adding uncertainty to a task will harm task performance on completion time, but not in response time.

RQ3: How are the users' subjective responses to uncertainty related to the objective responses?

Despite our expectations for finding strong relations between the subjective and objective measures, we found a small negative linear correlation between the scores of the system usability scale and time to answer the second call, small positive linear correlations between the presence score and both task completion time for Task 2 and time to answer the second call, and a small positive linear correlation between scores of the motion sickness questionnaire and time to answer the second call. We think with the increase in sample size, we can report stronger correlations between these variables.

RQ4: How does the user evaluate the quality of his/her experience through subjective measures?

Another purpose of the study was to report the results of the participants' evaluation of their experience with the system. The mean score of our results from the System Usability Scale (SUS) conveys a higher amount than the average SUS score which is 68. This gives an immediate insight into the overall good usability of the system and the need for minor improvements in the design [69]. In addition, based on the adjective rating scale introduced by [70], we also found that nearly 70 % of the participants' ratings of the usability of the system fit into "Best imaginable", "Excellent", and "Good" categories (See Figure 15). For the Immersive Experience questionnaire (IEQ), the Slater–Usoh–Steed presence (SUS) questionnaire, and the Motion Sickness questionnaire (MASQ), the average of the scores also falls into an acceptable range representing a good quality of the participants' experiences (See Figures 16–19).

In sum, we can conclude that the system with the help of the designed environment and story plot is able to create a pleasant virtual experience. In addition, with the help of tracing from the HTCVive pro controllers and headset we were able to successfully capture in real time the behavioral responses of the participants related to the time of actions, and user position to our variable of interest.

6. Conclusions and Future Works

In this paper, we investigated the effects of uncertainty level in a virtual office on participants' objective and subjective responses through a controlled human-subject study. We designed an experimental scenario inspired by a famous story name Amelia Bedelia written by Peggy Parish [62]. For the design of our system, we first investigated and carefully selected the virtual reality interfaces and environments that supported our research needs. In addition, we were inspired by previous games which applied uncertainty in their designs. The goal was to develop a system that supports a pleasant 3D immersive experience with real-world-like interactions and rich data-collecting techniques. In our usability study, participants were asked to complete two different tasks inside a virtual office where they were also involved in interpersonal communication with their boss on the first day of work. We measured the participants' objective responses through the log data captured from the tracing of HTCVive pro controllers and headsets as well as assessed their subjective experience through questionnaires. We determined that the two proposed versions of tasks received significantly different ratings from the participants for their perceived uncertainty after the experiment. In addition, our results supported that the time taken to submit different tasks differs significantly. In addition, results from the usability, immersion, presence, and motion sickness questionnaires conveyed that overall, the participants were satisfied with the experience by scoring the usability, presence, and immersion of the experience on average higher than 50% and the motion sickness of the experience less than 30%.

This paper suggested that our proposed VR system can manipulate the levels of uncertainty to study it. In the design of this system, we inspired ourselves from real-life situations. An example workplace scenario could be what happens regularly for one in the role of a manager. S/he may receive multiple unpredictable inputs at once and has to constantly monitor and choose what to do first, stay productive, and successfully monitor time allocations to be able to work with everyone involved [71]. To indicate how effectively our system replicates such real-life happenings under the same conditions, an evaluation of our proposed system against real-life baseline conditions is required. We decided not to consider this system evaluation in this paper because of our limitations in controlling the confounding factors coming from real-world settings that make it hard for us to have a valid measure of the effects of uncertainty. So, we leave it for future work. In addition, we plan to investigate more behavioral responses from the user in a future study and assess the feasibility of this application with a desktop-based version of it. Finally, a larger sample size helps us to report and study more powerful behavioral results of the study.

Author Contributions: Conceptualization, S.H.; Methodology, S.H.; Software, S.H.; Validation, S.H.; Investigation, S.H.; Writing—original draft, S.H.; Writing—review & editing, G.M.; Supervision, G.M.; Funding acquisition, G.M. All authors have read and agreed to the published version of the manuscript.

59. Farah, Y.A.; Dorneich, M.C.; Gilbert, S.B. Evaluating Team Metrics in Cooperative Video Games. In *Proceedings of the Human Factors and Ergonomics Society Annual Meeting*; SAGE Publications Sage CA: Los Angeles, CA, USA, 2022; Volume 66, pp. 70–74.
60. Xu, W.; Liang, H.N.; Yu, K.; Baghaei, N. Effect of gameplay uncertainty, display type, and age on virtual reality exergames. In Proceedings of the 2021 CHI Conference on Human Factors in Computing Systems, Yokohama, Japan, 8–13 May 2021; pp. 1–14.
61. Liu, Y.; Lin, Y.; Shi, R.; Luo, Y.; Liang, H.-N. Relicvr: A virtual reality game for active exploration of archaeological relics. In Proceedings of the Extended Abstracts of the 2021 Annual Symposium on Computer-Human Interaction in Play, Virtual, Austria, 18–21 October 2021; pp. 326–332.
62. Parish, P.; Siebel, F.; Thomas, B.S.; Canetti, Y. *Amelia Bedelia*; Harper & Row: New York, NY, USA, 1963.
63. Tabbers, H.; Martens, R.; van Merriënboer, J.J. Multimedia instructions and cognitive load theory: Split-attention and modality effects. In Proceedings of the National Convention of the Association for Educational Communications and Technology, Long Beach, CA, USA, 17 February 2000 ; Citeseer: Princeton, NJ, USA, 2000.
64. Brooke, J. SUS-A quick and dirty usability scale. *Usability Eval. Ind.* **1996**, *189*, 4–7.
65. Usoh, M.; Catena, E.; Arman, S.; Slater, M. Using presence questionnaires in reality. *Presence* **2000**, *9*, 497–503. [CrossRef]
66. Rigby, J.M.; Brumby, D.P.; Gould, S.J.; Cox, A.L. Development of a questionnaire to measure immersion in video media: The Film IEQ. In Proceedings of the 2019 ACM International Conference on Interactive Experiences for TV and Online Video, Salford, UK, 5–7 June 2019; pp. 35–46.
67. Gianaros, P.J.; Muth, E.R.; Mordkoff, J.T.; Levine, M.E.; Stern, R.M. A questionnaire for the assessment of the multiple dimensions of motion sickness. *Aviat. Space Environ. Med.* **2001**, *72*, 115. [PubMed]
68. Buhr, K.; Dugas, M.J. The intolerance of uncertainty scale: Psychometric properties of the English version. *Behav. Res. Ther.* **2002**, *40*, 931–945. [CrossRef]
69. Brooke, J. SUS: A retrospective. *J. Usability Stud.* **2013**, *8*, 29–40.
70. Bangor, A.; Kortum, P.; Miller, J. Determining what individual SUS scores mean: Adding an adjective rating scale. *J. Usability Stud.* **2009**, *4*, 114–123.
71. Zika-Viktorsson, A.; Sundström, P.; Engwall, M. Project overload: An exploratory study of work and management in multi-project settings. *Int. J. Proj. Manag.* **2006**, *24*, 385–394. [CrossRef]

MDPI

St. Alban-Anlage 66

4052 Basel

Switzerland

Tel. +41 61 683 77 34

Fax +41 61 302 89 18

www.mdpi.com

Sensors Editorial Office

E-mail: sensors@mdpi.com

www.mdpi.com/journal/sensors